Proceedings in Life Sciences

Insect Flight

Dispersal and Migration

Edited by
W. Danthanarayana

With 62 Figures

Springer-Verlag
Berlin Heidelberg New York
London Paris Tokyo

Professor Dr. WIJESIRI DANTHANARAYANA

Department of Zoology
The University of New England
Armidale, N.S.W. 2351
Australia

ISBN 3-540-16502-9 Springer-Verlag Berlin Heidelberg New York
ISBN 0-387-16502-9 Springer-Verlag New York Heidelberg Berlin

Library of Congress Cataloging in Publication Data. Insect flight. (Proceedings in life sciences) Based on a symposium entitled "Insect flight: dispersal and migration" held at the XVIIth International Congress of Entomology in Hamburg, Germany in August, 1984. Includes index. 1. Insects–Flight–Congresses. 2. Insects–Migration–Congresses. 3. Insects–Dispersal–Congresses. I. Danthanarayana, W. (Wijesiri), 1936– . II. International Congress of Entomology (17th: 1984: Hamburg, Germany) III. Series. QL496.7.I58 1986 595.7'056 86-6433

Typesetting, offsetprinting and bookbinding: Brühlsche Universitätsdruckerei, Giessen
2131/3130-543210

Preface

This volume is based on a Symposium entitled "Insect Flight: Dispersal and Migration" held at the XVIIth International Congress of Entomology in Hamburg, Germany in August, 1984. An earlier similar Symposium was held at the XVth International Congress of Entomology in 1976 in Washington DC, USA ("Evolution of Insect Migration and Diapause" edited by Hugh Dingle, Springer-Verlag, 1978). It is, therefore, appropriate that the present Symposium was held 8 years later in view of the intensified interest and developments in this field, particularly in the areas of evolution, genetics, endocrine/neuroendocrine influences, orientation, navigation and nocturnal flight activity.

It was intended that the Symposium should have a fairly representative, albeit non-complete, coverage of the various aspects of its topic, and the Congress facilitated the gathering of many specialists in the field, enabling the presentation and discussion of 18 papers. Fifteen of these are presented in this volume, mostly as modified and expanded versions. Also included is a contribution by Roger Farrow who presented his paper at another section of the Congress, and one by Wolfgang Stein who could not attend the Congress; an additional paper, by myself and Stuart Dashper, is included with the intention of supporting my presentation at the Congress. One of the most interesting contributions was that by Rudiger Wehner on "The role of polarized skylight in insect navigation". Much of what he said has been published elsewhere under the title "Astronavigation in Insects" (*Annual Review of Entomology 29*:277–298, 1984). Two of the papers presented at the Symposium, one by Christer Solbreck and Brigitta Sillen-Tullberg entitled "Role of migration in exploiting patchy and time-varying resources: dynamics of a seed-predator system" and the other by Robert F. Denno entitled "Wing polymorphism and migration in plant hoppers: the role of host plants" are not included in this volume, as their research is still in progress. The reader is referred to the Congress Abstract volume, pp. 344 and 345, for synopses of these works.

The majority of chapters in this volume give accounts of original research; these and the review articles include developments over the last several years. This material indicates that at least *some* of the older concepts are being questioned and new ones proposed. This indeed is a reflection of the dynamism and progress of the subject, and fulfils the primary aim of the Symposium. Chapters in this book have not been grouped into sections because of the overlapping nature and the broad range of topics covered, but the order followed after the Introduction is to begin with the opening address by Hugh Dingle on "Evolution and Genetics of Insect Migration" followed by those on physiology, behaviour, genetics/ecology, pest insects, and methodology respectively. A number of colleagues in Australia, Britain, and the USA to whom I extend my sincere thanks, kindly assisted in reviewing the papers.

I was very fortunate and privileged in obtaining the services of two distinguished and eminent researchers in the field, Hugh Dingle and Roy Taylor to give the opening and closing addresses respectively, and also to Chair the four sessions. I am very grateful to them for doing this onerous work, as well as for the encouragement given to me during the organization of the Symposium and the preparation of this volume. I also thank all those who participated in the Symposium and contributed to this volume, and to Springer-Verlag and especially Dr. Dieter Czeschlik for undertaking the task of publication so enthusiastically. Tacit support given by my wife, Sunimal, during the preparation of this work and the considerable typing assistance given by Rhonda McLauchlan are greatly appreciated. Financial support provided by the Alexander von Humboldt-Stiftung for my attendance at the Entomology Congress at Hamburg is gratefully acknowledged. L.R. Taylor holds an Emeritus Research Fellowship from the Leverhulme Trust and wishes to acknowledge their support.

Armidale, Spring 1986   W. Danthanarayana

Contents

List of Contributors

You will find the addresses at the beginning of the respective contribution

Belton, P.   60
Bodenhamer, J.E.   27
Danthanarayana, W.   1, 88, 120
Dashper, S.   120
Dingle, H.   11
Dixon, A.F.G.   145
Dohse, L.   235
Edwards, J.S.   196
Farrow, R.A.   185
Gatehouse, A.G.   128
Gibo, D.L.   172
Goldsworthy, G.J.   49
Henneberry, T.J.   253
Howard, M.T.   145
Lingren, P.D.   253
McAnelly, M.L.   27
Mikkola, K.   152

Pair, S.D.   204, 221
Pedraza Martinez, F.A.   204
Rankin, M.A.   27
Raulston, J.R.   204
Reynolds, D.R.   71
Riley, J.R.   71
Saks, M.   235
Sanchez Valdez, V.M.   204
Sparks, A.N.   204
Stein, W.   242
Stinner, R.E.   235
Taylor, L.R.   265
Truesdale, F.M.   221
Westbrook, J.K.   204
Wheeler, C.H.   49
Wolf, W.W.   221

1 Introductory Chapter

W. Danthanarayana

Major aspects of insect dispersal and migration by flight have been discussed in detail in several books (Baker 1978; Dingle 1978; Johnson 1969; Rabb and Kennedy 1979; Rainey 1976a; Williams 1958) and a large body of review papers (e.g. Dingle 1972, 1980; Johnson 1974; Kennedy 1975; Schneider 1962; Southwood 1962, 1975, 1981; Taylor L.R. and Taylor R.A.J. 1983). The purpose of this introduction is, therefore, mainly to draw attention to advances in this field over the last few years, in the light of work presented in this volume. In doing so an attempt is made to highlight areas in which more research effort is desirable. In an introduction such as this, it is necessary to emphasize the obvious in order to present an intelligible perspective of the topics covered and discuss their implications.

There have been many attempts to arrive at acceptable definitions for the terms dispersal and migration (e.g. Andrewartha and Birch 1954; Elton 1930; the references listed above). This has led, on occasion, to confusion and contention largely due to attempts to provide all embracing definitions covering the entire animal kingdom. Life-cycle strategies of insects in comparison to, say, vertebrates are often very different reflecting the vast differences between the two groups in their body size, generation times and longevity. Primarily because of the short life span, bicoordinate navigation does not occur in insect migrations except perhaps to a limited extent in a few (apparently rare) examples such as the monarch butterfly (Urquhart and Urquhart 1977, 1978, 1979). Also, homing associated with migration and based on memory, in the sense of that found in higher animals, is yet to be demonstrated for insects. Nevertheless, there is now a consensus on the terminology applicable to insect movements. Although none of the papers presented at the Symposium considered the definitions of terms, the question was raised in discussion, and a re-iteration of definitions appears to be appropriate at the outset.

'Dispersal' refers to scattering of a population, leading to an increase in the mean distance between individuals (Andrewartha and Birch 1954; Southwood 1962, 1981; Schneider 1962), and encompasses two types of movements: migratory and non-migratory. Non-migratory movements involve travel within the habitat associated with such activities as feeding, mating, and oviposition, and have been referred to as 'appetitive' (Provost 1952), 'vegetative' (Kennedy 1961a, 1975) or 'trivial' (Southwood 1962) movements. In contrast, migratory movements take insects beyond the habitat for the purposes of colonizing new habitats, re-colonizing old ones, aestivation or hibernation (Johnson 1969; Southwood 1962; Dingle 1972). Migration is thus an evolved adaptation for survival and reproduction (Johnson 1960), so promoting genotypic fitness.

Insect Flight: Dispersal and Migration
Edited by W. Danthanarayana
© Springer-Verlag Berlin Heidelberg 1986

The genetic ('gene flow') and ecological (colonization, escape from inhospitable environmental conditions) consequences of migration will in turn lead to the evolution of a range of migratory behaviours and the associated physiological and morphological characteristics. In the final analysis it is difficult to separate sharply the ecological, behavioural, physiological, and morphological factors of insect migration. Integration of genetics with the above factors is, therefore, of fundamental importance. Dingle (Chap. 2) deals with this aspect, and emphasizes that little is known of the genetics underlying migratory behaviour. He stresses the need to comprehend the genetic variance and genetic correlation so that evolutionary processes that connect migration to important life history and fitness components can be understood. Understanding of the genetics of migration is expected to make a significant contribution to modern evolutionary biology because of the polygenic nature of most migratory traits, the evolutionary implication of genetic variation, phenotypic expression, selection of life history patterns and the involvement of migration in maintaining life cycle flexibility. Dingle's overview is supported by Rankin, McAnelly, and Bodenhamer (Chap. 3) on some physiological aspects and Gatehouse (Chap. 9) on ecological aspects of migration. The above workers use clear-cut physiological and morphological characteristics to distinguish migrants from non-migrants, and the adaptive roles of the migratory traits under varying environmental conditions are then examined. Rankin and colleagues show that the individual cost of migration has been previously overestimated, for in populations where migration is a major life-history strategy selection maximizes the product of reproduction and flight, rather than acting on either of these parameters; the physiological mechanism involved is juvenile hormone (JH) stimulation of both migration and reproduction. Gatehouse considers the ecological factors that determine the migratory capacity of the African armyworm. These act via a density-dependent phase polymorphism that ensures rapid re-dispersal of dense populations resulting from meteorological disturbances. The resulting low densities are achieved by seasonally cycling selection acting on genetically determined variability in potential flight capacity. It is suggested that this strategy is shared by many other oligophagous and polyphagous tropical and subtropical noctuids which do not undergo diapause, and depend on ubiquitous host plants.

The behavioural factors of insect migration include: persistent, undistracted locomotor activity (i.e. increased ortho-kinesis) that becomes straightened out, in the sense of traversing new ground instead of frequent changes in direction (i.e. decreased klino-kinesis), temporary depression of vegetative responses, special take-off and alighting patterns (Kennedy 1961a, 1975) and temporarily dominating phototaxis (Johnson 1969) which takes the insects right out of their 'boundary layer' (Taylor 1958, 1960, 1974). The term 'boundary layer' refers to the relatively shallow layer of air immediately above the earth's surface in which the wind speed is less than the flight speed of the insect. Within this layer the insect can control its movement to a large extent, and control its course and track. The depth of the boundary layer is variable depending on the species' size and flight speed, the strength of the prevailing wind and the terrain. Above the boundary layer upwind flight is difficult, and the insect's course and track are to a large extent influenced by wind direction.

For many years it has been known that insects flying at night rise through a temperature inversion with no help from convection and attain high ceilings ranging from a few hundred metres to 1-2 km (Kennedy 1975). Typical ground tracks of these insects are downwind, but rather than being merely carried by the wind, insects actively fly up-wards and show a common orientation even under dark and completely overcast conditions (Johnson 1969; Kennedy 1975; Farrow, this volume; Riley 1975; Reynolds and Riley 1979, this volume; Schaefer 1976; Waloff 1972). In considering this riddle of orientation of night-flying insects, Kennedy (1975) stated that "we must, as C.B. Williams believed, start looking into some sensory guidance mechanism not yet considered for insects". Although an unambiguous answer to this question has not been found, there has been a considerable body of work during the last several years on insect orientation. It has been shown that insects use one or several cues available in our celestial system, something considered to be highly unlikely several years ago! These include: the use of the earth's magnetic field (Arendse 1978; Baker and Mather 1982; Dyer and Gould 1981; Gould 1980, 1982; Gould et al. 1978; Kirschvink 1983; Martin and Lindauer 1977), time-compensated stellar navigation (Sotthibandhu and Baker 1979), celestial skymarks (Wehner 1983, 1984), infra-red energy perception (Evans and Kusher 1980; Meyer 1977) and moon-compass reaction (Sottibandhu and Baker 1979). Riley and Reynolds review the nocturnal orientation problem in Chap. 6 and by a process of elimination conclude that time-compensated stellar navigation and perception of the terrestrial magnetic field might be used by the compass navigators, and accelerative anisotropies in the air motion could provide the necessary cues for the wind navigators, as both types of navigation have been detected by radar in high-flying nocturnal mi-grants. They expect nocturnally migrating insects to adopt wind-related cues or compass orientation in an hierarchical manner depending on wind conditions and on their individual flight strategies. The stage is now set for further experimental work on nocturnal orientation, preferably under natural conditions, for the elucidation of the circumstances under which insects use the various navigational strategies available to them.

The influence of the moon on flight and migratory activities are covered by Danthanarayana in Chap. 7. It is shown that there is lunar periodicity in these activities in two experimental insects and that this may extend to other species also. Lunar periodicity appears to be rhythmic, but it is not yet known with certainty whether insect lunar rhythms are of endogenous or exogenous origin, although the available evidence tends to support the former (Neumann 1981). Naylor (1982) and Cloudsley-Thompson (1980) point out that endogenous lunar rhythms in the behaviour and physiology of insects and other organisms, as well as the mechanism of the internal lunar clock have presumably been selected for and inherited. It is perhaps in this context that lunar periodicity of insect activities needs to be examined. Also there is a remarkable similarity between the pattern of lunar periodicity of insect flight described in this volume and the semi-lunar cyclicity of cardioactive neurotransmitterlike substances found in the central nervous system of the cockroach, human and mouse blood as well as in leaf extracts of several plant species (Rounds 1975, 1981, 1982). It has been proposed that some of these substances may be involved in "system" modulating organismic response to lunar-related forces (see Rounds 1982). As an operating hypothesis, Rounds suggested that these cardioactive materials are "modulators" or "enablers" and are released into the blood or tissue fluid to act on "exitability" of cells in general and make the target tissues more or less responsive to other triggers. This "general response system" is presumed to be responding, directly or indirectly, to the gravitational force changes associated with variations in the position of the moon, the sun, and the earth. There is thus a need to investigate whether the neurophysiological and hormonal activities that influence insect flight and migration are affected by lunar periodicity.

The influence of polarized light on some night-flying insects is the topic of Chap. 8 (Danthanarayana and Dashper). Waterman (1981) points out that with the exception of Wehner's work on *Cataglyphis* (Wehner 1972, 1976) and Wellington's observations on mosquito flight (Wellington 1974a,b, 1976), there are few recent field experiments on polarization sensitivity. He goes on to stress that experiments are needed not only to establish the adaptive utility of polarization sensitivity, but also to define the specific functions of the relevant detecting and processing mechanisms, and that "the rather elegant hypotheses of sensory mechanisms currently being elucidated or postulated, therefore, may seem a bit top heavy for the field's somewhat makeshift foundation!" (Waterman 1981). Work reported in this volume provides a modest, but significant, contribution to our knowledge of polarization sensitivity of several nocturnally active insects as it relates to species other than the social Hymenoptera. The significance of further research on these lines need not be stressed.

As migration involves specialized behaviours and sustained travel, the process is extremely demanding and requires considerable changes in the insect's physiology. Physiological factors involved in migration include: the control of movement (flight) and the utilization of fuels mediated by the neuroendocrine system and the "oogenesis flight syndrome" (Johnson 1963, 1969). In the oogenesis flight syndrome the locomotor drive is optimal in young pre-reproductive adults with the metabolism switched to increased flight muscle growth and fat accumulation, and as the ovaries mature the locomotor drive disappears, the flight muscle apparatus autolyses and migration terminates (Danthanarayana 1970; Johnson 1969, 1976). These migratory and reproductive changes are closely coordinated by the JH-system in response to environmental cues (Caldwell and Rankin 1972; Rankin 1974, 1978). In Chap. 3 Rankin et al. reconsider the oogenesis flight syndrome. They point out that in species of highly migratory insects, JH stimulates both flight and oogenesis, although in some species JH can cause flight-muscle histolysis or have no effect on flight behaviour. They see the JH control of migratory behaviour as a mechanism evolved in response to selection associated with colonizing life-styles in which, ideally, both flight and reproductive parameters would be maximized. The implications of the stimulation of both flight and oogenesis are considered in their discussion.

In addition to the key role played by JH, several other hormone systems, those that are essential for the mobilization of fuels, are part and parcel of the physiology of insect flight and migration. At the onset of flight the energy requirements are met by the breakdown of the endogenous fuel (glycogen and proline) found in the flight muscle. Thereafter trehalose from the blood is the main fuel, and the sugars from the crop may also contribute; most of the energy required during long-term and sustained flights is obtained from the main stores of glycogen in the fat body and intestinal wall (Mordue et al. 1980). Mobilization of stored fuel reserves in the fat body has received much attention during the last several years. In Chap. 4 Goldsworthy and Wheeler discuss fuel mobilization via adipokinetic hormones and octopamine and the coordination of the fuel supply and its utilization in locusts. They compare the chemical nature and actions of the two known adipokinetic peptides, and discuss the actions of octopamine on the fat body and flight muscle, suggesting that it may play a vital role in maintaining the oxidation of glucose in the early stages of flight.

A form of energy transfer that happens during flight, but not quite linked to the process of migration and dispersal is that small part of the energy transmitted to the wings that gets converted into sound. In Chap. 5 Belton covers this aspect and considers the effects of the major environmental factors that influence wing-beat frequency. The value of wing-beat sounds to insects is for sexual attraction as in the case of female biting Diptera using wing-beat sounds to attract males. These sounds have been of value in radar studies of migration because wing-beat frequencies of individual insects or swarms can be detected by radar using the Doppler effect or differences in intensity of the echos (Buchan and Stelle 1979; Riley and Reynolds, this volume; Schaefer 1976).

Turning to the ecological factors of insect migration, the primary ecological characteristic of migration is the displacement of the insect from one habitat (breeding site) so that it arrives at another (Johnson 1960, 1969). Simultaneous pre-reproductive flights are characteristic of most migrants, but there are species that migrate between bouts of oviposition (Johnson 1969, 1974). Since migration is linked closely to habitat variations, the evolution of the migratory capacity has reached its climax in species living in impermanent habitats (Southwood 1962) and is a characteristic life-cycle strategy among r-selected species (Dingle 1972, 1974, 1981a). The initiation of migration is evoked by extrinsic and intrinsic mechanisms that can anticipate the arrival of adverse conditions. Intrinsic timing mechanisms such as lunar rhythms will not only permit initiation of migration, but synchronize settlement of individuals at points of termination. Extrinsic environmental cues also have similar roles, and selection will favour those individuals that can develop such timing mechanisms (Lidicker and Caldwell 1982).

Where alternate habitats are geographically distant and their suitability is seasonal and predictive as in temperate environments, photoperiod provides a very precise cue for the initiation of migration (Dingle 1972, 1980; Johnson 1969; Vepsäläinen 1978). In situations where the geographical separation of habitats varies from short to long distances and the time factor is less relevant, as in migrations of tropical insects or within-season migrations of temperate ones, cues other than photoperiod may be used. These include temperature, food availability and quality (Danthanarayana 1976; Dingle 1972, 1981a,b; Dixon and Howard, this volume; Vepsäläinen 1978), rainfall, natural enemies, and disease pressure (Dingle 1981b, Gatehouse, this volume) and intraspecific density effects (Taylor L.R. and Taylor R.A.J. 1977, 1983). The seasonal synchronization of insect life histories by migration and diapause, evoked by the various environmental cues in different geographical situations has been discussed by Dingle (1981b). Intraspecific population pressure may operate through social behaviour as in locusts where crowding induces the migratory phase transformation (Kennedy 1956, 1961b; Uvarov 1966) and the development of the alate morphs in aphids (Dixon 1973, this volume; Hughes 1963, 1974; Kennedy 1972). Similar social effects are known to occur even at very low densities as in some Lepidoptera where mutual interference between even two individuals resulting from low-density larval crowding will make one individual develop into a more flightworthy adult (Danthanarayana et al. 1982). The impact of migration as a density-dependent process for population regulation has been stressed recently by Taylor and Taylor (Taylor L.R. and Taylor R.A.J. 1977, 1983; Taylor R.A.J. and Taylor L.R. 1979). Their concept is encompassed in a mathematical model, the Δ-model, the components of which are migration (or repulsion) and congregation (or attraction) which produces the Δ-response in a negative feedback loop, and a quantitative assessment of migration potential is possible because the model can be simulated. In their purely spatial concept, animal movement is not sub-divided into categories other than migration and attraction. In this balancing of the association between density-dependent aggregation and migration, all movements of animals in space have the common function of regulating populations by constant re-distribution of populations in space and time. This allows the colonizers and re-colonizers to exploit the shifting heterogeneity of habitats as exemplified by the movements of East African armyworm populations (Gatehouse, Chap. 9). In Chap. 10 Dixon and Howard show that the variability in reproductive investment between individuals within an aphid clone is programmed. This intraclonal tactical diversity in reproductive investment, characteristic of the Aphidinae, is shown to be associated with differential dispersal which is an adaptation to living in, rather than developing in, a heterogeneous environment. This, they say, may account for the success of most members of this sub-family.

In most insect species distance travel is achieved rapidly and economically by supplementing their own powers of flapping flight with the energy freely available from the wind (Johnson 1969; Pedgley 1982; Rainey 1951, 1976b; Taylor 1974). Mikkola, in Chap. 11, gives numerous examples of wind-dependent migrations into Finland. He shows that even among boundary-layer migrants wind seems to play a role in keeping the track of the migrant straight. Butterfly migration, often thought to be carried out without any wind assistance (Baker 1978; Walker and Riodan 1981), is shown to have an "above-country and resource-independent long-range flight" dependent on winds in addition to the frequently observed "cross-country progress" during which feeding and reproduction occurs. This is very evident from the work of Gibo (Chap. 12) on monarch butterflies, which use energy-saving mechanisms of soaring in thermals and flying at high altitudes. This enables them to escape from high latitudes as quickly as possible before they become trapped by the onset of deteriorating weather. It is suggested that selection will tend to favour energy-efficient displacement and rapid escape in the southern Ontario populations. Many insect migrations are thus synchronized to annual climatic cycles and the resulting seasonal changes in weather. Meteorological aspects of insect migrations have been well reviewed recently by Gauthreaux (1980), Pedgeley (1982), and Rainey (1976b, 1982). Farrow (Chap. 13) shows that a net southward displacement of micro-insects in the temperate latitudes of Australia is caused by the characteristic pattern of synoptic weather, and the rotation of winds' trajectories with the passage of weather systems ensures extensive re-distribution of aerial populations in all directions over inland Australia. The greatest potential for this displacement occurs in northerly airflows at night above the planetary boundary layer. Farrow says that this type of displacement and re-distribution of aerial populations obviates the necessity of requiring specific 'return' migrations in autumn to compensate for the poleward migrations of spring and at other times. Such partly misdirected and wasteful products of migration can also have beneficial effects on the ecosystem. Edwards (Chap. 14) shows how arthropod fallout on the snowfields of Pacific Northwest volcanoes plays a role in the establishment of vegetation, and thus in the diversification of the ecosystem. In this rare example of 'primary succession in reverse', the initial colonizers are scavengers and predators subsisting on wind-borne arthropods. This arthropod fallout also yields suitable nutrients for primary plant colonists, contributing to further diversification.

Papers by Gatehouse (Chap. 9), Raulston et al. (Chap. 15), and Wolf et al. (Chap. 16) also emphasize the significance of weather transport systems in insect migrations, particularly the spread of pest species in the African and North American continents. Pheromone trap sampling and monitoring of population build-up in corn fields (Raulston et al. Chap. 15) show that initial populations of the fall armyworm, the corn earworm and the tobacco budworm migrate from south (north-eastern Mexico) to north (north and west Texas) in the spring. Synchrony of weather systems with the emerging populations in the south is conducive to these migrations. Wolf et al. (Chap. 16), with the aid of blacklight traps on offshore oil platforms and radar observations, show that insects could move significant distances over water and pest species such as the fall armyworm can cross the widest part of the Gulf of Mexico with favourable winds. These studies thus improve our increasing knowledge (e.g. Dingle 1981b; Greenbank et al. 1980; Gunn and Rainey 1979; Joyce 1981; Rabb and Kennedy 1979; Rainey 1976a) of migrant agricultural pests. Stinner et al. (Chap. 17) stress the desirability of predicting pest outbreaks. Two mathematical models are presented, one based on a highly mobile species and the other on a species with limited mobility. In Chap. 18 Stein reviews the occurrence of dispersal among arthropods of public health importance and the influence of biotic, abiotic, and trophic factors on dispersal. He points out that the amount of published work on the dispersal of public health pests is scanty, particularly on Siphonaptera, Hemiptera, and Acarina. Finally, in Chap. 19 Lingren et al. provide information on night-vision equipment and its use in observing nocturnally active insects which constitute a greater portion of the pest types. In the concluding chapter (Chap. 20) Taylor discusses the meaning of the word migration with respect to different animal taxa. He proposes an interesting new classification that defines four different kinds of migration and provides an acceptable rationale for its adoption.

References

Andrewartha HG, Birch LC (1954) The distribution and abundance of animals. University of Chicago Press, Chicago

Arendse MC (1978) Magnetic field detection is distinct from light detection in the invertebrates *Tenebrio* and *Talitrus*. Nature 274:358–362

Baker RR (1978) The evolutionary ecology of animal migration. Holmes and Meier, New York

Baker RR, Mather JG (1982) Magnetic compass sense in the large yellow underwing moth, *Noctua pronuba* L. Anim Behav 30:543–548

Buchan PB, Stelle DB (1979) A radar-Doppler autocorrelation analysis of insect activity. Physiol Entomol 4:103–109

Caldwell RL, Rankin MA (1972) Effects of a juvenile hormone mimic on flight in the milkweed bug, *Oncopeltus fasciatus*. Gen Comp Endocrinol 19:601–605

Cloudsley-Thompson JL (1980) Biological clocks: their functions in nature. Weidenfeld and Nicholson, London

Danthanarayana W (1970) Studies on the dispersal and migration of *Sitona regensteinensis* (Coleoptera:Curculionidae). Entomol Exp Appl 13:236–246

Danthanarayana W (1976) Environmentally cued size variation in the light brown apple moth, *Epiphyas postvittana* (Walk.) (Tortricidae), and its adaptive value in dispersal. Oecologia (Berl) 26: 121–132

Danthanarayana W, Hamilton G, Khoul SP (1982) Low density larval crowding in the light brown apple moth and its adaptive value in dispersal. Entomol Exp Appl 31:353–358

Dingle H (1972) Migration strategies of insects. Science 175:1327–1335

Dingle H (1974) The experimental analysis of migration and life history strategies in insects. In: Barton Browne L (ed) Experimental analysis of insect behaviour. Springer, Berlin Heidelberg New York

Dingle H (ed) (1978) Evolution of insect migration and diapause. Springer, Berlin Heidelberg New York

Dingle H (1980) Ecology and evolution of migration. In: Gauthreaux SA Jr (ed) Animal migration, orientation, and navigation. Academic Press, New York

Dingle H (1981a) Geographic variation and behavioural flexibility in milkweed bug life histories. In: Denno RF, Dingle H (eds) Insect life history patterns: habitat and geographic variation. Springer, Berlin Heidelberg New York

Dingle H (1981b) Function of migration in the seasonal synchronization of insects. Entomol Exp Appl 31:36–48

Dixon AFG (1973) Biology of aphids. Arnold, London

Dyer FC, Gould JL (1981) Honey bee orientation: a backup system for cloudy days. Science 214: 1041–1042

Elton C (1930) Animal ecology and evolution. Clarendon, Oxford

Evans WG, Kusher JE (1980) The infrared perceptive fields of *Melanophila acuminata* (Coleoptera: Buprestidae). Can Entomol 112:211–216

Gauthreaux SA Jr (1980) The influence of long-term and short-term climatic changes on the dispersal and migration of organisms. In: Gauthreaux SA Jr (ed) Animal migration, orientation, and navigation. Academic Press, New York

Gould JL (1980) The case for magnetic sensitivity in birds and bees (such as it is). Am Sci 68:256–267

Gould JL (1982) Ethology: the mechanism of evolution and behaviour. Norton, New York

Gould JL, Kirschvink JL, Deffeyes KS (1978) Bees have magnetic remanence. Science 201:1026–1028

Greenbank DO, Schaefer GW, Rainey RC (1980) Spruce budworm (Lepidoptera:Tortricidae) moth flight and dispersal: new understanding from canopy observations, radar and aircraft. Mem Entomol Soc Can 110:1–49

Gunn DL, Rainey RC (1979) Strategy and tactics of control of migrant pests. Philos Trans R Soc Lond B Biol Sci 287:245–288

Hughes RD (1963) Population dynamics of the cabbage aphid, *Brevicoryne brassicae* (L.). J Anim Ecol 32:393–424

Hughes RD (1974) Living insects. Collins, Sydney

Johnson CG (1960) A basis for a general system of insect migration and dispersal by flight. Nature 186:348–350

Johnson CG (1963) Physiological factors in insect migration by flight. Nature 198:423–427

Johnson CG (1969) Migration and dispersal of insects by flight. Methuen, London

Johnson CG (1974) Insect migration: aspects of its physiology. In: Rockstein M (ed) The physiology of the insecta 3:279–476. Academic Press, New York

Johnson CG (1976) Lability of the flight system: a context for functional adaptation. In: Rainey RC (ed) Insect flight. Blackwell, Oxford

Joyce RJV (1981) The control of migrant pests. In: Aidley DJ (ed) Animal migration. Cambridge University Press, Cambridge

Kennedy JS (1956) Phase transformation in Locust biology. Biol Rev 31:349–370

Kennedy JS (1961a) A turning point in the study of insect migration. Nature 189:785–791

Kennedy JS (1961b) Continuous polymorphism in Locusts. Symp R Entomol Soc Lond 1:80–90

Kennedy JS (1972) The emergence of behaviour. J Aust Entomol Soc 11:168–176

Kennedy JS (1975) Insect dispersal. In: Pimental D (ed) Insects, science, and society. Academic Press, New York

Kirschvink JL (1983) Biomagnetic geomagnetism. Rev Geophys 21:672–675

Lidicker WZ Jr, Caldwell RL (1982) Editor's comments on papers 11 through 16. In: Lidicker WZ Jr, Caldwell RL (eds) Dispersal and migration. Hutchinson Ross, Pennsylvania

Martin H, Lindauer M (1977) The effect of the earth's magnetic field on gravity orientation in the honey bee, *Apis mellifera*. J Comp Physiol 122:145–187

Meyer JR (1977) Head capsule transmission of long wavelength flight in the Curculionidae. Science 196:524–525

Mordue W, Goldsworthy GJ, Brady J, Blaney WM (1980) Insect physiology. Blackwell, Oxford

Naylor E (1982) Tidal and lunar rhythms in animals and plants. In: Brady J (ed) Biological timekeeping. Cambridge University Press, Cambridge

Neumann D (1981) Tidal and lunar rhythms. In: Aschoff J (ed) Handbook of behavioural neurobiology, vol 4. Plenum, New York, pp 351–380

Pedgley D (1982) Windborne pests and diseases: meteorology of airborne organisms. Horwood, Chichester

Provost MW (1952) The dispersal of *Aedes taeniorhynchus* I. Preliminary studies. Mosq News 12: 174–190

Rabb RL, Kennedy GG (1979) Movement of highly mobile insects: concepts and methodology in research. North Carolina State University Press, Raleigh, NC

Rainey RC (1951) Weather and movements of locust swarms: a new hypothesis. Nature 168:1057–1173

Rainey RC (ed) (1976a) Insect flight. Symp R Entomol Soc Lond, vol 7. Blackwell, Oxford

Rainey RC (1976b) Flight behaviour and features of the atmospheric environment. In: Rainey RC (ed) Insect flight. Symp R Entomol Soc Lond 7:75–112

Rainey (1982) Putting insects on the map: spacial inhomogeneity and the dynamics of insect populations. Antenna 6:162–169

Rankin MA (1974) The hormonal control of flight in the milkweed bug, *Oncopeltus fasciatus*. In: Barton Brown L (ed) Experimental analysis of insect behaviour. Springer, Berlin Heidelberg New York

Rankin MA (1978) Hormonal control of insect migration. In: Dingle H (ed) Evolution of insect migration and diapause in insects. Springer, Berlin Heidelberg New York

Reynolds DR, Riley JR (1979) Radar observations of concentrations of insects above a river in Mali, West Africa. Ecol Entomol 4:161–174

Riley JR (1975) Collective orientation in night flying insects. Nature 253:113–114

Rounds HD (1975) A lunar rhythm in the occurrence of bloodborne factors in cockroaches, mice and men. Comp Biochem Physiol C Comp Pharmacol 50:193–197

Rounds HD (1981) Semi-lunar cyclicity of neurotransmitter-like substances in the CNS of *Periplaneta americana* (L.). Comp Biochem Physiol C Comp Pharmacol 69:293–299

Rounds HD (1982) A semi-lunar periodicity of neurotransmitter-like substances from plants. Physiol Plant 54:495–499

Schaefer GW (1976) Radar observations of insect flight. In: Rainey RC (ed) Insect flight. Symp R Entomol Soc Lond 7:157–197

Schneider F (1962) Dispersal and migration. Annu Rev Entomol 7:223–242

Sotthibandhu S, Baker RR (1979) Celestial orientation by the large yellow underwing moth, *Noctua pronuba*. Anim Behav 27:786–800

Southwood TRE (1962) Migration of terrestrial arthropods in relation to habitat. Biol Rev 37: 171–214

Southwood TRE (1975) The dynamics of insect populations. In: Pimental D (ed) Insects, science, and society. Academic Press, New York

Southwood TRE (1981) Ecological aspects of insect migration. In: Aidley DJ (ed) Animal migration. Cambridge University Press, Cambridge

Taylor LR (1958) Aphid dispersal and diurnal periodicity. Proc Linn Soc Lond 169:67–73

Taylor LR (1960) The distribution of insects at low levels in the air. J Anim Ecol 29:45–63

Taylor LR (1974) Insect migration, flight periodicity and the boundary layer. J Anim Ecol 43: 225–238

Taylor LR, Taylor RAJ (1977) Aggregation, migration, and population mechanics. Nature 265: 415–421

Taylor LR, Taylor RAJ (1983) Insect migration as a paradigm for survival. In: Swingland IR, Greenwood PW (eds) The ecology of animal movement. Clarendon, Oxford

Taylor RAJ, Taylor LR (1979) A behavioural model for the evolution of spacial dynamics. In: Anderson RM, Turner BD (eds) Population dynamics. Blackwell, Oxford

Urquhart FA, Urquhart NR (1977) Overwintering areas and migratory routes of the monarch butterfly (*Danaus p. plexippus*, Lepidoptera:Danaidae) in North America with special reference to the western population. Can Entomol 109:1583–1589

Urquhart FA, Urquhart NR (1978) Autumnal migration routes of the eastern populations of the monarch butterfly (*Danaus plexippus* L., Danaidae; Lepidoptera) in North American to the overwintering site in the Neovolcanic Plateau of Mexico. Can J Zool 56:1759–1764

Urquhart FA, Urquhart NR (1979) Vernal migration of the monarch butterfly (*Danaus p. plexippus*, Lepidoptera:Danaidae) in North America from the overwintering site in the neo-volcanic plateau of Mexico. Can Entomol 111:15–18

Uvarov B (1966) Grasshoppers and locusts, vol 1. Cambridge University Press, Cambridge

Vepsäläinen K (1978) Wing dimorphism and diapause in *Gerris*: determination and adaptive significance. In: Dingle H (ed) Evolution of insect, migration, and diapause. Springer, Berlin Heidelberg New York

Walker TJ, Riodan AJ (1981) Butterfly migration: are synoptic-scale wind systems important? Ecol Entomol 6:433–440

Waloff Z (1972) Orientation of flying locusts, *Schistocerca gregaria* (Forsk.) in migrating swarms. Bull Entomol Res 62:1–72

Waterman TH (1981) Polarization sensitivity. In: Autrum H (ed) Handbook of sensory physiology, vol 7/6. Springer, Berlin Heidelberg New York, pp 281–469

Wehner R (1972) Visual orientation performances of desert ants *(Catoglyphis bicolor)* towards astronomagnetic directions and horizontal landmarks. In: Galler SR, Schmidt-Koenig K, Jacobs GJ, Bellville RE (eds) Animal orientation and navigation. NASA, Washington

Wehner R (1976) Polarized-light navigation by insects. Sci Am 235:106–115

Wehner R (1983) Celestial and terrestrial navigation: Human strategies – insect strategies. In: Huber F, Markl H (eds) Behavioural physiology and neuroethology. Springer, Berlin Heidelberg New York

Wehner R (1984) Astronavigation in insects. Annu Rev Entomol 29:277–298

Wellington WG (1974a) Changes in mosquito flight associated with natural changes in polarized light. Can Entomol 106:941–948

Wellington WG (1974b) A special light to steer by. Nat Hist 83:46–53

Wellington WG (1976) Applying behavioural studies in entomological problems. In: Anderson JF, Kaya HK (eds) Perspectives in forest entomology. Academic Press, New York

Williams CB (1958) Insect migration. Collins, London

2 Evolution and Genetics of Insect Migration

H. DINGLE [1]

1 Introduction

The last few decades have seen the firm establishment of the notion that insect migration represents an adaptive syndrome. It was not always so. In the decade before the Second World War most entomologists thought large-scale movements of individuals between populations were rare. If such movements did occur, they were usually regarded as pathological attempts to relieve population pressure (see Lidicker and Caldwell 1982). There were of course exceptions which did not pass unnoticed, such as the spectacular movements of locusts and butterflies (Williams 1958), but on the whole most movement was thought to be a "rather quiet humdrum process" resulting from "the normal life of animals" (Elton 1927, p. 148). The perspective has changed dramatically with migration now seen as a fundamental element in the framework of an insect's life history and ecology (Johnson 1969; Taylor and Taylor 1983; Dingle 1984a).

There is enormous variation both within and among species in the way that natural selection has shaped this relation between migratory behavior and the structure of life histories. But running through the diversity is a common thread of adaptation to shifting environments (Southwood 1962, 1977; Vepsäläinen 1978; Dingle 1979, 1980, 1984a). As habitats deteriorate or become overcrowded, insects leave and move to new areas so that there is a shifting mosaic of departure from and invasion of habitats through time – the now famous "fern stele" of Taylor and Taylor (1977). Further, as Kennedy (1985) has again recently emphasized, these movements often involve specialized behavior that renders migrating individuals physiologically distinct from their nonmigrating counterparts. During migration, feeding and reproduction are suppressed by locomotory activity, but this suppression simultaneously primes them for action at the termination of migration. The ultimate relation between migration and life histories is that migratory behavior allows the individual choice of both place and time of breeding and so is a major component of fitness (Dingle 1984a).

We are thus beginning to grasp the critical features of the environmental "templet" (Southwood 1977) which mould the evolution of migratory patterns. What we need to fully understand this evolution, however, and what we have only just begun to assess, is information on the genetic basis for migration. The influence of genes is of fundamental importance, for if the templet of natural selection is to result in the evolution of migration and its appropriate "fit" to habitats and life histories, there must be a

---

[1] Department of Entomology, University of California, Davis, CA 95616, USA.

Insect Flight: Dispersal and Migration
Edited by W. Danthanarayana
© Springer-Verlag Berlin Heidelberg 1986

genetic basis for differences among individuals. But beyond simply knowing whether genes are involved, we need to know what sorts of genetic structure are available for selection and hence what the potential response to selection might be, and what constraints might derive from genetic correlations among traits or from genotype x environment interactions within traits across environments (Via 1984). All are important to comprehending the "substructure of fitness" (Istock 1978) with respect to migration. I shall explore these issues here and attempt to summarize briefly what we know of the genetic mechanisms underlying migratory behavior itself and of the genetic correlations between migration and suites of life-history traits. In doing so, I shall consider both theory and data and briefly discuss some of the methods available from genetics for analyzing continuous characters such as migratory behavior.

2 Genetic Variance and Genetic Correlation

In some cases that I shall discuss, gene influences on migration can be attributed to alleles at one or a few loci behaving in Mendelian fashion. In most cases, however, the relevant traits are polygenic and continuously varying and so do not segregate according to Mendel's classical laws. Rather, the appropriate analyses of such traits require the techniques of quantitative genetics. These are covered lucidly and in detail in Falconer (1981), so I shall briefly summarize here only some of those necessary to understand the particular cases I shall review.

The basic model of quantitative genetics assumes that individual differences in metric traits are the result of both genetic and environmental influences which can be expressed as

$$V_P = V_A + V_N + V_E \ ,$$

where V_A is the polygenic or additive genetic variance, V_N is the non-additive genetic variance, and V_E is the environmental variance. V_A arises from the average influences of all genes contributing to a character and is the variance involving specifically the resemblance between offspring and their parents (as distinct from resemblance to other members of the population); because of this, it determines sensitivity to selection and, hence, the maximum rate at which evolution can occur. V_N includes effects due to dominance and epistasis as well as effects resulting from the fact that genes can behave differently in different environments (genotype X environment interaction). V_E is the variance due to environmental fluctuations. If one is trying to estimate genetic variances in the laboratory, one usually attempts to maintain a constant environment so that $V_E = 0$ (and there are no G x E interactions). All of these variances sum to V_P which is the total or phenotypic variance in the population being considered.

We are particularly interested in V_A because of its bearing on rates of evolution. The relation between V_A and V_P is usually expressed as the heritability (h^2) of the trait, where $h^2 = V_A/V_P$. Estimating h^2 gives us an estimate of V_A and can be done in several ways. One is to take advantage of the fact that V_A involves offspring-parent resemblance and to estimate h^2 by offspring on parent regression, where the slope of the regression is a good estimator of h^2. Similarly one can estimate h^2 by selection where, again because of offspring-parent resemblance, the response to selection (R) is a function of the selection differential (S) and the heritability, so that $R = h^2 S$ and $h^2 = R/S$ (often called the "realized heritability"). Estimates of h^2 can also be derived from resemblances among kin, but are less accurate because they are confounded by V_N. (Note: I have been discussing what is sometimes called "narrow sense" heritability. The "broad sense" heritability is the ratio of the combined V_A and V_N to V_P.)

Metric characters may also be correlated (e.g., body size and clutch size) with the causes of correlation being both genetic and environmental. This is expressed as

$$r_P = h_x h_y r_A + e_x e_y r_E ,$$

where r_P is the phenotypic correlation between traits X and Y, r_A and r_E are the additive genetic and environmental correlations, respectively, h_x and h_y are the respective square roots of the heritabilities, and $e^2 = 1 - h^2$ (and so also includes nonadditive genetic variance). The genetic correlation indicates the extent to which genes influence the traits in common (usually due to pleiotropy) with the result that when selection influences one trait, it must influence the other as well. As a result, genetic correlations can constrain the outcome of selection (Lande 1976; Istock 1983). The parent-offspring regression and selection can also be used to estimate r_A (although with selection, the correlation can only be determined with respect to the trait under selection) with methods analogous to those discussed for estimating h^2. In this case, the appropriate traits are X in the parents and Y in the offspring. As will be evident in what follows, both genetic variance and genetic correlation are important to understanding the evolution of migration and its role in life histories.

3 Theoretical Models

Most models of migration do not consider genetics explicitly. For example, Cohen (1967) and Parker and Stuart (1976) consider migratory "strategies" from the perspective of optimization theory. Cohen examines migration as a problem of optimal choice between randomly varying alternatives with migration and remaining sedentary as possible "pure strategies" at opposite ends of an environmental continuum of varying risks. Mixed strategies are likely when survivorship probabilities of migrants and nonmigrants are independent and environmental variance is high. Both Cohen (1967) and Ziegler (1976) have suggested that genetic determinism for migration, as opposed to responding to environmental cues such as photoperiod, is favored where environments are more unpredictable, but do not consider what the genetic mechanisms might be. Parker and Stuart (1976) use the concept of evolutionarily stable strategies (ESS's) to consider migration as a function of optimal investment, accumulated gain, and between patch search costs. As in Cohen's model various pure and mixed strategies are possible depending on factors such as competition, resource sharing, and ability to assess resources.

Roff (1975) has used computer simulations to analyze migration specifically with respect to the underlying genetic mechanisms. Roff examined migration with respect to habitat stability which was a function of the mean and variance of the rate of increase of the population $[\mu (\lambda), \sigma^2 (\lambda)]$, and the mean and variance of the carrying capacity $[\mu (K), \sigma^2 (K)]$. Stability increased with increasing means or decreasing variances. Migration was assumed to be either: (1) not under genetic control; (2) determined by a

Table 1. Effects of increased "environmental stability" on the proportion of "dispersers" maintained in a population (Roff 1975)

Parameter altered	Genetic model[a]					
	"Simple"			Quantitative		
	D.I.	D.D.	D.K.D.	D.I.	D.D.	D.K.D.
Increase mean rate of increase (λ)	+	+	+	+	+	+
Decrease σ^2 (λ)	−	−	+	−	−	?
Increase carrying capacity (K)	−	−	−	−	+	?
Decrease σ^2 (K)	−	−	−	−	?	−

[a] D.I. = density independent; D.D. = density dependent; D.K.D. = density and carrying capacity dependent.

single gene with two alleles; or (3) determined polygenically (quantitatively). Case 1 gave little insight into the conditions permitting a stable dispersal polymorphism to exist. In the simple genetic model a stable polymorphism is possible if the allele for migration is dominant; quantitative inheritance can also lead to a stable polymorphism. As σ^2 (K) decreased so did migration, a result found in Southwood's (1962) survey of insect migration and by Vepsäläinen (1978) in his analysis of gerrids. Increasing environmental stability caused by raising μ (λ) did not, however, decrease migration, so the influence of stability on migration can vary according to circumstances. Some of the results are summarized in Table 1. One further interesting result was that if habitats remain stable for long periods, there will be a gradual loss of dispersing genotypes as migrants leave and are not replaced. This is consistent with the frequent observation of aptery or brachyptery in isolated habitats (reviewed in Dingle 1980, 1984b). An important conclusion coming from Roff's models is that the proportion of migrants maintained in a population depends on both the habitat parameter which varies and the mode of inheritance of the trait (Table 1).

4 Variation in Flight or Migratory Behavior

4.1 Population Variation

At the simplest level one can demonstrate that gene differences influence a trait if observed variation persists when the organisms in question are reared under identical conditions. As an example, the milkweed bug, *Oncopeltus fasciatus* (Dallas), displays considerable geographic variation both in migration and in the duration of tethered flight in the laboratory (Dingle et al. 1980; Dingle 1984a). Migratory northern populations display a relatively high proportion of long-duration laboratory flights with approximately 25% flying for 30 min or longer. Tropical bugs which do not migrate for long distances showed little long-duration tethered flight with less than 5% flying for 30 min or more. Since laboratory environments were the same for all bugs, differences were presumably the consequence of genotype.

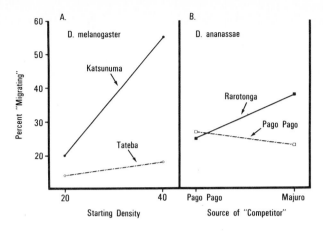

Fig. 1A,B. The "migration" responses of **(A)** two populations of *Drosophila melanogaster* from Japan at different densities and **(B)** two populations of *D. ananassae* when placed with a "competitor" species from the islands indicated. The index of migration was movement between population tubes in the laboratory. Note population (genotype) x environment interaction in each case. A after Sakai et al. (1957); B after Narise (1966)

Additional examples of population differences come from studies of *Drosophila* from Japan and various Pacific Islands (Fig. 1). Migration here was measured as the number of flies to move from one "population tube" to another in the laboratory which presumably reflected migratory tendency in the field. Sakai et al. (1957) examined migration in six wild strains of *D. melanogaster* (Meig.) from sites in Japan as a function of population density. Partial results for two strains are shown in Fig. 1A which indicates not only absolute differences between the populations, but also a population (genotype) x environment interaction with a much greater response in the Katsunuma population over the range of densities indicated. Figure 1B shows the responses of two populations of *D. ananassae* Bock and Wheeler when mixed with a "competitor" sibling species. Here again a population x environment interaction is evident with the Rarotonga strain responding strongly to the presence of the Majuro competitor, while the Pago Pago strain displays little if any change in behavior (Narise 1966).

4.2 Single-Locus Effects

In view of the complexity and multiplicity of factors which enter into migratory flight, it is hardly surprising that cases of single-locus influences are rare. In fact I know of no case where the primary influence of a Mendelian locus has been shown to be on some aspect of flight itself. There are, however, cases where single loci, usually of mutants rare in nature, influence migratory behavior via pleiotropic effects. One such is the yellow eye (y) mutant in the blowfly *Phaenicia sericata* Meigen (Chabora 1969). The measure of "emigration behavior" was the number of flies moving from one population cage to another via a narrow tube. Flies bearing either the y/y or y/+ genotype emigrated more than wild-type flies, apparently because they were photonegative and were more likely to seek out the darker exit hole where the tube left the cage (Table 2). Such is certainly the case in *Musca domestica* (L.) where y/y displayed higher emigration rates than heterozygotes or wild types only at high light intensities, showing a gene x environment interaction (Chabora and Kessler 1977). The influence on emigration is thus shown to be a pleiotropic effect of a gene which influences sensitivity to light.

Table 2. Influence of the yellow eye (Y) gene on the emigration behavior of the fly *Phaenicia sericata* (Chabora 1969)

Comparison	Hours after start	Mean number emigrating per replicate	
		y/y	+/+
y/y vs +/+	6	3.8	1.9[a]
	18	9.3	4.5[b]
	40	12.3	6.3[b]
		y/+	+/+
y/+ vs +/+	7	1.4	0.5
	19	5.9	2.3[a]
	38	8.4	3.5[b]

[a] $P < 0.05$; [b] $P < 0.01$.

A different and potentially interesting approach to the influences of single loci on migration has been the analysis of variation in the α-glycerophosphate dehydrogenase (α-Gpdh) locus. α-Gpdh is of particular importance because it is one of two major enzymes in a metabolic pathway which maintains glycolytic flux and produces ATP needed for flight (Sacktor 1974). In *D. melanogaster*, mutants deficient in α-Gpdh cannot sustain flight (O'Brien and MacIntyre 1972). Across the genus *Drosophila*, however, there is little variation at this locus; 173 of 175 species examined are monomorphic (Lakovarra and Keranen 1980). Oakeshott et al. (1982) did find some geographic variation in the frequency of "fast" and "slow" forms of the enzyme in *D. melanogaster*, but there was no obvious relation to flight behavior. Zera (1981) surveyed 11 species of waterstriders (Gerridae:Hemiptera) which showed considerable intra- and interspecific variation in the degree of wing polymorphism. Even though average number of observed electromorphs per species was high (5.36 ± 0.96), well above values for other insects, there was no significant association between α-Gpdh type and wing morph for any of the populations examined. Thus, any specific association between enzyme activity and migration has yet to be demonstrated.

4.3 Polygenic Influences

Even though only a few species have been studied, it is already evident that much of the genetic contribution to flight is a consequence of polygenic influences. Selection experiments are the most direct way to reveal these effects. Dingle (1968), for example, applied intense selection for long duration tethered flight to a migratory population of the milkweed bug, *Oncopeltus fasciatus*, and raised the proportion of bugs making flights of 30 min or longer from about 25% to over 60% in a single generation. Similarly, Sharp et al. (1983) increased the proportions of Caribbean and Mediterranean fruit flies, *Anastrepha suspensa* (Loew) and *Ceratitis capitata* (Wiedemann), respectively, flying from holding to catching containers from 1.7- to 3-fold in various laboratory and wild strains. Using offspring on parent regression, Caldwell and Hegmann (1969) estimated the heritabilities for duration of tethered flight in the lygaeid bug, *Lygaeus kalmii* Stål, at about 0.20 when offspring durations were regressed on male parents and about 0.40 when regressed on female parents, suggesting environmental influences due to maternal effects as well as additive gene influences.

In addition to duration, timing is also important to migration; when one departs can be as important as how far one travels (Ziegler 1976). Timing has been examined in an elegant selection experiment by Rankin (1978) in *Oncopeltus fasciatus*. Earlier work had demonstrated that intermediate titers of juvenile hormone promoted flight in this species and that some bugs delayed flight for several days relative to the rest of the population. By selecting for these delaying individuals, Rankin produced a population of late-flying individuals which also showed two correlated responses. First, oviposition was delayed, suggesting at least a partial reproductive diapause, and secondly, juvenile hormone titers remained at intermediate levels in the interval before the initiation of flight instead of rising to levels promoting egg development and oviposition as was usually the case. Hormone levels rose appropriately to promote reproduction following flight. Genes influencing migration thus appear to do so via their action on juvenile hormone. Rankin has thus demonstrated genetic and physiological relationships coordinating a set of life-history responses which include migration.

The genetics of flight physiology have also been examined by Curtsinger and Laurie-Ahlberg (1981) in *Drosophila melanogaster*. They estimated broad-sense heritabilities, i.e., without separating the contributions of V_A and V_N, and genetic correlations among various flight-related traits using isogenic lines of flies homozygous for second chromosomes derived from different geographic areas. Their results are briefly summarized in Table 3. Heritability estimates usually indicated between 30% and 40% genetic variance for traits such as wing-beat frequency or power output during flight. The major exception is wing-stroke amplitude which has a smaller estimated genetic variance, but this estimate is likely confounded by the fact that amplitude also varies as a consequence of attempts by the flies to steer. For similar reasons correlations of wing-stroke amplitude with other traits were not demonstrated. Genetic correlations were evident for wing-beat frequency with wing area and total power and for area and power. Frequency, area, and power were all correlated with live weight, but line differences in weight did not account for the other correlations observed. Chromosomes derived from different natural populations, therefore, contribute to genetic variation in flight characteristics. The robustness and consistency of measurements within lines indicate that the technique

Table 3. Broad-sense heritabilities (leading diagonal) and line (genetic) correlations (± SE) for flight-related traits in *Drosophila melanogaster* (Curtsinger and Laurie-Ahlberg 1981)

	WBF	Amp	A	Pt	Wt
Wing-beat frequency (WBF)	0.39	0.03 ± 0.24	0.48 ± 0.18	0.99 ± 0.01	0.57 ± 0.16
Wing-stroke amplitude (Amp)		0.16	−0.25 ± 0.23	−0.09 ± 0.24	−0.24 ± 0.23
Wing area (A)			0.33	0.45 ± 0.19	0.62 ± 0.14
Total power (P_t)				0.33	0.58 ± 0.16
Live weight (W_t)					—

of tethered flight can be used to analyze gene differences in flight physiology. Exactly what these differences might mean in terms of migration would seem a worthy subject for further exploration.

5 Variation in Wing Form

Wing polymorphism is one of the most conspicuous ways in which insects respond to environmental heterogeneity. In general, winglessness occurs in stable or isolated habitats, while macroptery occurs where habitats are transient or fluctuating (Southwood 1962; Vepsäläinen 1978; Dingle 1980). Since both forms occur in many species, there has been much interest in the interactions between environmental and genetic factors influencing the polymorphism. Among environmental influences, photoperiod, temperature, and crowding are all prominent (Harrison 1980; Dingle 1984b). But in addition to these environmental factors, genes and genotype x environment interactions also influence the appearance of the various wing morphs.

Aphids are among the most apparent of wing polymorphic insects, although problems with breeding and survival (as opposed to simply rearing parthenogenetic clones) have limited genetic studies (Blackman 1971). Nevertheless, it has been shown that different clones (genotypes) vary in their tendencies to produce alate morphs. Lamb and MacKay (1979, 1983) examined field-collected clones of *Acyrthosiphon pisum* Harris and demonstrated genetic differences in sensitivity to stimuli inducing alate production. Within clones across generations the responses were temporally stable with a highly significant correlation of independent measures of alate production among groups of sisters. Clonal genotype was thus a major contributor to frequency of alates. These results confirmed those of Sutherland (1969a,b, 1970) who demonstrated that laboratory clones varied in wing morph frequency in response to either crowding or host-leaf quality. In *Myzus persicae* L. short days induce a winged sexual generation. Where winters are mild, however, clones may vary; some remain parthenogenetic, some produce all sexual offspring, and others produce mixed broods of sexual and asexual females (Blackman 1971). Crosses between an intermediate clone with only a partial sexual generation and an all sexual clone displayed what seemed to be relatively simple Mendelian segregation for the sexual-asexual polymorphism.

Evidence for Mendelian segregation has also been found in wing polymorphic beetles. Jackson (1928) crossed individuals from macropterous and 95% brachypterous lines of the weevil *Sitona hispidula* (Fab.) with the proportions of the two morphs in the F_1 indicating segregation of Mendelian units with brachyptery dominant. A similar situation may be the case in the carabid beetle *Pterostichus anthracinus* Ill. (Lindroth 1946).

Single pair crosses also revealed a Mendelian unit influencing brachyptery in the milkweed bug, *Oncopeltus fasciatus* (Klausner et al. 1981). The brachypterous form was found in a laboratory population originally collected on the Caribbean island of Guadeloupe. Crosses between brachypters and normal bugs in two temperatures and two photoperiods revealed that the brachyptery was recessive in contrast to the situation in beetles, although polygenic modifiers also influenced the trait as there was considerable variation around the "typical" brachypterous form. This suggests that a single mutation can produce brachyptery, and that under appropriate conditions it could evolve quickly. Whether such single mutations are in fact of significance in the field is at present unknown. A further search in the natural population for brachypters or heterozygotes carrying the putative recessive allele failed to turn up any bugs bearing the trait.

Complexity in the genetic mechanisms underlying wing polymorphism seems to be the rule in waterstriders (Hemiptera:Gerridae) and in crickets. With respect to gerrids, Vepsäläinen (1974, 1978) suggested that wing form was Mendelian with short wings dominant, but the situation needs reevaluation since the mating histories of many of the females used in the test crosses were not known (reviewed in Harrison 1980). In a recent controlled study of the North American gerrid, *Limnoporus canaliculatus* (Say), Zera et al. (1983) found that in constant long days about 60% of the offspring of long-wing x long-wing crosses were macropterous, but that even with a short-wing x short-wing cross, there were about 6% macropterous offspring. If both sets of crosses were carried out in a short and decreasing photoperiod, the proportion of macropters increased to about 85% and 25%, respectively, demonstrating genotype by environment interactions. In the field long-winged individuals generally increase in frequency in the autumn (short days) in preparation for migration to diapause sites. Similar results have been found in crickets. In *Gryllus pennsylvanicus* Burmeister about 70% of the offspring are long-winged in long-wing crosses, while about 15% are long-winged in short-wing crosses (Harrison 1979). In contrast *G. firmus* Scudder produced only 20% long-winged offspring in long-wing crosses and reduced the frequency to only about 5% in short x short (Roff 1984).

There has as yet been little attempt to apply even the simplest methods of quantitative genetics to the problem of wing polymorphism. Mahmud (1980) did succeed in increasing the proportion of macropters in the planthopper *Laodelphax striatellus* (Fallèn) between the first and second progeny generations of long-wing crosses, but there was no further change in the third generation. In an interesting study Honek (1976a,b, 1979) selected for macroptery in the pyrrhocorid bug, *Pyrrhocoris apterus* (L.). In natural populations virtually 100% of the bugs are brachypterous in short days and enter diapause, while in long days about 20% are macropterus. With selection these proportions changed to about 90% macroptery in long days and 70% in short days after 29 generations. The short-day macropters still diapaused, so selection on wing length did not alter the diapause response. This is noteworthy because in *Oncopeltus* and *Dysdercus* there was also failure to find a correlation between age at first reproduction and other traits (see below). *P. apterus* is particularly interesting because macropters are flightless and in fact show little if any flight muscle development. The action of natural selection on wing form in this species is certainly not understood.

In fact our understanding of the respective roles of genes and environment in producing wing polymorphisms is cursory at best. It is apparent from the above studies that the genetic component of wing morph determination is complex. But the extent of the variation among different species and populations with respect to major gene effects or the influence of polygenes is simply not known at present. Because of their considerable potential for yielding insights into evolutionary mechanisms, the genetics, ecology, and physiology of wing polymorphisms should richly reward future study.

6 Migration and Life Histories

Migration allows insects to escape from unfavorable conditions and to colonize new habitats. It is thus intimately tied to life histories, and one would expect natural selection to produce a suite of life-history traits which co-evolve with migration and lead to successful colonizations. The traits that define colonizers are likely to involve rapid development, early reproduction, and high fecundities (Safriel and Ritte 1980; Dingle 1984a,b), and these characters should be associated with migratory capabilities. Of primary interest here is whether such associations exist as a colonizing syndrome involving genetic correlations.

Two independent sets of selection experiments have been used to assess colonizing syndromes in *Tribolium* (Ritte and Lavie 1977; Lavie and Ritte 1978; Wu 1981; reviewed in Dingle 1984a). In both cases beetles were selected for high and low "dispersal" rates and then assessed for correlated responses in life-history traits after several generations. Beetles dispersed either by moving between vials of flour on pipe cleaners (Wu) or along strings (Lavie and Ritte). Ritte and Lavie (1977) crossed high and low dispersing lines of *T. castaneum* Herbst and found that the dispersal trait behaved as a Mendelian dominant. More interesting was the fact that beetles of the high line developed more rapidly, produced more eggs during the first 4 days of reproduction, and produced a higher proportion of fertile eggs (Lavie and Ritte 1978). Wu (1981) compared *T. castaneum* with its congener, *T. confusum* Duval, which is a species displaying little dispersal behavior. Her results for *T. castaneum* confirmed those of Lavie and Ritte (1978) and also demonstrated that the correlations were absent in *T. confusum*. A colonizing life-history syndrome was thus present only in the more highly dispersing species.

The genetic architecture underlying the life history of a migratory population of *Oncopeltus fasciatus* also suggests a colonizing syndrome. Hegmann and Dingle (1982) used an analysis which incorporated both full-sibling and half-sibling families and calculated heritabilities and genetic correlations for a number of traits which included wing length, development time, age at first reproduction, and measures of fecundity. They found that long-winged (large) bugs developed more rapidly and produced larger egg clutches as a consequence of genetic correlations among these traits. An additional interesting finding was that age at first reproduction was not correlated with these other traits; Derr (1980) had similar results with the pyrrhocorid bug, *Dysdercus bimaculatus* (Stål). This absence of correlation with age at first reproduction means that the bugs retain considerable flexibility with respect to the timing of reproduction and migration because this flexibility is not constrained by genetic correlations with other life-history traits. Selection can act on age at first reproduction without generating correlated responses in other aspects of the life history (reviewed in Dingle 1984a).

Palmer (1985; see also Dingle et al. 1984) has assessed the genetic correlations of tethered flight and various life-history traits with wing length in migratory *O. fasciatus* via bidirectional selection. Some of the results are summarized in Fig. 2. First, there was a strong response to selection on wing length in both directions with both short- and long-winged lines significantly different from controls in both replicates. Since heritability for wing length was estimated at 0.55 by Hegmann and Dingle (1982), this result was expected. Secondly, there was a positive correlation of duration of tethered flight with long wings, as the bugs of the long-winged line flew significantly longer than

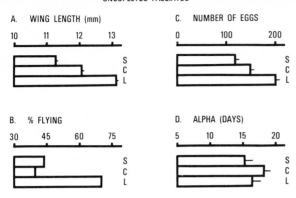

Fig. 2A–D. The direct response to selection on (**A**) wing length and correlated responses in (**B**) flight, (**C**) egg production in the first 5 days of oviposition, and (**D**) age at first reproduction (*alpha*) in the milkweed bug, *Oncopeltus fasciatus*. In **A, C,** and **D** means and standard errors are indicated in the short-wing (*S*), control (*C*), and long-wing (*L*) lines in the fifth generation of selection. In **B** the percent flying for 30 min or longer in tethered flight in the ninth generation of each line is indicated. Data from Palmer (1985) and Dingle et al. (1984)

controls. There was no demonstrable difference between controls and short-winged bugs. Thirdly, fecundity, as measured by the first 5 days of egg production, was also significantly greater in the long-winged lines. Finally, there was no correlation between wing length and age at first reproduction, again confirming predictions from the results of Hegmann and Dingle (1982). These results, as with those from the *Tribolium* experiments, strongly suggest a migration-colonization syndrome with a positive genetic correlation between flight and fecundity, but with flexibility of timing because age at first reproduction remains free of a "cost of correlation" (Dingle 1984a).

The genetic basis of life-history differences among various populations of *O. fasciatus* has been examined by Leslie (1986) using crosses among bugs of differing geographic origin. The populations were derived from individuals collected in California, Iowa, Maryland, Texas, Florida, and Puerto Rico. Two results are of particular interest here. First, multivariate analysis of an array of life-hsitory traits revealed that the six populations divided into four sets: California, Puerto Rico, Maryland and Iowa, and Texas and Florida. Given the geographic distances, it was not surprising that California and Puerto Rico separated from the remaining continental locations. The particular configuration of the other two subsets was, however, somewhat surprising. The implication is that rather than division into eastern and midwestern migratory populations moving north and south along their respective "flyways", the division in *O. fasciatus* is into a subset of more migratory genotypes which invade northern areas in the spring or even into a more or less separate migrant genetic unit which interbreeds little if at all with more southerly populations. Only much further research can distinguish among these possibilities.

The second result of interest was that for most traits the offspring of any cross involving a Puerto Rican parent tended to show dominance deviations toward that parent. In other words, the hybrids resembled the Puerto Rican mother *or* father more than the parent from the continental population in question. These dominance effects disappeared in the offspring of crosses between hybrid parents (these offspring were close to the mean between the original source populations) implying that the genetic basis of the traits was polygenic with averaging effects which became predominant in the second generation. The population in Puerto Rico may have been on the island for only about 450 years since the introduction of its primary host plant, *Asclepias curassavica* L. (Blakley and Dingle 1978). If this is true, it has evolved quite different migratory and reproductive habits in a relatively short time (Dingle 1981; Dingle et al. 1982; Dingle and Baldwin 1983; Leslie and Dingle 1983). Apparently, the divergence has resulted in a general "dominance" of the Puerto Rico genotype, although it should be noted that isolation has not as yet resulted in any loss of viability or fertility in the progeny of crosses between Puerto Rican and mainland bugs (Dingle et al. 1982; Leslie 1984).

7 Discussion

Even in the early embryonic form in which they now exist, studies of the genetic substructure of insect migration indicate complex systems of genetic variation. As might have been expected from a behavior subject to so many causes, gene involvement in regulation and control seems to be primarily polygenic. Those cases where clear and distinct phenotypes exist, such as wing polymorphism, display characteristics indicating that if major genes are involved, they are still subject to extensive regulation by greater or lesser sets of modifiers. The polygenic systems also provide much genetic variation on which natural selection can act, enough in fact to suggest that were selection to be unidirectional for more than a generation, evolutionary responses would be relatively rapid. The genetic correlations connecting migration to life histories further indicate that with some notable exceptions, like age at first reproduction, the entire life cycle structure will be influenced. This raises, in particular, profound questions concerning the nature of selection for wing polymorphism given the life history "costs" in terms of later reproduction and lowered fecundity often present in the macropterous forms (Dingle 1984b; Roff 1984; Zera et al. 1984). It also merits noting that as Roff (1975 and Table 1) has shown, the type of genetic mechanism, i.e., whether Mendelian or polygenic, influences the course of evolution when environments fluctuate.

When we recognize how integral genetic variation, phenotypic expression, and selection of life-history patterns are to the process of evolution and how important migration is to maintaining life-cycle flexibility, it becomes readily evident that an understanding of the genetics of migration will make a significant contribution to modern evolutionary biology (Istock 1983; Dingle 1984a; Palmer 1985, and Fig. 2). The distribution of birth and death over space and time across seasons and generations combined with the variable contribution of individuals to those events provide the fundamental basis for understanding evolution as a process. Studies of migration de facto have a rightful place at center stage since migration so often determines where births and deaths occur. Nor are these studies likely to be merely esoteric academic exercises. Many of the most economically important insects are migrants. Their populations *will* evolve when we apply management techniques, rendering an understanding of their ecology and evolution absolutely mandatory (Dingle 1979; Barfield and O'Neil 1983).

To take a case in point, many migrant Lepidoptera in the southeastern United States are apparently "pied piper" migrants, migrating north in the spring, but caught by winter before they can migrate south (Rabb and Stinner 1978; Dingle 1980). Common evolutionary sense supported by Roff's (1975) models predicts that the migrants should be selected out of the population. Evidently they are not, and we need to know why. Similar examples abound.

Critical to progress in understanding the evolution and genetics of insect migration are laboratory assays which adequately index migratory behavior in the field. Are *Drosophila* moving between laboratory population tubes (Fig. 2) really migrants in any meaningful ecological sense? Some results of attempts to "calibrate" laboratory performance with behavior in the field give us encouragement. Migrant *Oncopeltus fasciatus* do fly more in tethering experiments (Dingle et al. 1980) and armyworm moths [*Spodoptera exempta* (Wlk.)] take-off in the laboratory at the same time as they do in the field (Gatehouse and Hackett 1980). The very careful experiments of Curtsinger and Laurie-Ahlberg (1981) convincingly demonstrate that tethered flight in *Drosophila melanogaster* is a robust measure of differences among genetic lines. These hints of success, however, should spur further efforts, not contribute to false confidence. I have reviewed in detail elsewhere the issue of whether laboratory and field behavior are comparable with respect to flight (Dingle 1984b).

A further problem requiring attention is the genetic basis of the relation between migration and diapause. That the two behaviors are related both ecologically and physiologically is now well established (Dingle 1982, 1984a,b), but the contributions of genes and the extent to which genes are shared still largely awaits exploration. As a major contributor to the timing of all events in insect life cycles, including migration, diapause is likewise a fundamental element in insect life-history evolution. The repeated demonstration of extensive genetic variation in diapause (Hoy 1978) and the observed absence of correlation between age at first reproduction and other fitness traits, at least in some of the few cases where it has been looked for, serve to stress again the relevance of understanding the genetic architecture of life cycles.

The above summary reveals all too clearly just how sketchy our knowledge is of the contributions of genes to migration and colonizing syndromes. I think enough is revealed, however, to encourage the hope that there are exciting and interesting possibilities in further investigation. Perhaps we can even say that in crude form we have the opening lines of a Genesis for the field. Further lines and chapters only await our insight and ingenuity.

Acknowledgments. I am grateful to Ken Evans, Peter Frumhoff, Francis Groeters, Larry Harshman, Beth Jakob, Susan Scott, and Randy Snyder for their useful and critical comments on the manuscript. My own research on insect migration was funded by grants from the U.S. National Science Foundation.

References

Barfield CS, O'Neil RJ (1983) Is an ecological understanding a prerequisite for pest management? Fl Entomol 67:42–49
Blackman RL (1971) Variation in the photoperiodic response within natural populations of *Myzus persicae* (Sulz.). Bull Entomol Res 60:533–546

Blakley NR, Dingle H (1978) Competition: butterflies eliminate milkweed bugs from a Caribbean island. Oecologia (Berl) 37:133–136

Caldwell RL, Hegmann JP (1969) Heritability of flight duration in the milkweed bug, *Lygaeus kalmii*. Nature 223:91–92

Chabora PC (1969) Mutant genes and the emigration behavior of *Phaenicia sericata* (Diptera, Calliphoridae). Evolution 23:65–71

Chabora PC, Kessler ME (1977) Genetic and light intensity effects on the activity and emigratory behavior of the house fly, *Musca domestica* (L.). Behav Genet 7:281–290

Cohen D (1967) Optimization of seasonal migratory behavior. Am Nat 101:5–17

Curtsinger JW, Laurie-Ahlberg CC (1981) Genetic variability of flight metabolism in *Drosophila melanogaster* I. Characterization of power output during tethered flight. Genetics 98:549–564

Derr JA (1980) The nature of variation in life history characters of *Dysdercus bimaculatus* (Heteroptera:Pyrrhocoridae), a colonizing species. Evolution 34:548–557

Dingle H (1968) The influence of environment and heredity on flight activity in the milkweed bug *Oncopeltus*. J Exp Biol 48:175–184

Dingle H (1979) Adaptive variation in the evolution of insect migration. In: Rabb RJ, Kennedy GG (eds) Movement of highly mobile insects: concepts and methodology in research. North Carolina State University Press, Raleigh, NC, pp 64–87

Dingle H (1980) Ecology and evolution of migration. In: Gauthreaux SA Jr (ed) Animal migration, orientation, and navigation. Academic Press, New York, pp 1–101

Dingle H (1981) Geographic variation and behavioral flexibility in milkweed bug life histories. In: Denno RF, Dingle H (eds) Insect life history patterns: habitat and geographic variation. Springer, Berlin Heidelberg New York, pp 57–73

Dingle H (1984a) Behavior, genes, and life histories: complex adaptations in uncertain environments. In: Price PW, Slobodchikoff CN, Gaud WS (eds) A new ecology: novel approaches to interactive systems. Wiley, New York, pp 169–194

Dingle H (1985) Migration, chapter 8, vol 9. In: Kerkut GA, Gilbert LI (eds) Comprehensive insect physiology, biochemistry and pharmacology. Pergamon, Oxford

Dingle H, Baldwin JD (1983) Geographic variation in life histories: a comparison of tropical and temperate milkweed bugs *(Oncopeltus)*. In: Brown VK, Hodek I (eds) Geographic variation in insect life cycle strategies. Academia Junk, The Hague, pp 143–165

Dingle H, Blakely NR, Miller ER (1980) Variation in body size and flight performance in milkweed bugs *(Oncopeltus)*. Evolution 34:371–385

Dingle H, Blau WS, Brown CK, Hegmann JP (1982) Population crosses and the genetic structure of milkweed bug life histories. In: Dingle H, Hegmann JP (eds) Evolution and genetics of life histories. Springer, Berlin Heidelberg New York, pp 209–229

Dingle H, Leslie JF, Palmer JO (1986) Behavior genetics of flexible life histories in milkweed bugs *(Oncopeltus fasciatus)*. In: Huettel M (ed) Evolutionary genetics of invertebrate behavior. Plenum, New York

Elton C (1927) Animal ecology. Sidgwick and Jackson, London

Falconer DS (1981) Introduction to quantitative genetics, 2nd edn. Longman, London

Gatehouse AG, Hackett DS (1980) A technique for studying flight behaviour of tethered *Spodoptera exempta* moths. Physiol Entomol 5:215–222

Harrison RG (1979) Flight polymorphism in the field cricket *Gryllus pennsylvanicus*. Oecologia (Berl) 40:125–132

Harrison RG (1980) Dispersal polymorphisms in insects. Annu Rev Ecol Syst 11:95–118

Hegmann JP, Dingle H (1982) Phenotypic and genetic covariance structure in milkweed bug life history traits. In: Dingle H, Hegmann JP (eds) Evolution and genetics of life histories. Springer, Berlin Heidelberg New York, pp 177–185

Honěk A (1976a) Factors influencing the wing polymorphism in *Pyrrhocoris apterus* (Heteroptera, Pyrrhocoridae). Zool Jahrb Abt Syst Oekol Geogr Tiere 103:1–22

Honěk A (1976b) The regulation of wing polymorphism in natural populations of *Pyrrhocoris apterus* (Heteroptera, Pyrrhocoridae). Zool Jahrb Abt Syst Oekol Geogr Tiere 103:547–570

Honěk A (1979) Independent response of 2 characters to selection for insensitivity to photoperiod in *Pyrrhocoris apterus*. Experienta (Basel) 35:762–763

Hoy MA (1978) Variability in diapause attributes of insects and mites: some evolutionary and practical implications. In: Dingle H (ed) Evolution of insect migration and diapause. Springer, Berlin Heidelberg New York, pp 101–126

Istock CA (1978) Fitness variation in a natural population. In: Dingle H (ed) Evolution of insect-migration and diapause. Springer, Berlin Heidelberg New York, pp 171–190

Istock CA (1983) The extent and consequences of heritable variation for fitness characters. In: King CR, Dawson PS (eds) Population biology: retrospect and prospect. Oregon State University Colloq Pop Biol. Columbia University Press, New York, pp 61–96

Jackson DJ (1928) The inheritance of long and short wings in the weevil, *Sitona hispidula*, with a discussion of wing reduction among beetles. Trans R Soc Edinb 55:665–735

Johnson CG (1969) Migration and dispersal of insects by flight. Methuen, London

Kennedy JS (1985) Migration, behavioral and ecological. In: Rankin MA (ed) Migration: mechanisms and adaptive significance. Contrib Mar Sci (Suppl) 27:1–20

Klausner E, Miller ER, Dingle H (1981) Genetics of brachyptery in a lygaied bug island population. J Hered 72:288–289

Lakovaara S, Keranen (1980) Variation at the α-Gpdh locus of drosophilids. Hereditas 92:251–258

Lamb RJ, Mac Kay PA (1979) Variability in migratory tendency within and among natural populations of the pea aphid, *Acyrthosiphon pisum*. Oecologia (Berl) 39:289–299

Lamb RJ, Mac Kay PA (1983) Micro-evolution of the migratory tendency, photoperiodic response and developmental threshold of the pea aphid, *Acyrthosiphon pisum*. In: Brown VK, Hodek I (eds) Diapause and life cycle strategies in insects. Junk, The Hague, pp 209–218

Lande R (1976) The maintenance of genetic variability by mutation in a polygenic character with linked loci. Genet Res 26:221–234

Lavie B, Ritte U (1978) The relation between dispersal behavior and reproductive fitness in the flour beetle *Tribolium castaneum*. Can J Genet Cytol 20:589–595

Leslie JF (1986) Life history variation and population structure in a seasonally migratory insect (Hemiptera:Lygaeidae:*Oncopeltus*). Evolution (in press)

Leslie JF, Dingle H (1983) A genetic basis for oviposition preference in the large milkweed bug, *Oncopeltus fasciatus*. Entomol Exp Appl 34:215–220

Lidicker WZ JR, Caldwell RL (eds) (1982) Dispersal and migration. Hutchinson Ross, Stroudsburg, Pennsylvania

Lindroth CH (1946) Inheritence of wing dimorphism in *Pterostichus anthracinus* Ill. Hereditas 32: 37–40

Mahmud FS (1980) Alary polymorphism in the small brown planthopper *Laodelphax striatellus* (Homoptera: Delphacidae). Entomol Exp Appl 28:47–53

Narise T (1966) The mode of migration of *Drosophila ananassae* under competitive conditions. Stud Gen Univ Texas 4:121–131

Oakeshott JG, Gibson JB, Anderson PR, Knibb WR, Anderson DG, Chambers GK (1982) Alcohol dehydrogenase and glycerol-3-phosphate dehydrogenase clines in *Drosophila melanogaster* on different continents. Evolution 36:86–96

O'Brien SJ, MacIntyre RJ (1972) The α-glycerophosphate cycle in *Drosophila melanogaster* II. Genetic aspects. Biochem Genet 7:127–138

Palmer JO (1985) Ecological genetics of wing length, flight propensity, and early fecundity in a migratory insect. In: Rankin MA (ed) Migration: mechanisms and adaptive significance. Contrib Mar Sci (Suppl) 27:653–663

Parker GA, Stuart RA (1976) Animal behavior as a strategy optimizer: evolution of resource assessment strategies and optimal emigration thresholds. Am Nat 110:1055–1076

Rabb RL, Stinner RE (1978) The role of insect dispersal and migration in population processes. In: Vaughan CR, Wolf W, Klassen W (eds) Radar, insect, population ecology, and pest management. NASA Conf Publ No 2070. NAA Wallops Flight Center, Wallops Island, Virginia, pp 3–16

Rankin MA (1978) Hormonal control of insect migratory behavior. In: Dingle H (ed) Evolution of insect migration and diapause. Springer, Berlin Heidelberg New York, pp 5–32

Ritte U, Lavie B (1977) The genetic basis of dispersal behavior in the flour beetle *Tribolium castaneum*. Can J Genet Cytol 19:717–722

Roff DA (1975) Population stability and the evolution of dispersal in a heterogeneous environment. Oecologia (Berl) 19:217–237

Roff D (1984) The cost of being able to fly. Oecologia (Berl) 63:30–37

Sactor B (1974) Biological oxidations and energetics in insect mitochondria. In: Rockstein M (ed) The physiology of insects, vol 4. Academic Press, London, pp 271–353

Safriel UN, Ritte U (1980) Criteria for the identification of potential colonizers. Biol J Linn Soc 13:287–297

Sakai K, Narise T, Iyama S (1957) Migration studies in several wild strains of *Drosophila melanogaster*. Nat Inst Genet Jpn Annu Rep No 7:73–75

Sharp JL, Boller ET, Chambers (1983) Selection for flight propensity of laboratory and wild strains of *Anastrepha suspensa* and *Ceratitis capitata* (Diptera: Tephritidae). J Econ Entomol 76:302–305

Southwood TRE (1962) Migration of terrestrial arthropods in relation to habitat. Biol Rev 37:171–214

Southwood TRE (1977) Habitat, the templet for ecological strategies? J Anim Ecol 46:337–365

Sutherland ORW (1969a) The role of crowding in the production of winged forms by two strains of the pea aphid, *Acyrthosiphon pisum*. J Insect Physiol 15:1385–1410

Sutherland ORW (1969b) The role of the host plant in the production of winged forms by two strains of the pea aphid, *Acyrthosiphon pisum*. J Insect Physiol 15:2179–2201

Sutherland ORW (1970) An intrinsic factor influencing alate production by two strains of the pea aphid, *Acyrthosiphon pisum*. J Insect Physiol 16:1349–1354

Taylor LR, Taylor RAJ (1977) Aggregation, migration and population mechanics. Nature 265:415–421

Taylor LR, Taylor RAJ (1983) Insect migration as a paradigm for survival by movement. In: Swingland IR, Greenwood PW (eds) The ecology of animal movement. Clarendon, Oxford, pp 181–214

Vepsäläinen K (1974) Determination of wing length and diapause in waterstriders (*Gerris* Fabr., Heteroptera). Hereditas 77:163–176

Vepsäläinen K (1978) Wing dimorphism and diapause in Gerris: determination and adaptive significance. In: Dingle H (ed) Evolution of insect migration and diapause. Springer, Berlin Heidelberg New York, pp 218–253

Via S (1984) The quantitative genetics of polyphagy in an insect herbivore I. Genotype-environment interaction in larval performance on different host plant species. Evolution 38:881–895

Williams CB (1958) Insect migration. Collins, London

Wu A-C (1981) Life history traits correlated with emigration in flour beetle populations. PhD Thesis, University of Illinois at Chicago Circle

Zera AJ (1981) Extensive variation at the α-glycerophosphate dehydrogenase locus in species of waterstriders (Gerridae: Hemiptera). Biochem Genet 19:797–812

Zera AJ, Innes DJ, Saks ME (1983) Genetic and environmental determinants of wing polymorphism in the waterstrider, *Limnoporus canaliculatus*. Evolution 37:513–522

Ziegler JR (1976) Evolution of the migration response: emigration by *Tribolium* and the influence of age. Evolution 30:579–592

3 The Oogenesis-Flight Syndrome Revisited

M. A. RANKIN[1], M. L. MCANELLY, and J. E. BODENHAMER

1 Introduction

The question of how suites of interrelated characters evolve to produce an optimal life-history strategy is central to the development of life-history theory. Dingle (1985), Lande (1979, 1982), and others have emphasized the need for genetic analysis of such character suites. An understanding of the physiological relationships between life-history traits is equally important. Physiological mechanisms may link fitness characters in unsuspected ways and may put at least short-term constraints on response to selection. Understanding such mechanisms can allow predictions concerning correlated responses when selection is imposed on a particular trait and may yield insight into phenotypic links between characters.

Migration is an important component of the life histories of many organisms. Certainly insects display a wide variety of movement patterns which can have great impact on fitness. Among insects, migration may be a means of avoiding adverse conditions − escape in space − but it is also an important colonizing device that allows exploitation of temporary or patchy habitats (Southwood 1962, 1977; Southwood et al. 1974; Denno 1976, 1979, 1983, 1985; Solbreck 1978, 1985; Dingle 1982, 1985).

Migration has been considered a "costly" strategy in terms of reproductive output, and (without taking into account the potential for improved opportunities in a new habitat) it was frequently assumed that migrants in general have decreased reproductive fitness relative to nonmigrants either due to reduction in total egg production, increased mortality of dispersers or some combination of these effects (Grinnell 1922; Elton 1936; Dingle 1972; Roff 1975, 1977; Denno and Dingle 1981; Lidicker and Caldwell 1982). Migration was thought to be favored by individual selection only when local conditions were adverse or were about to deteriorate. Group selection arguments were invoked by some authors (Wynne-Edwards 1962; Van Valen 1971; Myers 1976) to explain the frequent occurrence of migration when individual selective advantage to migrants was obscure.

Indeed, when apterae or nonmigrants are compared with their migratory conspecifics or members of closely related nonmigratory species, negative effects of migration on reproduction are often evident. In species that display morphological differences between nonfliers and potential migrants, the energy and material required to build the flight apparatus has reasonably been assumed to diminish reserves available for repro-

---

[1] Department of Zoology, University of Texas, Austin, Texas 78712, USA.

Insect Flight: Dispersal and Migration
Edited by W. Danthanarayana
© Springer-Verlag Berlin Heidelberg 1986

duction and to prolong the developmental period. Many examples can be cited in which a decrease or delay in reproduction is observed in presumed migrants compared to non-migrants. Zera (1984, 1985) has shown that age to first reproduction is less and reproductive output (especially early output) is significantly greater in wingless as compared to winged waterstriders *Limnoporus canaliculatus*. In other insects as well the sacrifice extracted by migration or possession of wings has been measured in increased time to first reproduction, decreased numbers of developed ovarioles, or decreased numbers of eggs per clutch (Dingle and Arora 1973; Roff 1977, 1984; Walters and Dixon 1983; Wratten 1977).

Observing that migration in many insects is restricted to the post-teneral, pre-reproductive period, i.e., the period after the cuticle has hardened following adult emergence, but prior to (or between) oviposition(s) (such that the presence of fully developed or developing oocytes is somehow inhibitory to migratory behavior), C.G. Johnson (1969) and others described the "oogenesis-flight syndrome" as one characteristic of insect migration. The assumption inherent in this term was that migration and reproduction are alternate physiological states. This idea was reinforced by the fact that migration is often associated with and induced by conditions which produce adult diapause (delay in reproduction) such as short photoperiod, low temperatures or poor food quality, and migrants are often physiologically similar to diapausing insects in having hypertrophied fat bodies and immature ovaries. Thus, physiologically as well as ecologically, migration was thought to be an alternative to immediate reproduction in the young adult.

If migration results in reproductive costs to the individual and/or is the physiological adverse to reproduction, these facts must be taken into account in theories of the evolution of migration and life-history strategies. However, one must not overlook in such calculations the long-term benefits which might accrue to the successful migrant, or the consequences of not migrating. Furthermore, migration is an important colonizing device and has probably been very important in the spread of insects into the many and diverse habitats which they occupy today. As Simberloff (1981) and others have written – a good colonist needs not only to travel efficiently, it also requires high fecundity and rapid reproductive development in the new habitat. Thus, a good colonist might be expected to display a suite of life-history characters which involve both adaptations for long flight (such as longer wings, larger body size, well-developed flight musculature) and adaptations for rapid and prolific reproduction. With the view that migratory flight involves significant energetic cost to the migrant and that flight and reproduction are physiologically antagonistic, severe limitations would seen to exist on the extent to which both aspects of such a colonization strategy might evolve.

We suggest that such constraints may not in fact exist to the extent previously supposed, at least in highly migratory insects. A growing body of theory and empirical evidence suggests that migration can often be part of a suite of covarying traits that evolve as a consequence of adaptation for successful colonization (Simberloff 1981; Parsons 1982; Palmer 1985). Many migrant species may in fact be highly evolved in terms of the balance of the reproductive costs versus the benefits of migration, and this balance very likely involves neuroendocrine mechanisms that link migratory behavior with reproductive development (Caldwell and Rankin 1972; M.A. Rankin 1974, 1978, 1980; Rankin and Rankin 1980a).

In several species, flight has no *observed* deleterious effects on reproduction (Cockbain 1961) and may actually stimulate reproduction (B. Johnson 1958; Highnam and Haskell 1964; Rygg 1966; Lavie and Ritte 1978; Slansky 1980). In *Oncopeltus* subjected to tethered flight, for example, age to first oviposition and the inter-clutch interval were significantly decreased, and the mean number of eggs produced tended to be greater, in flown than in unflown bugs (Slansky 1980). *Oncopeltus* selected for long wings (a trait correlated with flight propensity) for nine generations show greater early fecundity than unselected controls or those selected for short wings (Palmer 1985).

The decreased ovariole number in gregarious *Locusta migratoria* (Norris 1950) has often been cited as evidence of the trade-off between migration and reproduction (e.g., Dempster 1963; Dingle 1972). However, crowded *Schistocerca gregaria* do not show decreased reproductive capacity (Norris 1952); indeed, migratory behavior is not restricted to the gregarious phase in either species (Ramchandra 1942; Davey 1953, 1959; Roffey 1963). Performance of flight greatly accelerates oocyte growth in the locusts, *Locusta migratoria* and *Schistocerca gregaria*, although in the former the effect is dependent on whether the animals are isolated or crowded (Highnam and Haskell 1964).

When migration is associated with adult diapause, environmental cues elicit a delay in reproduction which is correlated with the occurrence of migration to the diapause site. However, such a delay is not necessarily the result of performance of the migratory flight itself. For example, Brower and his co-workers have shown that monarch butterflies actually enhance their lipid reserves on the southward migratory flight by heavy feeding on late-blooming composites in Texas (Brower 1985). Furthermore, as female monarch butterflies leave the winter hibernacula in March, they do so mated and carrying rapidly developing oocytes. Similarly, female *Hippodamia convergens* (Coccinellidae) mate before leaving their mountaintop hibernacula in the spring and leave the diapause site with rapidly developing ovaries. The negative relationship between ovarian development and flight which is observed in pre-diapause *H. convergens* is much less marked in post-diapause animals (Rankin and Rankin 1980b). In leaving the hibernacula, females continue to fly until ovaries are completely developed (see below). Indeed, if collected at hibernacula and distributed in gardens heavily infested with aphid prey, these beetles frequently will neither feed nor lay eggs until they make a flight (Williams 1958; Hagen 1962; Rankin and Rankin, unpublished observations), a behavior that has reduced the efficacy of these beetles as biotic control agents.

Another insect in which long-duration flight is not necessarily associated with reproductive costs is the grasshopper, *Melanoplus sanguinipes* (Fabr.). Using a tethered flight assay (as per Dingle 1965; Rose 1972; Rankin and Rankin 1980b; Davis 1980) which had been validated in field tests of animals from known migratory and sedentary populations (McAnelly and Rankin, in press, a; McAnelly 1984, 1985), we examined the effect of numerous environmental variables on flight behavior of *M. sanguinipes*. The investigation included three populations [referred to as CO (Colorado), AZ (Arizona), and NM (New Mexico)] which showed markedly different tendencies for movement in the field and in laboratory flight tests, Colorado being relatively sedentary, Arizona and New Mexico highly migratory (Fig. 1).

The behavioral differences between these populations were found to be primarily genetically controlled rather than environmentally induced (McAnelly 1984; Tables 1

M.A. Rankin et al.

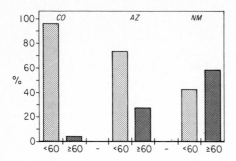

Fig. 1. Flight performance of field-caught insects tested within 24 h of capture. Insects from a highly migratory population (AZ) made significantly more long-duration flights than those from a relatively sedentary population (CO). $x^2 = 53.64, P < 0.001. n = $(CO) 289; (AZ) 182; (NM) 50. (From McAnelly 1985). (NM tests were done in a different year from AZ and CO and were not used for statistical analysis.)

Table 1. Comparisons of the number of "migrants" (individuals that made at least one long flight in four trials) in 2nd, 3rd, and 4th generation lab-reared progeny of NM and CO field parents (1st generation progeny were not available for simultaneous comparison)

	2nd Generation[a]		3rd Generation		4th Generation	
	NM	CO	NM	CO	NM	CO
No long flights	5	41	1	25	4	11
One or more long flights	82	45	32	52	19	35
Migrants (%)	94.3	52.3	98.1	43.2	89.7	59.3
	$x^2 = 38.81$		$x^2 = 36.94$		$x^2 = 8.59$	
	$P < 0.001$		$P < 0.001$		$P < 0.01$	

[a] Reared under a variety of photoperiods. No effect of any photoperiod on flight, therefore all data were combined.

Table 2. Comparison of flight performance of the progeny of crosses between CO and NM grasshoppers

	CO × CO	NM × CO	CO × NM	NM × NM
No long flights	11	12	34	4
One or more long flights	16	16	51	35
Migrants (%)	59.3	57.1	60.0	89.7

$x^2 = 12.71, P < 0.01.$

and 2). The effect of numerous environmental parameters (photoperiod, rearing temperature, food quality, and density) on flight were investigated and found to have little or no influence on migratory behavior of grasshoppers from any of the populations. In contrast, the marked differences in flight behavior observed in the field between CO (sedentary) and NM (migratory) animals were retained in the second, third, and fourth

Table 3. Progeny of grasshoppers from parents that made no long flights (n = 12) versus those from parents that flew 3 or 4 (n = 15) times in 4 trials

	NF × NF	F × F
No long flights	7	1
At least one long flight	5	14
Total	12	15
Migrants (%)	42	93

χ^2 = 8.54, P < 0.01

generation lab progeny of CO and NM field parents (Table 1), and crosses between NM and CO lab animals (which were as viable as within population matings) confirm that these differences between populations were not due to a maternal effect (McAnelly 1984). Progeny of both randomly constituted reciprocal crosses were virtually identical to those of the CO × CO cross and markedly different from those of the NM × NM cross (Table 2), suggesting that long-duration flight behavior in this species is heritable (McAnelly 1984). Preliminary results from within-population crosses of fliers and non-fliers show considerable separation of flight performance in the F_1 generation (Table 3), and suggest that there may be a strong genetic influence on the within-population variability as well.

Like *Oncopeltus fasciatus, Danaus plexippus*, and numerous other insect migrants, *M. sanguinipes* shows a clear negative relationship between the tendency to make a long flight and the ratio of reproductive tract weight to total body weight in females (Fig. 2), though not in males. Thus, a classical oogenesis-flight syndrome exists. However, *M. san-*

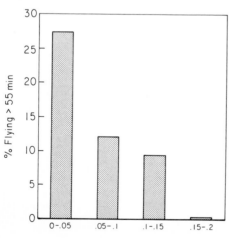

Reproductive Tract Wt/Total Body Wt

Fig. 2. The oogenesis-flight syndrome in *M. sanguinipes*. Ovarian development was assessed as the ratio of reproductive tract weight relative to total body weight. Increasing weight of ovaries corresponded to increasing egg size and vitellogenesis in oocytes (n = 41). (From McAnelly and Rankin, in press, b)

guinipes females are also capable of inter-reproductive flight. In a sample of male-female pairs, of the females that made at least one long flight prior to first oviposition, 12.5% (*n* = 8) of the CO, 35.7% (*n* = 14) of the AZ, and 20% (*n* = 25) of the NM females flew after ovipositing. Inter-reproductive flights occurred as early as 24 h after egg laying (McAnelly and Rankin, in press, b).

In all three lab-reared groups there was a trend that was significant at $p < 0.01$ for NM grasshoppers toward decreasing age at first oviposition in insects that were flight tested repeatedly (in 60 min flight tests) relative to those that were flight tested only twice (Fig. 3). When NM females which were 14-days-old or more and had not oviposited were given one flight trial in which they were allowed to fly for several hours, the flown females began ovipositing within 2 days and laid significantly more pods per female than controls (Fig. 4). Numbers of egg pods per female among tethered controls (that were subjected to the same handling and exposure to flight-testing temperatures as the flown animals) were identical to those from controls not handled in any manner.

Thus, the *M. sanguinipes* populations examined in this study showed highly significant heritable differences in migratory behavior. The differences in flight tendency appear to be due to selection in habitats which favor different frequency of migration (high frequency in NM and AZ, low frequency in CO). In the populations which dis-

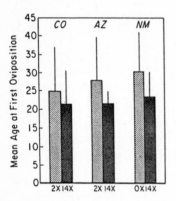

Fig. 3. Effect of flight on age at first reproduction. Comparison of insects flown 14 times (*dark hatching*) with those flown only twice (*light hatching*). *n* = (CO 14X) = 6; (CO 2X) = 5; (AZ 14X) = 7; (AZ 2X) = 8; (NM 14X) = 15; (NM 2X) = 12. Significant at $p < 0.01$ for NM (f = 2.65, df = 28) (From McAnelly and Rankin, in press, b)

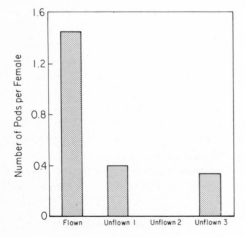

Fig. 4. Comparison of NM females flown once (*n* = 20) with unflown controls. *Unflown 1* = females tethered and suspended under floodlights, but not flown (*n* = 10). *Unflown 2 and 3* = no handling (*n* = 20). χ^2 of the total number of egg pods (not the number of pods/female) produced by each group = 29.12, SD = 3, $P < 0.001$. (From McAnelly and Rankin, in press, b)

played a high frequency of migratory behavior, performance of long-duration flight apparently caused acceleration of reproduction and enhanced reproductive output. Our working hypothesis is that migration and reproduction are evolving in *M. sanguinipes* populations such that where migration has been heavily selected, its cost in terms of early reproduction is low. We suspect that the basis for this evolution is a physiological link between the two processes through the neuroendocrine system (see below), but we have no evidence to this point in this species as yet. Where there has been less selection for migration in the population (as in McAnelly's CO population) we may find that reproductive costs of flight will be higher, although further investigation will be necessary to specifically address this point.

2 Endocrine Influence on Flight Behavior

Three endocrine or neuroendocrine factors have been shown to directly or indirectly influence flight behavior in insects: (1) the adipokinetic hormone of locusts (and probably also in monarch butterflies), (2) most recently the indole amine octopamine in locusts and Lepidoptera, and (3) JH in *Oncopeltus fasciatus* (Dallas), *Hippodamia convergens* (Guerin-Meneville), and the monarch butterfly, *Danaus plexippus* (L.).

Of the three hormones that have been shown to affect flight, the action of the locust adipokinetic hormone has been the most clearly demonstrated. Its effect is probably not directly on flight behavior per se, but rather on the flight fuel system. AKH controls lipid mobilization from the fat body, formation of the primary lipid transport protein in the hemolymph, and probably uptake and utilization of lipid by the flight muscle. It is necessary for long-duration flight in locusts and is produced in the corpus cardiacum (CC), its release is under control of the NCCI and II, probably via octopaminergic synapses (Goldsworthy 1985). Removal of the CC decreases flight performance (speed on a roundabout) of locusts. (See Goldsworthy, this volume, for detailed discussion of the locust AKH system.)

AKH activity has also been reported in a number of other insects including the beetle *Tenebrio molitor* (Goldsworthy et al. 1972), the monarch butterfly *Danaus plexippus* (Dallmann and Herman 1978), and the tobacco hornworm moth *Manduca sexta* (Beenakkers et al. 1978). Extracts of *Periplaneta* CC cause AKH effects in locusts, but they decrease hemolymph lipid in the roach itself (as does locust AKH) (Downer 1972; Downer and Steele 1975). *Carausius morosus* CC extracts have no effect on lipid or carbohydrate levels in that insect, but have AKH effects in locusts and a hypolipemic effect in roaches (Gäde 1979). Thus, a small family of similar peptides seem to elicit a variety of effects in different insects.

In several species of insects, juvenile hormone has been shown to affect flight muscle development or histolysis. For example, in *Ips confusus* (Borden and Slater 1968) and *Dysdercus fulvoniger* (Davis 1975), high juvenile hormone levels presumably associated with the onset of oogenesis, induce flight muscle degeneration. In contrast, in the Colorado potato beetle *Leptinotarsa decemlineata* (de Wilde and de Boer 1961, 1969; de Wilde et al. 1968), juvenile hormone seems to be necessary for flight muscle regeneration after diapause and is probably involved in migratory behavior. In the cockchafer *Melolontha melolontha* the state of activity of the corpus allatum (CA) affects the orientation of flight (Stengel 1974) either towards or away from oviposition and feeding sites.

The apparent relationship of diapause and migration, and Johnson's (1966) suggestion that migration was pre-reproductive because it was due to a lack of JH, led to an investigation of the effect of JH on flight behavior in a typical insect migrant, *Oncopeltus fasciatus* (Caldwell and Rankin 1972). These studies have since been extended to two other migrants, *H. convergens* and *Danaus plexippus*. We briefly review these results, giving more details for monarch work since it has not previously been published.

2.1 Oncopeltus fasciatus – large milkweed bug

The summer range of *Oncopeltus fasciatus* extends from Canada to Central America, but it does not survive winter conditions in temperate areas. Dingle (1968) has shown, on the basis of laboratory flight tests and some highly suggestive field evidence; that a portion of the resident southern population migrates north in the spring and colonizes the very abundant milkweed fields in the north. It appears in Iowa milkweed patches in late June or early July and the population rapidly expands, reaching a peak in early September. By October, *O. fasciatus* has nearly disappeared from this area, apparently due to southward flights of much of the population.

Long-duration tethered flight behavior has been used as a criterion for migratory behavior in this species (Dingle 1965, et seq; Rankin 1974, et seq; Palmer 1985; Slansky 1980), this type of flight behavior being largely restricted to the post-teneral, pre-reproductive period. In response to short photoperiods, *Oncopeltus* undergoes an adult reproductive diapause which provides a considerably longer period of time prior to egg maturation during which females will make long flights and also increases the percentage of individuals of both sexes in laboratory populations that make a long flight at some time during that period (Caldwell and Rankin 1972). The combination of short photoperiod and low temperatures in the fall probably induce migratory behavior in most of the northern population. As noted above, however, although *Oncopeltus* displays a classical oogenesis-flight syndrome, performance of long-duration flight does not elicit a reduction in total reproductive output or longevity, nor does it induce a delay in first reproduction (Slansky 1980). Furthermore, Palmer (1985) has shown that selection for long wings results not only in increased frequency of long-duration flight in the population, but also in a correlated increase in reproductive output.

Food quality or availability can also greatly affect flight and reproduction. When fed on suboptimal food, green pods or flowers, female *Oncopeltus* delay reproduction and will make long flights even in long day and relatively high-temperature conditions which would normally induce reproduction. Since the first northern colonizers typically arrive in Iowa in late June or early July when temperatures are high and photoperiods are approaching the longest of the year, it is likely that food quality (flowers and green pods at this time) is an important factor in stimulating northward flights in the spring. By this same mechanism, flight can also be stimulated among reproductively mature residents by a food shortage, thus greatly increasing the probability of finding new food sources.

Detailed experiments involving gland extirpations and implantations showed that juvenile hormone (JH) has a dual role in adult development of this species. The hormone stimulates migratory behavior in both sexes and ovarian development in the female (Caldwell and Rankin 1972; Rankin 1974 et seq).

As the ovaries attain their full development in the female, further migratory flight is inhibited, although short appetitive flights still occur. Although it would appear that the two effects of JH are thus somewhat antagonistic, later experiments have shown them actually to be part of a closely coordinated program of adult development which seems to be directed primarily by changing titers of this hormone in response to environmental cues.

JH titer determinations on pooled hemolymph samples from adult *Oncopeltus* exposed to different photoperiod regimens indicated that long photoperiods result in immediate increase in hemolymph titers of the hormone (associated with rapid onset of oviposition behavior), while short photoperiods bring on a more gradual increase in JH levels following adult emergence. Short photoperiods enhance migratory flight and delay onset of reproduction (Rankin and Riddiford 1978). In any photoperiod regime, migratory flight typically occurs during periods of intermediate hormone titers. It would appear that lower titers of JH or shorter exposure times are necessary to stimulate flight before reproductive development, which ultimately inhibits long flights. Selection for four generations for late onset of migratory flight behavior resulted in a correlated delay in onset of reproduction and a delay in the associated rise in JH titers (Rankin 1978).

Treatment with the chemical allatecomizing agents precocene I and II (Bowers 1976; Bowers et al. 1976) not only caused cessation of oviposition and resorption of oocytes, but also produced a decrease in long-duration flight behavior in the population which could be restored to high levels by JH replacement therapy (M.A. Rankin 1980). Like starvation, precocene treatment stimulates a transient increase in frequency of migratory behavior and in the duration of shorter flights. This increased flight behavior after precocene treatment can be quite prolonged under some circumstances and seems to be associated with an incomplete inhibition of JH production (as determined by CA size and state of ovarian development, Rankin, unpublished results). The ultimate inhibition of long-duration flight after precocene treatment and its restoration by JH replacement therapy indicate, however, that JH is necessary for both long-duration flight behavior and reproduction in *Oncopeltus fasciatus*.

No effect of the corpus cardiacum (CC) on flight has been observed in *Oncopeltus fasciatus*. When the equivalent of half a CC was injected as a saline homogenate, no effect on flight behavior was observed. Similarly CC extracts injected into precocene-treated animals who were displaying no long-duration flight activity had no restorative effect on flight (Rankin and Riddiford 1977) nor did it elevate hemolymph lipid levels, while JH injection both stimulated flight behavior and caused a rapid increase in hemolymph lipid (Rankin, unpublished results).

2.2 Hippodamia convergens

Another insect in which the role of JH in migratory flight has been examined is the ladybeetle, *Hippodamia convergens*. This insect is one of the most widespread of the American coccinellids, occurring throughout the western, central, and southern United

States. It is an economically important predator of aphid and mite pests on cultivated crops and in fact is commercially available as a natural control agent for such pests.

Adults emerge from "hibernation" sites in the spring and migrate up to hundreds of kilometers to areas of aphid infestation (Hagen 1962) where they may develop high densities (Dickson et al. 1955). As aphid populations decrease, the coccinellids move to neighboring habitats. Such movements may occur several times in a season and, depending on food abundance, the species may be uni- or multivoltine. If prey is unavailable or scarce, young adults enter reproductive diapause and an extended migratory phase during which they move to mountaintop aggregation sites. Diapausing beetles may remain at the hibernation sites from 6 to 9 months until they migrate back to lower altitudes (Hagen 1962; Rankin and Rankin 1980b). Some type of diapause development seems to be involved in the return of the migratory response at this time (Rankin and Rankin 1980a,b).

Laboratory tethered-flight tests can be used as an assay for migratory behavior. Beetles that fly 30 min on a tether will nearly always fly much longer. It appears that migration to the aestivo-hibernation sites is accomplished primarily by newly emerged (pre-reproductive female) beetles. As is true for many insect migrants, long-duration tethered flights tend to occur most frequently among pre-reproductive *H. convergens* females. Long daylength and high-quality food result in a sharp decline in flight after 1 week, by which time oviposition has begun. Under short photoperiod and/or poor food quality (frozen aphids) long flights continue for some time in the population. Photoperiod seems only to influence the duration of the long-flight phase when food quality is optimal. When this is the case, short photoperiod greatly lengthens the period of time in which beetles will make long flights, although the percentage of long fliers in the population is lower when food is optimal than when it is not. When food quality is poor, flight activity is enhanced under either photoperiod regime to quite high levels. Thus, the beetles do respond to photoperiod under some conditions, but food quality seems to be the primary cue which triggers migratory behavior. Similarly, ovarian development is dependent upon food quality, photoperiod having little, if any, effect; *H. convergens* can be reproductive under both long- and short-day conditions (Rankin and Rankin 1980b).

Under long days there seems to be a clear oogenesis-flight syndrome in this species, i.e., ovarian development is inversely correlated with flight activity (Fig. 5a). Under short photoperiods, peak flight activity occurs in females with partially developed, rather than completely undeveloped ovaries. Furthermore, some flight activity continues in the short-day population until ovarian development is complete (3.0 mg; Fig. 5a), while it virtually ceases in the long-day animals by the time the reproductive tract has attained 1.5–2.0 mg fixed wet weight. Thus, the oogenesis-flight syndrome is less pronounced and flight continues longer in the population under short photoperiods (Rankin and Rankin 1980b).

Juvenile hormone is necessary for ovarian development in this species (Fig. 6) although, as with many beetles, a brain factor is also necessary for completion of oogenesis (S.M. Rankin 1980). Topical application of a JH mimic to *H. convergens* stimulated a significant increase in long-term flight behavior in both males and females and ovarian development in females (Fig. 7). Topical application of precocene II to *H. convergens* inhibited flight activity in treated animals of both sexes and oogenesis in females for

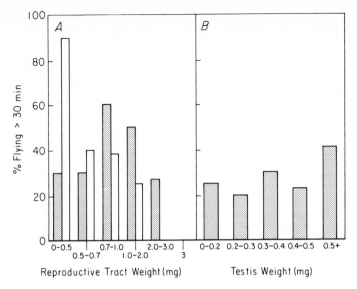

Fig. 5A,B. Relationship of reproductive tract development to long-duration, tethered-flight behavior in *Hippodamia convergens*. **A** Flight activity of females collected in March (short-day photoperiod; *shaded bars*; n = 38) and in mid-summer (long-day photoperiod; *open bars*; n = 50) around Austin, Texas, with respect to weight of the fixed reproductive tract. **B** Flight activity of males collected in mid-summer (long-day photoperiod; n = 35), with respect to testis weight. (From Rankin and Rankin 1980b)

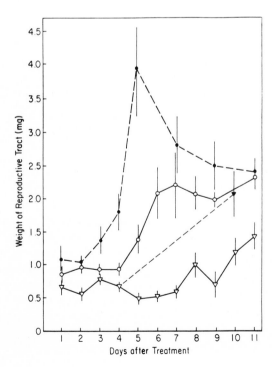

Fig. 6. Effects of ten applications of precocene II and JHM on weight of the entire fixed-reproductive tract of female *H. convergens*. *Closed circles* indicate JHM-treated animals; *open circles*, acetone-treated controls; *open triangles*, precocene-treated animals; the *solid triangle* connected to the precocene-treatment group indicates precocene-treated animals given JHM repacement therapy on day 4. *Bars* = SEM. (From Rankin and Rankin 1980a)

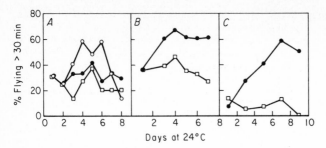

Fig. 7A–C. Effect of JHM on flight behavior of *H. convergens*. **A** Females (*open squares* = acetone-treated, *n* = 15; *closed circles* = 5 μg JHM day 2, *n* = 15; *open circles* = 10 μg JHM day 2, *n* = 15). **B** Males (*open squares* = acetone-treated, *n* = 30; *closed-circles* = 10 μg JHM on day 2, *n* = 30). **C** Females collected in June from the hibernacula, given repeated applications of JHM (*open squares* = acetone treated days 2, 4, 6, and 8, *n* = 30; *closed circles* = 10 μg JHM on days 2, 4, 6, and 8, *n* = 30). (From Rankin and Rankin 1980a)

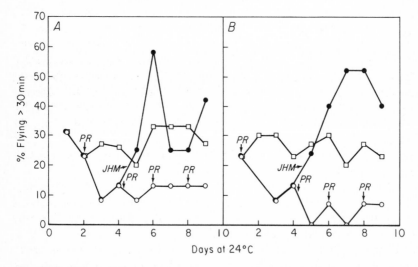

Fig. 8A,B. Effects of 10 μg applications of precocene II administered every 2 days or of precocene II followed by JHM treatment, on flight behavior of *H. convergens* males (**A**) or females (**B**)

about 10 days; JHM (altosid) treatment to precocene-treated beetles significantly increased their migratory behavior over that of precocene-treated or acetone-treated controls and also stimulated oogenesis in females (Fig. 8). These results indicate that JH stimulates migratory flight behavior in this species along with reproductive development. It is likely that the hormone serves to coordinate migration with reproduction in the young adult. Higher doses of JHM (10 μg/animal) seemed to have a dual effect in females, increasing flight and stimulating ovarian development. The sharp decline in flight on day 8 among females given 10 μg JHM may have been due to an ovarian-based inhibitory effect (Fig. 7; Rankin and Rankin 1980a).

In one group of females collected shortly after their arrival at the hibernacula and treated immediately and repeatedly with JHM, the flight response to JH treatment was

Table 4. Effects of JHM treatment on the mean weight of the reproductive tract in *H. convergens*

Treatment	Weight of tract (mg ± SD)
Tested in March, treated day 2 only with 10 μg JHM/female	2.40 ± 0.18
1 μl acetone/female	0.82 ± 0.09
Tested in July, treated on days 2, 4, 6, and 8 with 10 μg JHM/female	1.25 ± 0.24
1 μl acetone/female	0.65 ± 0.01

sustained and did not involve cessation of flight after several days (Fig. 7C), nor was there substantial ovarian development (Table 4). In contrast, all other animals used had been stored at low temperatures in diapause for several months prior to testing, and the two groups were significantly different in their flight activities ($P < 0.001$, Mann-Whitney U). We have found that among animals collected from aggregation sites in Texas, flight activity is very low when beetles are newly arrived at the aggregation sites. There seems to be a gradual increase of the migratory response during diapause. The increase in responsiveness to migratory stimuli may be due to a gradual rise in JH titers or a change in sensitivity to the hormone similar to that observed in diapausing monarch butterflies (Herman 1981). The relatively moderate increase in ovarian weight among newly aggregated females in response to high and repeated applications of JHM (Table 4) tends to support this suggestion. As noted above, this seems to be a situation in which the delay in reproduction associated with migration is due to factors other than the performance of flight per se. Juvenile hormone stimulates both flight and reproduction in the post-diapause beetle, thereby ensuring that reproductive delay associated with colonization will be minimized.

The increased flight behavior of JHM-treated males occurred without a subsequent depression in flight activity (Fig. 7B), suggesting that in the absence of ovaries, flight stimulation by JH may be more prolonged. This is similar to the effect of JH in male and ovariectomized female *Oncopeltus fasciatus* (Rankin 1974). As with *O. fasciatus*, precocene treatment inhibited both oogenesis and flight behavior in *H. convergens*, and JH replacement therapy restored both (Fig. 8). We have not investigated the possible involvement of the corpus cardiacum in flight in this species, nor have we pursued the mechanism of the JH effect on flight. However, like *Oncopeltus fasciatus*, there is clearly a coordinated stimulation of both long-flight behavior and reproductive development in this species. This, in combination with the "relaxation" of the oogenesis-flight syndrome in short photoperiods suggests that selection has acted to maximize both flight and reproduction in the migrant colonizer and has done so through the neuroendocrine system.

2.3 Danaus plexippus

Populations of monarch butterflies in North America undergo dramatic migrations to overwintering areas each fall. These butterflies travel from Canada and northern United States to overwintering sites in southern California, Florida, and Mexico and migrate back again to the north the following spring. In the fall, decreasing day lengths and lower temperatures appear to affect the neuroendocrine system in such a way that the activity of the CA and possibly the brain neurosecretory cells is altered (Herman 1973; Barker and Herman 1976; Herman 1985). It is clear that monarchs of the fall generation are physiologically and behaviorally different from those of the summer generations. Herman (1973) has shown that winter animals from the California populations are reproductively inactive owing at least in part to inactive CA. Juvenile hormone titer determinations done on field-collected animals (Lessman and Herman 1983) indicate a decrease in JH titers associated with the southward flight of the monarch and an increase associated with northward journeys. If lower titers of JH stimulate migratory flight, while higher titers or longer exposure to the hormone are necessary for ovarian development, it is possible that changing titers of JH during the migratory and reproductive periods govern which activity will predominate as is the case with *Oncopeltus fasciatus*.

Dallmann and Herman (1978) found evidence of an adipokinetic hormone in monarch CC extracts which appears to elevate hemolymph diglycerides during flight and may act in much the same way the locust AKH does. It is possible that the monarch AKH may act along with JH to stimulate migratory flight. Yet another possibility would be that the effect of JH on flight could be by way of an effect on the CC hormone. These possibilities were explored in two series of experiments.

Monarchs were either collected around Austin during their fall or spring migration, collected from Mexican aggregation sites, or reared in the laboratory from eggs laid by field-collected animals. Several types of experiments were done to determine (1) whether an oogenesis-flight syndrome exists in this species, (2) the effect of retrocerebral gland removal and JH replacement therapy on flight, and (3) the effect of AKH and JH on flight behavior of intact animals.

Since many insect migrants display an oogenesis-flight syndrome in which migratory flight behavior diminishes as the ovaries mature, and since *D. plexippus* has frequently been reported to migrate at a time when the ovaries are immature or regressed, it was of interest to determine whether our flight assay would identify an oogenesis-flight syndrome in this species.

In the first experiment, animals were lab-reared under L:D 16:8 photoperiod, 23 °C as caterpillars and switched to 12 °C, L:D 11.5:12.5 as adults; in a second experiment animals were field collected as spring migrants. Decrease in flight activity with ovarian development was much more striking in lab-reared animals (Fig. 9A) than in spring migrants (Fig. 9B). Possibly spring migrants, which are emerging from the winter diapause, tolerate more ovarian development as an adaptation for rapid exploitation of the northern breeding grounds.

In contrast, in males, greater reproductive development is correlated with increased flight activity (Fig. 9C). Since reproductive-tract development in males is stimulated by JH, this correlation along with the results outlined below may indicate that longer exposure or higher titers of JH results in both enhanced flight activity and development of the reproductive tract.

The influence of juvenile hormone (JH) and adipokinetic hormone (AKH) treatment on flight of intact animals was examined using animals that were either field-collected during their migration or collected from diapause aggregation sites in Mexico. In the

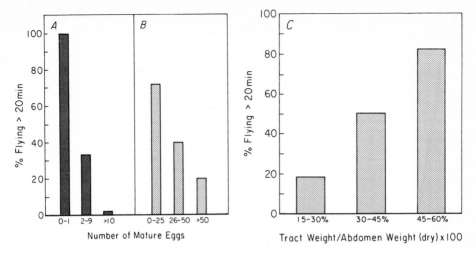

Fig. 9A–C. Oogenesis-flight syndrome in monarch butterflies. **A** Percent of a lab-reared population that made a continuous flight longer than 20 min, graphed with respect to ovarian development as measured by number of mature (chorionated) eggs in the reproductive tract ($n = 14$). **B** Percent of a population of field-collected spring migrants making at least one continuous flight during a single flight test, with respect to ovarian development as measured by number of chorionated eggs in the reproductive tract ($n = 21$). **C** Percent of field-collected and lab-reared males making 20 min flights, with respect to testis development ($n = 147$)

first experiment, animals were collected as migrants in late October, 1982, held at $10°$–$12\ °C$, 11.5:12.5 L:D for about 1 month. All butterflies were flight tested once before treatment to determine flight proclivity and then divided into four groups, each of which received an equal number of fliers and nonfliers as determined by the initial flight test. Five to 7 days after the initial flight test, animals received hormone or carrier injections and were flight tested. A second treatment and flight test were given 2 days later.

Animals in Experiment 2 were collected in late January from Mexican aggregation sites, returned to Austin by airplane and maintained for 2 weeks as above. The treatment-flight test protocol was the same as in Experiment 1, as were the results; data for the two experiments were pooled and are shown in Fig. 10.

AKH alone had a moderate stimulatory effect on flight in both post-treatment flight tests in both experiments, while JH injections caused a similar increase in flight behavior, but only in the second flight test of each experiment (Fig. 10) and the response to JH given alone was more variable than to AKH given alone or in combination with JH. Further, the effect of AKH on flight behavior was immediate, while response to JH required some time (at least more than 2 h) to develop. Both hormone treatments appeared to stimulate flight activity to an equal degree and their effects were not additive.

In order to further investigate the effect of JH on flight behavior, monarchs were allatectomized and given JH replacement therapy, and flight behavior was monitored. Animals were maintained at $10°$–$12\ °C$ in short-day photoperiods and were flight tested 7 days after surgery and every 3 days thereafter. Hormone or carrier treatments began on day 19 and continued at 3-day intervals until the termination of the experiment.

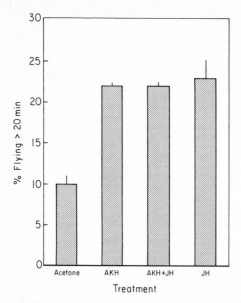

Fig. 10. Effect of AKH and JH on flight behavior of intact *Danaus plexippus*: in experiments with locust adipokinetic hormone, AKH (Peninsula Biochemicals) (0.5 μg in 1 μl saline) and/or 20 μg JH I (in 0.5 μl acetone) was injected 1 h prior to flight testing; controls received 0.5 μl acetone. Injections were given through the intersegmental membrane between the 2nd and 3rd abdominal tergites using a calibrated 5 μl micropipette pulled to a fine tip. A tether-attachment device (strapping tape doubled so that fibers in the two pieces of tape were at right angles to one another) was affixed to the dorsal mesothoracic cuticle with rubber cement and remained in place for the lifetime of the animal. Butterflies were attached to the tether via a small hole made in the tape; the tether consisted of a fishhook swivel which passed through the hole in the tape at one end and which was suspended from its other end by a light wire which was in turn attached to a flight rack. The tether device allowed the butterfly to orient in any direction during flight (n = 50, acetone; 54, AKH; 55, AKH + JH; 54, JH; *bars* = SEM)

The CC-CA-X group was divided into three groups on day 19, one of which received 20 μg JH I in hexane, the second 0.5 μl of hexane, and the third remained untreated.

Allatectomy was seen to significantly decrease and JH replacement therapy to restore flight activity in these populations (Fig. 11). Animals were dissected at the termination of the experiment and examined for amount of reproductive-tract development and for possible regeneration of the retrocerebral complex. No evidence of regeneration was observed. Number of mature eggs in sham-operated animals ranged from 0–118 with a mean of 23.6, and ranged in unoperated controls from 0–75 with a mean of 34.6. None of the allatectomized, untreated animals had mature oocytes, while two of the six hexane-treated females did, with a mean of 3.5 eggs. Allatectomized, JH-treated animals had from 3–59 mature eggs with a mean of 21.3. Why hexane treatment stimulated oogenesis (and flight) to a slight degree is not clear. Many precautions were taken against solvent contamination, but the data suggest that we may have had some JH contamination of the hexane. Alternatively, hexane may act as a weak JH mimic in this species.

These experiments suggest that the monarch butterfly, like *Oncopeltus fasciatus*, increases flight activity in direct response to JH. We should add, however, that because

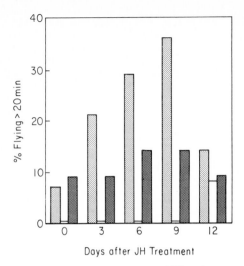

Days after JH Treatment

Fig. 11. Effect of allatectomy and JH replacement therapy on flight in *Danaus plexippus*.
CC-CA removal was done with very fine forceps after shaving and opening the neck membrane to
expose the posterior brain and retrocerebral complex. Wounds were sealed with low-melting point
wax. Fiber optic microscope illuminators were used in preference to incandescent lights in order to
reduce heat stress during surgery. Animals were allowed to recover from surgery for 1 week before
flight testing. JH was applied in 0.5 μl acetone or hexane to CC-CA-ectomized animals on day 19
and every 3 days thereafter prior to flight testing. Hormone or carrier was applied topically to the
abdominal sternites. Experimental animals were reared as adults in a L:D 11.4:12.5 10 °C regime
and were fed a solution of honey and water (1:6) at 16 °C every 3 days just prior to treatment and
flight testing. If they were not being tested, animals were fed only once a week and allowed to exercise in large net cages for 30 min at 16 °C after feeding.
Hormone-treated animals were fed separately, handled with gloves, and reared in containers which
were clearly marked as hormone contaminated and which were cleaned and stored separately from
control containers. Animals were flight tested for 1 h at 3-day or 1-week intervals in front of a low-
speed fan, with or without a spotlight for added heat. During the 1-h test, animals were monitored
at 2-min intervals; notation was made as to whether the animal was flapping its wings, quivering the
wings, or stopped. Minutes of continuous flight and % flying longer than 20 min were noted. Flight
testing apparatus as for Fig. 10. (n = 14, Ca-CCX + hexane or nothing, *open bars*; n = 21, sham-
operated; n = 23 unoperated controls — graphed together as *darker shaded bar*; n = 15 CA-CCX +
JH, *lighter shaded bar*)

of difficulties in identifying migratory as opposed to trivial flight in a tethered-flight
assay with these butterflies, these results must be considered preliminary. It would also
be useful to know the effect of AKH injections on flight of CA-CC-ectomized animals.
Further, because of the length of time our animals were held in the cold, all experiments were done on the equivalent of diapausing winter or early spring animals, i.e.,
all had experienced short-day length as young adults and a prolonged period at 10°–12 °C
before flight testing. The question of JH stimulation of the southward flight when
JH titers are low or decreasing and the ovaries are undeveloped has not been addressed
by these experiments. However, it may be a situation similar to that observed in *Oncopeltus fasciatus* in response to starvation; JH titers rapidly decrease after food is with-
drawn, but flight activity increases for a short period of time. The mechanism for this
effect is not clear. It may be that the threshold for stimulation of flight by JH is lower

than that required to stimulate oogenesis and in the absence of developing ovaries, flight is stimulated. Alternately starvation in *Oncopeltus fasciatus* may stimulate flight behavior by a mechanism not involving JH; in monarchs other factors such as AKH may be involved in flight stimulation, at least in the fall, and JH may simply add to that effect during the spring migration.

3 Discussion

Thus, in three species of highly migratory insects, JH has been shown to stimulate both flight and oogenesis. This apparently dual action of the hormone is by no means universal; as noted above, in some insects, JH can cause flight-muscle histolysis or have no obvious effect on flight behavior (Rankin, unpublished). If JH evolved originally as a hormone controlling metamorphosis, its effect on reproductive development in the adult probably evolved secondarily. Tertiary effects of the hormone such as control of caste determination, aggressive behavior or flight behavior seem to have evolved separately in many insect groups. JH control of migratory behavior may have developed as a response to selection associated with a colonizing life-style in which, ideally, both flight and reproductive parameters would be maximized. The stimulation of both flight and oogenesis by the same hormone would be an efficient mechanism for production of a migrant in which at least early reproductive output would be maximized in the newly colonized habitat. The argument gets more complex, however, when a period of diapause, involving low JH and cessation of oogenesis is involved. In this case one must propose that either (1) a second factor is necessary for oogenesis and in its absence JH only stimulates flight (this may well be the situation for *H. convergens*) and/or (2) that higher titers of (or perhaps different periods of exposure to) the hormone are necessary to stimulate reproduction than are necessary to induce migratory flight. This may be the case for *Oncopeltus fasciatus* in its response to starvation, for example. (3) A third possibility is that migratory behavior can be stimulated by another hormone, such as AKH, as well as by JH, as is perhaps true in the monarch butterfly in its southward flight. However, more precise information regarding effects of endocrine manipulation on clearly identified migratory behavior in the monarch is necessary before the actual role of JH in long-distance movement in this species can be determined.

In acridids as well, the neuroendocrine system may act to coordinate both flight and reproduction. The increase in oocyte growth with flight observed by Highnam and Haskell (1964) in *S. gregaria* and *L. migratoria* was accompanied by a decrease in paraldehyde fuchsin staining material in the corpora cardiaca and in neurosecretory cells of the pars intercerebralis. There were also cyclic changes in the volume of the corpora allata associated with oocyte growth. These results suggest that in locusts in the absence of flight, release of neuroendocrine factors necessary for oocyte growth may be delayed. The possibility that the acceleration and enhanced reproductive output observed among New Mexican *Melanoplus sanguinipes* may involve a similar effect of flight on the neuroendocrine system is currently being investigated.

References

Barker JF, Herman WS (1976) Effect of photoperiod and temperature on reproduction of the monarch butterfly, *Danaus plexippus*. J Insect Physiol 22:1565–1568

Beenakkers AMT, Van der Horst DJ, Van Marrewijk WJA (1978) Regulation of release and metabolic function of the adipokinetic hormone in insects. In: Gillard PJ, Boer HH (eds) Comparative endocrinology. Elsevier/North Holland, Amsterdam, pp 445–448

Borden JH, Slater CE (1968) Induction of flight muscle degeneration by synthetic juvenile hormone in *Ips confusus* (Coleoptera: Scolytidae). Z Vgl Physiol 61:366–368

Bowers WS (1976) Discovery of insect antiallatotropins. In: Gilbert LI (ed) The juvenile hormones. Plenum Press, New York, pp 394–408

Bowers WS, Ohta R, Cleere JH, Marsella PA (1976) Discovery of antijuvenile hormones in plants. Science 193:542–548

Brower LP (1985) New perspectives on the migration biology of the monarch butterfly, *Danaus plexippus* L. In: Rankin MA (ed) Migration: mechanisms and adaptive significance. Contrib Mar Sci Suppl, vol 27, pp 748–785

Caldwell RL, Rankin MA (1972) Effects of a juvenile hormone mimic on flight in the milkweed bug, *Oncopeltus fasciatus*. Gen Comp Endocrinol 19:601–605

Cockbain AJ (1961) Viability and fecundity of alate aliencolae of *Aphis fabae* Scop. after flights to exhaustion. J Exp Biol 38:181–187

Dallmann SH, Herman WS (1978) Hormonal regulation of hemolymph lipid concentration in the monarch butterfly, *Danaus plexippus*. Gen Comp Endocrinol 36:142–150

Davey JT (1953) Possibility of movements of the African Migratory Locust in the solitary phase and the dynamics of its outbreaks. Nature 172:720–721

Davey JT (1959) The African Migratory Locust (*Locusta migratoria migratorioides* Rch. and Frm., Orth.) in the Central Niger Delta, part 2. The ecology of Locusta in the semi-arid lands and seasonal movements of populations, *Locusta* no 7

Davis MA (1980) Variation in flight duration among individual *Tetraopes* beetles: implications for studies of insect flight. J Insect Physiol 26:403–406

Davis NT (1975) Hormonal control of flight muscle histolysis in *Dysdercus fulvoniger*. Ann Entomol Soc Am 68:710–714

Dempster JP (1963) The population dynamics of grasshoppers and locusts. Biol Rev Camb Philos Soc 38:490–529

Denno RF (1976) Ecological significance of wing-polymorphism in Fulgoroidea which inhabit salt marshes. Ecol Entomol 1:257–266

Denno RF (1979) The relation between habitat stability and the migration tactics of planthoppers, *Miscellaneous publications*. Entomol Soc Am 11:41–49

Denno RF (1983) Tracking variable host plants in space and time. In: Denno RF, McClure MS (eds) Variable plants and herbivores in natural and managed systems. Academic Press, New York, pp 291–341

Denno RF (1985) Fitness, population dynamics, and migration in planthoppers: the role of host plants. In: Rankin MA (ed) Migration: mechanisms and adaptive significance. Contrib Mar Sci Suppl, vol 27, pp 613–630

Denno RF, Dingle H (1981) Considerations for the development of a more general life history theory. In: Denno RF, Dingle H (eds) Insect life history patterns: habitat and geographic variation. Springer, Berlin Heidelberg New York, pp 1–8

Dickson RC, Laird EF, Pesho GR (1955) The spotted alfalfa aphid (*Theriaphis maculata*; Predator relationship: Hemiptera, Cocinellidae, Syrphidae, Neuroptera). Hilgardia 24:93–118

Dingle H (1965) The relation between age and flight activity in the milkweed bug *Oncopeltus*. J Exp Biol 42:269–283

Dingle H (1968) The influence of environments and heredity on flight activity in the milkweed bug *Oncopeltus*. J Exp Biol 48:175–184

Dingle H (1972) Migration strategies of insects. Science 175:1327–1335

Dingle H (1982) Function of migration in the seasonal synchronization of insects. Entomol Exp Appl 31:36—48

Dingle H (1985) Migration and life history patterns. In: Rankin MA (ed) Migration: mechanisms and adaptive significance. Contrib Mar Sci Suppl, vol 27, pp 27—42

Dingle H, Arora G (1973) Experimental studies of migration in bugs of the genus *Dysdercus*. Oecologia (Berl) 12:112—140

Downer RGH (1972) Interspecificity of lipid-regulating factors from insect corpus cardiacum. Can J Zool 50:63

Downer RGH, Steele JE (1972) Humoral stimulation of lipid transport in the American cockroach, *Periplaneta americana*. Gen Comp Endocrinol 19:1341

Elton CS (1936) Animal ecology, 2nd ed. MacMillan, London

Gade G (1979) Studies on the influence of synthetic adipokinetic hormone and some analogs on cyclic AMP levels in different arthropod systems. Gen Comp Endocrinol 37:122—130

Goldsworthy GJ (1985) The endocrine control of flight muscle metabolism in locusts. Adv Insect Physiol 17:147—204

Goldsworthy GJ, Mordue W, Guthkelch J (1972) Studies on insect adipokinetic hormones. Gen Comp Endocrinol 18:545—551

Grinnell J (1922) The role of the "accidental." Auk 39:373—380

Hagen KS (1962) Biology and ecology of predacious Coccinellidae. Annu Rev Entomol 7:289—326

Herman WS (1973) The endocrine basis of reproductive inactivity in monarch butterflies overwintering in central California. J Insect Physiol 19:1883—1887

Herman WS (1981) Studies on the adult reproductive diapause of the monarch butterfly, *Danaus plexippus*. Biol Bull (Woods Hole) 160:89—106

Herman WS (1985) Hormonally mediated events in adult monarch butterflies. In: Rankin MA (ed) Migration: mechanisms and adaptive significance. Contrib Mar Sci Suppl, vol 27, pp 799—815

Highnam KC, Haskall PT (1964) The endocrine system of isolated and crowded *Locusta* and *Schistocerca* in relation to oocyte growth, and the effects of flying upon maturation. J Insect Physiol 10:849—864

Johnson B (1958) Factors affecting the locomotor and settling responses of alate aphids. Anim Behav 6:9—26

Johnson CG (1966) A functional system of adaptive dispersal by flight. Annu Rev Entomol 11:233—260

Johnson CG (1969) Migration and dispersal of insects by flight. Methuen, London

Kennedy JS (1961) A turning point in the study of insect migration. Nature 189:785—791

Lande R (1979) Quantitative genetic analysis of multivariate evolution, applied to brain: body size allometry. In: Dingle H, Hegmann J (eds) Evolution and genetics of life histories. Springer, Berlin Heidelberg New York

Lande R (1982) A quantitative genetic theory of life history evolution. Ecology 63:607—615

Lavie B, Ritte U (1978) The relationship between dispersal behavior and reproductive fitness in the flour beetle *Tribolium castaneum*. Can J Genet Cytol 20:589—595

Lessman CA, Herman WS (1983) Seasonal variation in hemolymph juvenile hormone of adult monarchs (*Danaus p. plexippus*: Lepidoptera). Can J Zool 61:88—94

Lidicker WZ Jr, Caldwell RL (1982) Dispersal and migration. Hutchinson Ross, Stroudsberg, PA

McAnelly ML (1984) The role of migration in the life history of the grasshopper, *Melanoplus sanguinipes*. PhD Dissertation, University of Texas at Austin

McAnelly ML (1985) Variation in migratory behavior and its control in the grasshopper, *Melanoplus sanguinipes*. In: Rankin MA (ed) Mechanisms and adaptive significance. Contrib Mar Sci Suppl, vol 27, pp 687—703

Myers JH (1976) Distribution and dispersal in populations capable of resource depletion. Oecologia (Berl) 23:255—269

Norris MJ (1950) Reproduction in the African migratory locust (*Locusta migratora migratorioides* R. and F.) in relation to density and phase. Anti-Locust Bull 6:1—10

Norris MJ (1952) Reproduction in the Desert Locust (*Schistocerca gregaria* Forsk.) in relation to density and phase. Anti-Locust Bull 13

Palmer JO (1985) Ecological genetics of wing length, flight propensity, and early fecundity in a migratory insect. In: Rankin MA (ed) Migration: mechanisms and adaptive significance. Contrib Mar Sci Suppl, vol 27, pp 653—663

Parsons PA (1982) Adaptive strategies of colonizing animal species. Biol Rev 57:117–148

Ramchandra R (1942) Some results on the Desert Locust (*Schistocerca gregaria* Forsk.) in India. Bull Entomol Res 33:241–265

Rankin MA (1974) The hormonal control of flight in the milkweed bug, *Oncopeltus fasciatus*. In: Barton Brown L (ed) Experimental analysis of insect behaviour. Springer, Berlin Heidelberg New York, pp 317–328

Rankin MA (1978) Hormonal control of insect migration. In: Dingle H (ed) Evolution of migration and diapause in insects. Springer, Berlin Heidelberg New York, pp 5–32

Rankin MA (1980) Effects of precocene I and II on flight behaviour in *Oncopeltus fasciatus*, the migratory milkweed bug. J Insect Physiol 26:67–76

Rankin MA, Rankin SM (1980b) Some factors affecting presumed migratory flight activity of the convergent ladybeetle, *Hippodamia convergens* (Coccinellidae: Coleoptera). Biol Bull 158: 336–369

Rankin MA, Riddiford LM (1977) Hormonal control of migratory flight in *Oncopeltus fasciatus*: the effects of the corpus cardiacum, corpus allatum, and starvation on migration and reproduction. Gen Comp Endocrinol 33:309–321

Rankin MA, Riddiford LM (1978) The significance of haemolymph juvenile hormone titer changes in the timing of migration and reproduction in adult *Oncopeltus fasciatus*. J Insect Physiol 24: 31–38

Rankin SM (1980) The physiology of migration and reproduction in the ladybird beetle, *Hippodamia convergens* Guerin-Meneville. PhD Dissertation, University of Texas, Austin

Rankin SM, Rankin MA (1980a) The hormonal control of migratory flight behaviour in the convergent ladybird beetle, *Hippodamia convergens*. Physiol Entomol 5:175–182

Roff DA (1975) Population stability and the evolution of dispersal in a heterogeneous environment. Oecologia (Berl) 19:217–237

Roff DA (1977) Dispersal in dipterans: its costs and consequences. J Anim Ecol 46:443–456

Roff DA (1984) The cost of being able to fly: a study of wing polymorphism in two species of crickets. Oecologia (Berl) 63:30–37

Roffey J (1963) Observations on night flight in the Desert Locust (*Schistocerca gregaria* Forskal). Anti-Locust Bull no 39

Rose DJW (1972) Dispersal and quality in populations of *Cicadulina* species (Cicadellidae). J Anim Ecol 41:589–609

Rygg TD (1966) Flight of *Oscinella frit* L. (Diptera, Chloropidae) females in relation to age and ovary development. Entomol Exp Appl 9:74–84

Simberloff D (1981) What makes a good island colonist? In: Denno RF, Dingle H (eds) Insect life history patterns: habitat and geographic variation. Springer, Berlin Heidelberg New York, pp 195–206

Slansky F Jr (1980) Food consumption and reproduction as affected by tethered flight in female milkweed bugs *(Oncopeltus fasciatus)*. Entomol Exp Appl 28:277–286

Solbreck C (1978) Migration, diapause, and direct development as alternative life histories in a seed bug *Neacoryphus bicrucis*. In: Dingle H (ed) The evolution of insect migration and diapause. Springer, Berlin Heidelberg New York, pp 195–217

Solbreck C (1985) Insect migration strategies and population dynamics. In: Rankin MA (ed) Migration: mechanisms and adaptive significance. Contrib Mar Sci Suppl, vol 27, pp 631–652

Southwood TRE (1962) Migration of terrestrial arthropods in relation to habitat. Biol Rev 37: 171–214

Southwood TRE (1977) Habitat, the templet for ecological strategies? J Anim Ecol 46:337–465

Southwood TRE, May RM, Hassell MP, Conway GR (1974) Ecological strategies and population parameters. Am Nat 108:791–804

Stengel M (1974) Migratory behaviour of the female of the common cockchafer *Melolontha melolontha* L. and its neuroendocrine regulation. In: Barton Browne L (ed) Experimental analysis of insect behaviour. Springer, Berlin Heidelberg New York

Taylor LR, Taylor RAJ (1977) Aggregation, migration and population mechanics. Nature, London 265:415–421

Van Valen L (1971) Group selection and the evolution of dispersal. Evolution 25:591–598

Walters KFA, Dixon AFG (1983) Migratory urge and reproductive investment in aphids: variation within clones. Oecologia (Berl) 58:70–75 de Wilde J, de Boer JA (1961) Physiology of diapause in the adult Colorado potato beetle II. Diapause as a case of pseudoallatectomy. J Insect Physiol 6:152–161

de Wilde J, de Boer JA (1969) Humoral and nervous pathways in photoperiodic induction of diapause in *Leptinotarsa decemlineata* Say. J Insect Physiol 15:661–675

de Wilde J, Staal G, de Kort C, DeLoof A, Baard G (1968) Juvenile hormone titer in the hemolymph as a function of photoperiodic treatment in the adult Colorado potato beetle (*Leptinotarsa decemlineata* Say). Proc K Ned Akad Wet Ser C Biol Med Sci 71:321–326

Williams CB (1958) Insect migration. Macmillan, New York

Wratten SD (1977) Reproductive strategy of winged and wingless morphs of the aphids *Sitobion avenaa* and *Metapolohium dirhodum*. Ann Appl Biol 85:319–331

Wynne-Edwards VC (1962) Animal dispersion in relation to social behavior. Oliver and Boyd, Edinburgh

Zera AJ (1984) Differences in survivorship, development rate and fecundity between the long-winged and wingless morphs of the waterstrider, *Limnoporus canaliculatus*. Evolution 38:1023–1032

Zera AJ (1985) Wing polymorphism in waterstriders (Gerridae: Hemiptera): mechanism of morph determination and fitness differences between morphs. In: Rankin MA (ed) Migration: mechanisms and adaptive significance. Contrib Mar Sci Suppl, vol 27, pp 664–686

4 The Endocrine Control of Flight Metabolism in Locusts

G. J. GOLDSWORTHY and C. H. WHEELER [1]

Reserves of respiratory fuels in insect flight muscles are usually only sufficient to meet energy requirements at the initiation of flight, and are too small to sustain prolonged flight activity. The flight muscles must thus take up fuels from the haemolymph. In locusts, the carbohydrate content of the haemolymph is greater than that stored in the tissues, but the opposite is true for the fat reserves (see Goldsworthy 1983). Hormones play an important role in coordinating the supply and utilisation of fuels during sustained flight in locusts, and this will be the major theme of this review.

1 Adipokinetic Hormones

Adipokinetic activity in locust corpora cardiaca was first demonstrated simultaneously by Beenakkers (1969) and Mayer and Candy (1969). Subsequent investigations of the nature and action of the active materials present in the corpora cardiaca were facilitated by the development of a simple quantitative bioassay, and showed that the majority of the biological activity is present in the glandular lobes (Goldsworthy et al. 1972). It is now clear that there are at least two adipokinetic peptides in the corpora cardiaca of locusts. To distinguish between the two known peptides, we will use the abbreviation AKH-I for the peptide characterised by Stone et al. (1976), and AKH-II for compound-II of Carlsen et al. (1979). Where no particular peptide is specified, or where tissue extracts have been used as a hormone source, we will use AKH as a general term for 'adipokinetic hormone(s)'.

AKH-I is a blocked decapeptide with the structure Glu—Leu—Asn—Phe—Thr—Pro—Asn—Trp—Gly—Thr—NH_2 (Stone et al. 1976; Broomfield and Hardy 1977). The two known AKH-II peptides are also blocked, but are octapeptides (Carlsen et al. 1979; Gäde et al. 1984) whose structure varies between locust species: they are both analogues of the red pigment concentrating hormone (RPCH) of prawns (Fernlund and Josefsson 1972), namely (Ala^6)–RPCH in *Locusta* and (Thr^6)–RPCH in *Schistocerca gregaria* (Siegert et al. 1985; Gäde et al. 1986) and *Schistocerca nitans* (Gäde et al. 1986).

1.1 The Actions of AKH

AKH acts on the fat body to increase the release of stereospecific diacylglycerols (Tietz and Weintraub 1980; Lok and Van der Horst 1980) into the haemolymph. The mecha-

---

[1] Department of Zoology, University of Hull, Hull HU6 7RX, United Kingdom.

Insect Flight: Dispersal and Migration
Edited by W. Danthanarayana
© Springer-Verlag Berlin Heidelberg 1986

nism of its action probably involves activation (phosphorylation) of a lipase by a cAMP-dependent-protein kinase. AKH increases the levels of cAMP in the fat body (Spencer and Candy 1976; Gäde 1979) and, recently, Pines et al. (1981) have shown that crude extracts of glandular lobes of the corpora cardiaca activate a cAMP-dependent-protein kinase. In addition, these authors showed that diacylglycerol lipases in the fat body are activated by cAMP and cGMP in vitro. The effect of either AKH-I or AKH-II on these lipases was not investigated, but RPCH activated locust fat body protein kinase in vitro, and monoacylglycerol transferase and an unspecified lipase in vivo.

Neutral lipids such as diacylglycerols are water-insoluble and need to be carried in the haemolymph as part of macromolecular complexes called lipoproteins. Mwangi and Goldsworthy (1977) proposed that during AKH action the resting haemolymph lipoprotein, A$yellow$, and the C_L-proteins in the haemolymph combine together with extra diacylglycerol from the fat body to form a new lipoprotein, A$^+$ (see Goldsworthy 1983; Wheeler et al. 1984a). The C_L-proteins bind reversibly to A$^+$ lipoprotein; they can be pulled off the lipoprotein during polyacrylamide electrophoresis, and bound [^{3}H]-labelled C_L-proteins are displaced from A$^+$ in vitro. We believe that these C_L-proteins play a part in directing the interactions of the lipoproteins with the tissues (see below). Evidence for the model proposed by Mwangi and Goldsworthy (1977) has been discussed recently by Goldsworthy (1983) and Goldsworthy and Wheeler (1984) and will not be dealt with here. It should however be noted that there is no evidence for a direct effect of AKH on haemolymph protein or lipoprotein interactions in the formation of lipoprotein A$^+$, although this possibility cannot be excluded.

A second site of action of AKH is in the flight muscles where, by increasing the oxidation of lipids, it indirectly inhibits trehalose utilisation by an inhibition of glycolysis (see Sect. 3). The evidence for a direct action of AKH on the flight muscles derives from experiments in vivo and in vitro and has been discussed at length recently (Goldsworthy 1983), but the exact mechanism of this direct effect on the flight muscles remains uncertain. We have however shown that AKH also exerts an indirect effect on flight muscle metabolism via its effect on the transformation of lipoprotein A$yellow$ to lipoprotein A$^+$. Robinson and Goldsworthy (1977) found that the quality and quantity of lipoprotein preparations are important for the effect of AKH on lipid and carbohydrate utilisation in perfused flight muscles: progressive dilution of haemolymph preparations containing lipoprotein A$^+$ produces a graded attenuation of the inhibition of glycolysis, even in the presence of hormone, whereas a fourfold increase in the concentration of haemolymph containing A$yellow$ (with or without AKH) does not further suppress the utilisation of trehalose by perfused flight muscles. Wheeler and his colleagues (1984b) have demonstrated the presence in the flight muscles of a lipoprotein lipase, and we have recently shown that the activity of this enzyme is indirectly controlled by AKH.

AKH has no direct effect on lipoprotein lipase activity in flight muscle in vitro, but when flight muscles from animals injected with synthetic AKH-I are assayed using artificial lipid emulsions, the lipase activity is higher than that in flight muscles from non-injected locusts. This effect of AKH-I is dose-dependent (C.H. Wheeler, unpublished observations) and reversible; injection of C_L-proteins prior to removal of the flight muscles abolishes the AKH-dependent increase in enzymatic activity (see Fig. 1). In

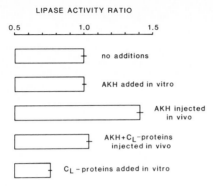

LIPASE ACTIVITY RATIO

0.5   1.0   1.5

no additions

AKH added in vitro

AKH injected
in vivo

AKH+C$_L$-proteins
injected in vivo

C$_L$ - proteins added in vitro

Fig. 1. The effect of sAKH-I and C$_L$-proteins on flight-muscle lipase activity in vitro and in vivo. Lipase activities are expressed as an activation ratio in which the measured lipase activity of the various treatment groups was divided by the activity of the untreated control group; values > 1 indicate activation, whilst those < 1 signify inhibition of the enzyme. In the groups indicated, sAKH-I (2 pmol) and C$_L$-proteins (5 mg/locust) were injected 90 min and 10 min respectively before muscle removal. sAKH-I and C$_L$-proteins, when added in vitro, gave respective concentrations of 2 pmol/250 μl and 9 μg/μl. Lipase activity was determined by measurement of [^{3}H]-glycerol release from locust haemolymph lipids (radiolabelled after the method of Wheeler et al. 1984b) sonicated with glycerol and phosphatidylcholine to form stable emulsions in buffer (pH 7.65) with 10% albumin. Product and substrate were separated by addition of 10% TCA (see Wheeler et al. 1984b). C$_L$-proteins were prepared from 2 g of lyophilised resting haemolymph taken up in 25 ml distilled water and mixed with 3 vol of dextran sulphate solution (0.66 g/l in 0.05 M MnCl$_2$). After centrifugation, solid ammonium sulphate was added to the decanted supernatant to reach 100% saturation. The proteinaceous precipitate was dissolved in 25% ammonium sulphate before separation on Sephadex G-75 in 0.1 M ammonium acetate. The peak of C$_L$-proteins was freeze dried and stored at $-15\,^\circ$C until needed

addition, C$_L$-proteins inhibit lipase activity in vitro (Fig. 1). These results suggest that in resting locusts, high concentrations of free C$_L$-proteins exert a direct inhibitory effect on lipase activity, but that during AKH action the binding of C$_L$-proteins to lipoprotein A$^+$ leads to 'activation' by a reduction of apoprotein binding to the enzyme.

Furthermore, the lipoprotein lipase in flight muscle shows a remarkable substrate specificity. Diacylglycerols associated with lipoprotein A$^+$ are hydrolysed at four to five times the rate of those associated with lipoprotein A*yellow* at all concentrations tested (Fig. 2). This discrimination is further emphasised when mixed substrates are used; in the presence of lipoprotein A$^+$ (at concentrations found normally in flying locusts), hydrolysis of A*yellow* lipids is reduced to ca. 10% of the values when A*yellow* is present as the sole substrate (C.H. Wheeler, unpublished observations). This provides a molecular basis for the observation that only lipoprotein A$^+$ in the presence of AKH is able to provide lipids to the flight muscles during long-term flight (Mwangi and Goldsworthy 1981) or contraction of the flight muscles in vitro (Robinson and Goldsworthy 1977).

1.2 AKH-I Versus AKH-II

One intriguing question relates to the need for two adipokinetic peptides in the locust. Do they perform different functions? AKH-II may be responsible for some lipid mobi-

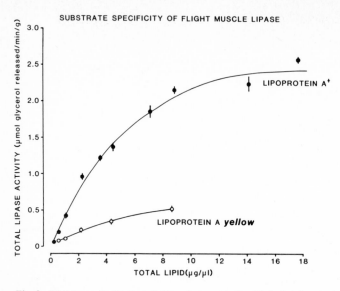

Fig. 2. Flight-muscle lipase activity when assayed with increasing concentrations of two different lipoprotein substrates. Lipoprotein A*yellow* was prepared from haemolymph of control resting locusts in which the majority of lipids are carried as part of lipoprotein A*yellow*. Preparations of lipoprotein A⁺ were obtained from locusts injected with 2 pmol sAKH-I (90 min before bleeding) in which the majority of lipid is carried as part of lipoprotein A⁺. Radiolabelling and preparation of substrates and enzyme assay were performed by the method of Wheeler et al. (1984b).

lisation during flight; although it is present in the corpora cardiaca in lesser amounts than AKH-I, it has a similar lipolytic potency to AKH-I on a molar basis at low doses of hormone (G. Gäde and G.J. Goldsworthy, unpublished observations), but with AKH-II the maximum lipid increase reaches only ca. 60% of that with AKH-I (Fig. 3). AKH-II does stimulate lipoprotein A⁺ formation, but produces significantly smaller amounts than does AKH-I, even when very high doses of AKH-II are used (K. Mallison and G.J. Goldsworthy, unpublished observations).

It has been suggested (Orchard and Lange 1983a) that, because it causes hypertrehalosemia in ligated locusts, AKH-II may contribute to the supply of carbohydrate to the flight muscles during flight. Glycogen phosphorylase in the fat body does increase in activity during flight (Van Marrewijk et al. 1980), and AKH-I is a potent activator of this enzyme (Gäde 1981). The relative activities of the two locust adipokinetic peptides (see Fig. 3) in activating glycogen phosphorylase are, however, similar to their hyperlipaemic potencies (G. Gäde, R. Ziegler, G.J. Goldsworthy, unpublished observations). But surprisingly, their relative effects in causing cAMP accumulation in locust fat body in vivo are in complete contrast to their hyperlipaemic and glycogen phosphorylase-activating properties (Fig. 3). The physiological significance of these differences has yet to be determined. Nevertheless, because of the minimal levels of glycogen stored in the fat body of laboratory locusts (Goldsworthy and Gäde 1982), neither peptide is likely to make a quantitative contribution to carbohydrate homeostasis during the early stages of flight (cf. Orchard and Lange 1983a). Indeed, the levels of haemo-

DOSE RESPONSE CURVES FOR NATURAL LOCUST ADIPOKINETIC HORMONES

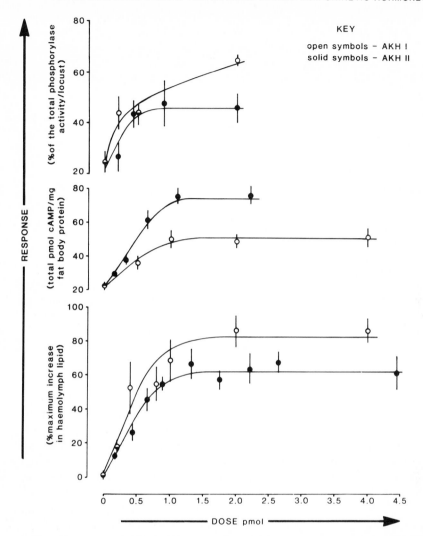

Fig. 3. A comparison of the activities of AKH-I and II separated on Sephadex LH20 in 6% aqueous butanol (Carlsen et al. 1979) and assayed in three different systems. In all cases hormones were injected in vivo before assay. In the *upper* graphs, glycogen phosphorylase activity (% of the total in the active form/locust fat body) is shown. The *middle* graphs show the cAMP/mg fat-body protein measured (by a competitive protein-binding assay) 15 min after hormone injection. The *lower* graphs show the haemolymph total lipid measured 90 min after hormone injection. In all cases, measurements were made at the time of peak response, and values (mean ± SE) for at least 10 (cAMP and lipid assays) or 6 (phosphorylase assay) observations are corrected for any changes obtained by injection of saline. The doses of hormone are estimated on the basis that on average ca. 95 ng of AKH-I and ca. 11 ng of AKH-II is present in methanolic extracts of one pair of glandular lobes from *Locusta* (see Goldsworthy and Wheeler 1984)

lymph trehalose fall dramatically during the early stages of flight and only reach a new steady state after ca. 30 min (see Goldsworthy 1983) when lipid is the predominant fuel for the flight muscles. At this time either or both of the locust adipokinetic peptides could play an important role in maintaining a steady (but low level) supply of carbohydrate to the flight muscles, but this is difficult to reconcile with the rapid (within 5 min) activation of glycogen phosphorylase during flight shown by Van Marrewijk et al. (1980).

1.3 The Release of AKH During Flight

Release of AKH during flight is under the control of secretomotor centres in the lateral areas of the protocerebrum (Rademakers 1977), and this control is probably octopaminergic (Orchard and Loughton 1981). Although the concentration of carbohydrates in the haemolymph can modulate AKH release, it seems unlikely that their decreasing concentration during the early stages of flight is directly responsible for the release of the hormones (see Goldsworthy 1983). Indeed, according to Orchard and Lange (1983b) the times of release of AKH-I and AKH-II appear to be out of phase. Orchard and Lange (1983b) propose that adipokinetic peptides are not released during the initial 15 min of flight. They believe that lipid mobilisation is biphasic, with the increase in lipid during the first 15 min phase being effected by octopamine (see below), and the subsequent, more prolonged phase being due to release of AKH-I and AKH-II. Nevertheless, a rapid release of AKH is suggested by Rademakers and Beenakkers (1977) who found a significant increase in the number of exocytotic profiles in the glandular lobe tissue after only 5 min of flight. When release of hormones is assessed by bioassay of haemolymph from flown locusts, however, AKH is not detected in measurable amounts until 10 or 15 min of the commencement of flight. The inability to measure AKH titers during the early minutes of flight could be due to lack of sensitivity in the methodology (see Goldsworthy and Wheeler 1984) and this is clearly an important question for future research.

2 Octopamine

Octopamine is chemically related to the vertebrate catecholamines, and is present in large amounts in insect nervous tissue (Evans 1978). In the haemolymph of resting locusts only the D-isomer is present (Goosey and Candy 1980).

2.1 Changes in Haemolymph Titre of Octopamine

The concentration of octopamine in the haemolymph increases dramatically during the first 10 min of flight in *Schistocerca*, and then declines rapidly towards the resting level as flight continues (Goosey and Candy 1980). The site of release or origin of this octopamine is uncertain, but Goosey and Candy (1982a) suggest that it may be released from nerves, originating in the thoracic ganglia, which innervate the flight muscles. In

Locusta, excitation by handling produces a rapid increase in haemolymph octopamine within 15 min, which persists for some time, but falls appreciably between 1 and 2 h after excitation (Orchard et al. 1981). In *Schistocerca* the turnover of octopamine appears to be more rapid both during flight (Goosey and Candy 1980) and rest (Goosey 1981), and in response to mechanical stress (Davenport and Evans 1984a), than it is in *Locusta*.

2.2 Actions of Haemolymph Octopamine

Circulating octopamine may act both on fat body, and flight and skeletal muscles (see Goldsworthy 1983; Evans and Siegler 1982). Orchard and his colleagues (1981) describe a small, but rapid and relatively long-lived octopamine-mediated hyperlipaemic response to stress in *Locusta*. This appears to be a direct effect of the octopamine on the fat body (Orchard et al. 1982). The rapid rise in the haemolymph titre of octopamine during flight may, therefore, initiate the mobilisation of diacylglycerol during the earliest moments of flight (Orchard et al. 1981; Orchard and Lange 1983b).

Octopamine stimulates the oxidation in vitro of a variety of substrates in locust flight muscle; it increases also the size of muscle contractions (Candy 1978). The effects on glucose oxidation occur at concentrations of octopamine comparable to those found in the haemolymph after a few minutes flight and are specific to the naturally occurring D-isomer (Goosey and Candy 1980). According to Candy (1978), octopamine doubles the rate of flight muscle glucose oxidation in vitro, but stimulation of butyrate or diacylglycerol oxidation is less pronounced. This may be significant in the overall control of flight muscle metabolism in the presence of several hormones during the early minutes of flight (see below). Octopamine may act on flight muscle metabolism via a stimulation of adenylate cyclase (Worm 1980); theophylline and dibutyryl cAMP have a similar, although slightly weaker, effect than octopamine on flight muscle glucose oxidation in vitro (Candy 1978).

3 An Overview of the Hormonal Control of Locust Flight Muscle Metabolism

In previous reviews from this laboratory (Goldsworthy and Cheeseman 1978; Goldsworthy 1983), parallels have been drawn between hormonal and metabolic events occurring during locust flight and those in some exercising vertebrates: muscular activity leads initially to an increased utilisation of 'blood' carbohydrate by the active muscles, and mobilisation of fat body or liver glycogen. But as exercise is prolonged, lipid mobilisation from the fat body or the adipose tissue is stimulated, and muscle metabolism is increasingly powered by fatty acid oxidation. Consequently, during exercise changes in tissue and 'blood' metabolites of a wide variety of animals can be superficially similar (Goldsworthy and Cheeseman 1978). Indeed, these considerations apply also to fasting animals in which changes in metabolites and metabolism can be quite similar between disparate animals but, more importantly, can resemble qualitatively those observed during sustained exercise: their requirements in both situations are identical, because they need to prevent too rapid a depletion of their limited stores of carbohydrate fuel (at least in mammals, some tissues have a continuous requirement for glucose). To achieve this, they switch to an alternative (and more ergonomic) fuel in the form of lipid.

An essential component of this strategy is that glycolysis in the muscles is inhibited during periods of increased fatty acid oxidation. In locusts, the point at which glycolysis is inhibited is not the phosphofructokinase reaction (as it is in mammals), because this enzyme in locusts and a variety of other insects is insensitive to citrate (Goldsworthy 1983). How then is the inhibition of glycolysis and the consequent reduction in the rate of utilisation of haemolymph trehalose brought about? Robinson and Goldsworthy (1977) found that the oxidation of trehalose by the flight muscles was reduced during diacylglycerol oxidation, but that of glycerol was unaffected. Ford and Candy (1972) have identified the aldolase reaction as a possible regulatory point for glycolysis when butyrate is present in flight muscle perfusions, and Storey (1980) has shown that locust flight muscle aldolase is sensitive to inhibition by both citrate and palmitoyl-carnitine. Because the concentration of acyl-carnitine in the flight muscles increases fivefold within the first 30 min of flight (Worm et al. 1980), it seems likely that accumulation of this intermediate (during increased oxidation of fatty acids in the flight muscles) could regulate glycolytic flux and offer an explanation for the observed decrease in trehalose oxidation which is characteristic of prolonged flight (see Goldsworthy 1983).

Within the last few years it has become clear that a number of hormones may operate to control locust flight metabolism. The role of AKH is well documented, at least for AKH-I, but there is the possibility that AKH-II could play an additional and separate role from AKH-I. This remains, however, to be determined. The changes in the haemolymph titre of octopamine, and its effects on fat body and flight muscle in vitro, are also clearly important. Finally, it is possible that 'insulin-like' hormones (in their action, if not in their chemical nature) such as the hypolipaemic factor (Orchard and Loughton 1980) and the storage lobe factor of Goosey and Candy (1982b) could be involved in the control of locust flight metabolism.

A particularly intriguing aspect of recent developments in this area is an analogy between octopamine in insects, and catecholamines in vertebrates. Octopaminergic cells in the locust metathoracic ganglia innervate the extensor tibiae muscles, where octopamine plays a neuromodulator role comparable with the role of noradrenaline in the mammalian sympathetic nervous system (O'Shea and Evans 1979; Evans and Siegler 1982). Octopamine which is released into the haemolymph could play an important role in directing energy metabolism during the first few minutes of flight. Orchard and his colleagues (1981) suggest that it may mediate a 'flight or fight' response. Indeed, Davenport and Evans (1984a,b) have shown that octopamine is released rapidly not only during mechanical stress, but also during the metabolic stress of starvation.

The full importance of octopamine in affecting flight muscle metabolism has yet to be elucidated, but it is tempting to suggest that it may play a vital role in maintaining the oxidation of glucose in the early stages of flight. Thus, at a time when increasing concentrations of diacylglycerol in the haemolymph (brought about by the release of AKH) compete with decreasing concentrations of trehalose, octopamine may favour glucose oxidation in the flight muscles. The increase in diacylglycerol oxidation in the flight muscles would therefore be gradual (see Mayer and Candy 1969; Van der Horst et al. 1980) and allow a smooth transition from mainly carbohydrate to mainly lipid oxidation when, eventually, sufficient diacylglycerol has been made available in the haemolymph to support flight muscle metabolism adequately. In this way, the flight muscle machinery would be prevented from 'stalling' due to inadequate or inappropriate levels of fuel in the haemolymph.

Acknowledgements. We thank Ms. A. Street for typing the manuscript. The original research described in this paper was supported by a grant from the S.E.R.C.

References

Beenakkers AMT (1969) The influence of corpus cardiacum on lipid metabolism in *Locusta migratoria*. Gen Comp Endocrinol 13, abstract 12

Broomfield CE, Hardy PM (1977) The synthesis of locust adipokinetic hormone. Tetrahedron Lett 25:2201–2204

Candy DJ (1978) The regulation of locust flight muscle metabolism by octopamine and other compounds. Insect Biochem 8:177–181

Carlsen J, Herman WS, Christensen M, Josefsson L (1979) Characterization of a second peptide with adipokinetic and red-pigment concentrating activity from the locust corpora cardiaca. Insect Biochem 9:497–501

Davenport AP, Evans PD (1984a) Stress-induced changes in the octopamine levels of insect haemolymph. Insect Biochem 14:135–143

Davenport AP, Evans PD (1984b) Changes in haemolymph octopamine levels associated with food deprivation in the locust, *Schistocerca americana gregaria*. Physiol Entomol 9:269–274

Evans PD (1978) Octopamine distribution in the insect nervous system. J Neurochem 30:1009–1013

Evans PD, O'Shea M (1978) The identification of an octopaminergic neurone and the modulation of a myogenic rhythm in the locust. J Exp Biol 73:235–260

Evans PD, Siegler MVS (1982) Octopamine mediated relaxation of maintained and catch tension in locust skeletal muscle. J Physiol (Lond) 324:93–112

Fernlund P, Josefsson L (1972) Crustacean color-change hormone: amino acid sequence and chemical synthesis. Science 177:173–175

Ford WCL, Candy DJ (1972) The regulation of glycolysis in perfused locust flight muscle. Biochem J 130:1101–1112

Gäde G (1979) Studies on the influence of synthetic adipokinetic hormone and some analogs on cyclic AMP levels in different arthropod systems. Gen Comp Endocrinol 37:122–130

Gäde G (1981) Activation of fat body glycogen phosphorylase in *Locusta migratoria* by corpus cardiacum extract and synthetic adipokinetic hormone. J Insect Physiol 27:155–161

Gäde G, Goldsworthy GJ, Kegel G, Keller R (1984) Single step purification of locust adipokinetic hormones I and II by reversed phase high-performance liquid chromatography, and amino acid composition of the hormone II. Hoppe-Seyler's Z Physiol Chem 365:391–398

Gäde G, Goldsworthy G, Schaffer MH, Carter Cook J, Rinehart KL Jr (1986) Sequence analysis of adipokinetic hormones II from corpora cardiaca of *Schistocerca nitans, Schistocerca gregaria,* and *Locusta migratoria* by fast atom bombardment mass spectrometry. Biochem Biophys Res Commun 134:723–730

Goldsworthy GJ (1983) The endocrine control of flight metabolism in locusts. In: Berridge MJ, Treherne JE, Wigglesworth VB (eds) Advances in insect physiology, vol 17. Academic Press, London, pp 149–204

Goldsworthy GJ, Cheeseman P (1978) Comparative aspects of the endocrine control of energy metabolism. In: Gaillard PJ, Boer HH (eds) Comparative endocrinology. Elsevier, Amsterdam, pp 423–436

Goldsworthy GJ, Gäde G (1982) The chemistry of hypertrehalosemic factors. In: Downer RGH, Laufer H (eds) Insect endocrinology. Liss, New York, pp 109–119

Goldsworthy GJ, Wheeler CH (1984) Adipokinetic hormones in locusts. In: Hoffman JK, Porchet M (eds) Biosynthesis, metabolism and mode of action of invertebrate hormones. Springer, Berlin Heidelberg New York, pp 126–135

Goldsworthy GJ, Mordue W, Guthkelch J (1972) Studies on insect adipokinetic hormones. Gen Comp Endocrinol 18:545–551

Goosey MW (1981) The regulation of insect flight muscle metabolism by octopamine. PhD Thesis, University of Birmingham

Goosey MW, Candy DJ (1980) The D-octopamine content of the haemolymph of the locust *Schistocerca americana gregaria* and its elevation during flight. Insect Biochem 10:393–397

Goosey MW, Candy DJ (1982a) The release and removal of octopamine by tissues of the locust *Schistocerca americana gregaria*. Insect Biochem 12:681–685

Goosey MW, Candy DJ (1982b) The regulation of substrate oxidation in locust flight muscle by a factor from the corpus cardiacum. Biochem Soc Trans 10:276

Lok CM, Van der Horst DJ (1980) Chiral 1,2-diacylglycerols in the haemolymph of the locust, *Locusta migratoria*. Biochim Biophys Acta 618:80–87

Mayer RJ, Candy DJ (1969) Control of haemolymph lipid concentration during locust flight: an adipokinetic hormone from the corpora cardiaca. J Insect Physiol 15:611–620

Mwangi RW, Goldsworthy GJ (1977) Diglyceride-transporting lipoproteins in *Locusta*. J Comp Physiol B Metab Transp Funct 114:177–190

Mwangi RW, Goldsworthy GJ (1981) Diacylglycerol-transporting lipoproteins and flight in *Locusta*. J Insect Physiol 27:47–50

Orchard I, Lange AB (1983a) Release of identified adipokinetic hormones during flight and following neural stimulation in *Locusta migratoria*. J Insect Physiol 29:425–429

Orchard I, Lange AB (1983b) The hormonal control of haemolymph lipid during flight in *Locusta migratoria*. J Insect Physiol 29:639–642

Orchard I, Loughton BG (1980) A hypolipaemic factor from the corpus cardiacum of locusts. Nature 286:494–496

Orchard I, Loughton BG (1981) Is octopamine a transmitter mediating hormone release in insects? J Neurobiol 12:143–153

Orchard I, Loughton BG, Webb RA (1981) Octopamine and short-term hyperlipaemia in the locust. Gen Comp Endocrinol 45:175–180

Orchard I, Carlisle JA, Loughton BG, Gole JWD, Downer RGH (1982) In vitro studies on the effects of octopamine on locust fat body. Gen Comp Endocrinol 48:7–13

O'Shea M, Evans PD (1979) Potentiation of neuromuscular transmission by an octopaminergic neurone in the locust. J Exp Biol 79:169–190

Pines M, Tietz A, Weintraub H, Applebaum SW, Josefsson L (1981) Hormonal activation of protein kinase and lipid mobilisation in the locust fat body in vitro. Gen Comp Endocrinol 43:427–431

Rademakers LHPM (1977) Identification of a secretomotor centre in the brain of *Locusta migratoria*, controlling the secretory activity of the adipokinetic hormone producing cells of the corpus cardiacum. Cell Tissue Res 184:381–395

Rademakers LHPM, Beenakkers AMT (1977) Changes in the secretory activity of the glandular lobe of the corpus cardiacum of *Locusta migratoria* induced by flight. A quantitative electron microscope study. Cell Tissue 180:155–171

Robinson NL, Goldsworthy GJ (1977) Adipokinetic hormone and the regulation of carbohydrate and lipid metabolism in a working flight muscle preparation. J Insect Physiol 23:9–16

Siegert K, Morgan P, Mordue W (1985) Primary structures of locust adipokinetic hormones II. Biol Chem Hoppe-Seyler 366:723–727

Spencer IM, Candy DJ (1976) Hormonal control of diacylglycerol mobilization from fat body of the desert locust, *Schistocerca gregaria*. Insect Biochem 6:289–296

Stone JV, Mordue W, Batley KE, Morris HR (1976) Structure of locust adipokinetic hormone, a neurohormone that regulates lipid utilization during flight. Nature 263:207–211

Storey K (1980) Kinetic properties of purified aldolase from flight muscle of *Schistocerca americana gregaria*. Role of the enzyme in the transition from carbohydrate to lipid-fuelled flight. Insect Biochem 10:647–655

Tietz A, Weintraub H (1980) The stereospecific structure of haemolymph and fat body 1,2-diacylglycerol from *Locusta migratoria*. Insect Biochem 10:61–63

Tietz A, Weintraub H, Peled Y (1975) Utilization of 2-acyl sn-glycerol by locust fat body microsomes: specificity of the acyltransferase system. Biochim Biophys Acta 388:165–170

Van der Horst DJ, Houben NMD, Beenakkers AMT (1980) Dynamics of energy substrates in the haemolymph of *Locusta migratoria* during flight. J Insect Physiol 26:441–448

Van Marrewijk WJA, Van Den Broek ATM, Beenakkers AMT (1980) Regulation of glycogenolysis in the locust fat body during flight. Insect Biochem 10:675–679

Wheeler CH, Mundy JE, Goldsworthy GJ (1984a) Locust haemolymph lipoproteins visualised in the electron microscope. J Comp Physiol 154:281–286

Wheeler CH, Van der Horst DJ, Beenakkers AMT (1984b) Lipolytic activity in the flight muscles of *Locusta migratoria* measured with haemolymph lipoproteins as substrates. Insect Biochem 14:261–266

Worm RAA (1980) Involvement of cyclic nucleotides in locust flight muscle metabolism. Comp Biochem Physiol 67C:23–27

Worm RAA, Luytjes W, Beenakkers AMT (1980) Regulatory properties of changes in the contents of coenzyme A. carnitine and their acyl derivatives in flight muscle metabolism of *Locusta migratoria*. Insect Biochem 10:403–408

5 Sounds of Insects in Flight

P. BELTON [1]

1 Introduction

Insects in flight displace air in time with their wing beat and therefore by definition produce sounds. The sounds range from the inaudibly low frequencies of 5-8 Hz produced by saturniid moths with wings 10 cm long (Bienz-Isler 1968) to the almost impossibly high frequencies above 1000 Hz of small biting midges with wings 1 mm long (Sotavolta 1947). Frequency is correlated with wing length, but despite painstaking studies by Sotavolta and others, no simple relationship has emerged between frequency and size that applies to all insects.

The last comprehensive review of the factors affecting frequency was made by Chadwick in 1953. Since then there has been a significant increase in knowledge about insect flight and its associated metabolism (Kammer and Heinrich 1978; Rainey 1976) yet there is still much to learn about the frequency and amplitude of the wing beat of free-flying insects. The present review covers the effect of the major environmental and some physical factors that influence the flight sounds of insects. It is illustrated by several experiments on mosquitoes done at Simon Fraser University and, in the space available, cannot be comprehensive.

These sounds are relevant to migration because the wing-beat frequencies (WBF) of individuals or swarms of insects can be detected by radar using the Doppler effect (Buchan and Satelle 1979) or differences in intensity of the echoes (Schaefer 1976). Although mosquitoes are not noted for their ability to migrate they fit one definition of this term by showing "persistent directional movement" (Kennedy 1961); they can fly continuously for 20 or more hours and are known to travel 150 km or more in nature (Hocking 1953). My primary interest is in the use of wing-beat sounds by female biting Diptera as an attractant for males and I will also consider them briefly from this viewpoint.

Experiments carried out by R.A. Costello and I used free-flying yellow fever mosquitoes, *Aedes aegypti* (L.) in a small soundproof chamber. Frequencies were determined from sonograms as described by Belton and Costello (1979). Statistical significance was tested in most cases with a binomial expansion, i.e. if eight of ten mosquitoes either increased or decreased their frequency, we were 95% confident that this did not occur by chance ($P = 0.044$). Some of the mosquitoes used in the feeding experiments were weighed as a group and the masses reported in Table 1 are therefore only representative.

[1] Centre for Pest Management, Biological Sciences, Simon Fraser University, Burnaby, British Columbia, Canada V5A 1S6.

Insect Flight: Dispersal and Migration
Edited by W. Danthanarayana
© Springer-Verlag Berlin Heidelberg 1986

2 Measuring Flight Sounds

2.1 Frequency

A microphone above or below a flying insect will measure some component (pressure or pressure gradient) of the air movement, with *opposite* signs, on the up- and down-stroke of the wings. From the side, front or rear a microphone will pick up the pressure or pressure gradient of air movement, with the *same* sign, on both the up- and down-stroke. Depending on its position therefore a microphone would record either the fundamental (WBF) or its first harmonic (2 × WBF) or more probably some combination of the two (Fig. 1a).

Analyses have shown that many other harmonics are present in the flight sound of several insects (Belton and Costello 1979) evidently caused by more complex movements of the wing (Nachtigall 1976) and possibly also by the mechanical characteristics of the thoracic box as it is deformed by the indirect flight muscles (Fig. 1b). The same

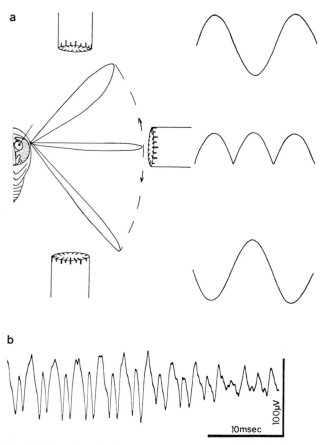

Fig. 1. a Frequency doubling in flight sound. Traces at the *right* represent theoretical oscillograms of the sounds recorded by microphones shown at the *left*. **b** Oscillogram of the flight sound of a free-flying mosquito

argument would apply to radar echoes from the wings of flying insects, but harmonics are not prominent in radar signatures. This is consistent with Schaefer's (1976) statement that radar "reflections from the wings are minute".

2.2 Intensity

The acoustic power output of a flying insect can be estimated by measuring the mean sound pressure level (SPL) around it. For example, $A.$ $aegypti$ has a mean SPL of approximately 50 dB at 2.54 cm. This represents an intensity of 10^{-11} W cm^{-2} or a total at the surface of a 2.54 cm sphere of 8.1×10^{-10} W and is a measure of the acoustic power dissipated in flight. Calculations of the metabolic rates of flying mosquitoes made by Hocking (1953) are about 12–13 W per Newton weight. $A.$ $aegypti$ weighs about 20 μN (mass about 2 mg) so that an estimate of its total metabolic power is 25×10^{-5} W. One-fifth of this power may be transferred to the wings (Weis-Fogh 1972) giving an estimate of 5×10^{-5} W. The fraction of the total power supplied to the wings that is dissipated as sound is therefore $8.1 \times 10^{-10}/5 \times 10^{-5}$, a microscopic 16-millionths of the total, although errors could be compounded in this estimate. Because mosquitoes make such sophisticated use of wing-beat sounds to attract mates (Belton 1974), this estimate makes it very unlikely that *any* insect devotes a significant amount of its metabolic energy toward producing sound from its wings in flight. These calculations also make it clear that wings are very inefficient radiators of the wavelengths they generate and this seems to be true for the whole span of wing lengths mentioned earlier.

2.3 Tethering

Insects must be held in fixed positions if sounds are to be recorded at a particular direction or distance and in many species this increases the load on the flight mechanism and results in higher WBFs and amplitudes (Chadwick 1953). The term flight mechanism is used here to describe the combination of wing muscles, thoracic skeleton and wings. Heinrich (1971), however, pointed out that the vertical loading on the wings was reduced in a tethered sphingid moth and observed a 20% (112°–89°) reduction in wing-beat amplitude and, although he does not discuss it, his results at 30 °C, show a 15% (27–23 Hz) reduction in frequency.

Costello and I found a consistent increase in frequency of about 3% when free-flying $A.$ $aegypti$ females were placed on a flight-mill similar to that used by Hocking (1953) and a further increase of 2% when the mill was stopped, consistent with a response to increased loading of the flight mechanism. The behaviour of tethered insects evidently depends on visual clues and on input from sense organs that detect air movement, consequently differences in response to tethering may vary with the species. The sphingids used by Heinrich may have been hovering when tethered but, in any case, the change in frequency brought about by tethering is small enough to be of little consequence for all but the most critical investigations.

3 Physiological Factors Affecting Flight Sound

3.1 Age

It seems to be normal for flying insects to emerge before the flight muscles and their associated enzymes are completely developed. This may be the reason why the WBF, of those insects that have been studied, increases to a steady level over several days or sometimes weeks. Many of the factors involved have been reviewed in detail by Johnson (1976). The extent of the change in frequency and its duration varies between species. In males of four species of cockroach, Farnworth (1972) found an increase of about 40% (22–31 Hz) over 14 days. In *A. aegypti* we found that the WBF of both sexes increased to a maximum in 3 days, males from 450 to 650 Hz and females from 370 to 500 Hz or 44% and 35% respectively. In *Drosophila*, however, Chadwick (1953) found an increase of only 16% in males and 18% in females over 10 days so it is not possible to generalise even within the same order of insects.

It is unlikely that these changes in frequency are important in the context of this Symposium because insects evidently will not fly over long distances before the WBF has stabilised.

3.2 Sex

In most insects that have been investigated there also seems to be a general tendency for males to have higher WBFs than females. To some extent this may be related to differences in size, but size alone does not explain the differences measured in mosquitoes and related Nematocera where sound is used to distinguish the sexes.

Schaefer (1976) found differences of 10% between male and female desert locusts with large females flapping at about 18 Hz and found that several species of acridids can be sexed from their radar signatures. Farnworth (1972) found smaller differences in *Periplaneta americana* with the females which cannot sustain flight some 3% lower than the males which flap at about 33 Hz at 30 °C.

We have commonly found differences in frequency close to an octave (100%) between the sexes of free-flying mosquitoes of the same age (Belton and Costello 1979) and Sotavolta (1947) found differences of between 40% and 127% in chironomid and ceratopogonid midges all of which use sound to attract males. Farnworth (1972) cites several studies that demonstrate greater activity of metabolic enzymes in male than in female insects and it seems quite likely that the females of many species would obtain a selective advantage by conserving energy in this way.

3.3 Weight

In those nematocerous flies, mosquitoes and midges, that use wing-beat sounds to attract males, it is possible that a blood meal or a heavy load of eggs might change the WBF enough to make them less attractive. Costello (1974) tested this in *A. aegypti* by attaching a length of nylon fishing line (with a mass of 1 mg) to the abdomen of females. The WBF of all ten females tested increased, but the mean was only 5% (537–564 Hz)

Table 1. Physical and physiological factors that affect WBF, measurements in Hz from 10 *Aedes aegypti* females (mean ± SE)

Tethering	Free	Flight-mill	Fixed[a]
	486 ± 7	506 ± 7	523 ± 5 s*
Feeding	Before		After
	477 ± 11	Sugar	488 ± 11 s*
	481 ± 8	Blood	487 ± 8 s*
(Mass)	1.2 mg		1.9 mg
Added weight	537 ± 10		564 ± 10 s*
(Mass)	1.2 mg		2.2 mg

[a] s* = Significant with 95% confidence (binomial expansion).

– small compared with the approximately 100% increase in weight. He found that when the mosquitoes fed on sugar or blood there was a smaller increase in WBF in proportion to the increase in weight – 2.3% for sugar and 1.2% for blood for a mean increase in weight of 0.75 mg (Table 1).

All of these treatments increased SPL slightly but, as the insects were in free flight, no measurements were made of wing-beat amplitude. We also noted slight increases in WBF in gravid females followed by a decrease after they oviposited. The effect of increased weight on the abdomen may be compensated by an increase in aerodynamic lift caused by a greater angle between the thorax and abdomen as Hocking (1953) demonstrated. The nylon line probably increased the drag on the females considerably more than a swollen abdomen would, but even so it did not change the WBF enough to influence their attraction to males.

4 Environmental Factors Affecting Flight Sound

4.1 Temperature

The effect of temperature on WBF is complex and evidently can differ in insects with different flight mechanisms. The distinction between the two recognised types of mechanism lies in the physiology of the indirect flight muscles. In the synchronous type that is found in most orders of flying insects there is a 1:1 relationship between the electrical nerve and muscle activity and the twitches that cause wing movement. In the asynchronous type, found in some Hemiptera, and all Coleoptera, Hymenoptera and Diptera, there is only a general correlation between electrical activity and twitches. The muscles twitch more frequently than the nerves fire, so that there is evidently a requirement for an elastic and mechanically resonant thorax that enables the indirect wing muscles to contract between electrical impulses when stretched.

In sphingids *(Hemaris)* which have a synchronous flight mechanism, Sotavolta (1947) found a slight but linear increase in WBF from 15°–35 °C with a slope of 0.5 Hz °C^{-1} (Fig. 2). In locusts, with a similar flight mechanism, Weis-Fogh (in Chadwick 1953) found no change in the WBF of tethered *Schistocerca* between 22° and 33 °C and this is also true for migrating locusts (R.C. Rainey, personal communication at this Sympo-

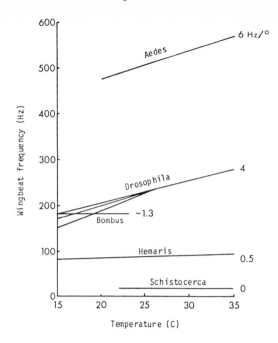

Fig. 2. Effect of temperature on wing-beat frequencies. Measurements of *Aedes* original, others adapted from authors mentioned in text

sium). In *P. americana*, which is anatomically similar to the locust, but does not migrate on the wing, the WBF increases slightly from $19°-27°C$ (0.6 Hz $°C^{-1}$), levelling off between $27°$ and $35°C$ (Farnworth 1972).

In insects with asynchronous muscles where wing loading, resonance and clicks of the thoracic box probably have a much greater influence on WBF, the effect of temperature varies with the size of the insect. This is evidently because larger insects have a smaller surface area relative to volume which may, by itself, influence temperature regulation. Unwin and Corbet (1984) showed that the WBF of insects of this type, with a mass of around 100 mg, changed WBF very little with temperature, whereas in *Drosophila* it increased by about 4 Hz $°C^{-1}$ and in heavier insects it decreased with temperature. The heavier insects they studied were bumblebees which are known to regulate body temperature (Heinrich 1976) and, at lower ambients, probably have a higher body temperature than other bees and flies. At temperatures above those found in their normal environment ($30°-40°C$ depending on species) Sotavolta (1947) found a negative correlation between WBF and temperature in Coleoptera, Lepidoptera and most Diptera he observed. He found one exception in a female mosquito (unidentified) whose WBF increased until it stopped flying at $39°-40°C$. Our experiments with female mosquitoes show a positive correlation between WBF and temperature between $20°$ and $35°C$, corresponding very closely in slope with Sotavolta's figures for *Drosophila*.

4.2 Humidity

As might be expected from the relationship between WBF and temperature, most investigators have found a positive correlation between WBF and relative humidity (RH)

Table 2. Environmental factors that affect WBF, measurements in Hz from 10 *Aedes aegypti* females (mean ± SE)

Relative humidity	5%	50%	90%[a]
26 °C	499 ± 7	502 ± 8	501 ± 8 n.s.
34 °C	547 ± 8	557 ± 6	562 ± 8 s*
Atmospheric pressure	89	101	114 kPa
	499 ± 7	504 ± 8	501 ± 7 s*
Wind speed	0–3		6 km h⁻¹
Swarm of ♂ *Aedes*	517		440
Illumination	0		700 lx
	491 ± 5		493 ± 6 n.s.

[a] n.s. = Not significant; s* = significant with 95% confidence (binomial expansion).

that, at least at high temperatures, is evidently related to the heat lost by evaporation. At high metabolic rates oxygen must be exchanged rapidly. The spiracles open and the properties of cuticle may also change. Sotavolta (1947) observed a 10% increase in WBF for a 60% increase in RH in three dipterous species. In *A. aegypti* we found significant changes in WBF only above 30 °C when there was a 2% change, from 547–562 Hz, with a 90% increase in RH (Table 2).

Farnworth (1972) found that increases in RH produced significant increases in the WBF of *P. americana* only above 27 °C and found that the thoracic temperature was indeed higher at 95% RH than at 50%.

4.3 Atmospheric Pressure

Some of the first experiments on flight sounds investigated the effects of changes in pressure because it altered the density of the air and therefore the loading on the wings (Chadwick and Williams 1949). These investigators observed a decrease of 17% in the WBF of *Drosophila* when the pressure was increased from 101 kilopascals (1 atm) to 506 kPa and an increase of 10% at 13 kPa.

Sotavolta (1952) found that reduced pressure (down to 10 kPa) had little or no effect on the WBF of social Hymenoptera, calyptrate and syrphid flies but, in confirmation of Chadwick and William's findings, increased it by up to 10% in *Drosophila*, nematoceran flies and in Lepidoptera. Our experiments with *A. aegypti* used modest changes in pressure (of ± 13 kPa) and found small, but consistent decreases in frequency when pressure was either increased or decreased.

At sea level, atmospheric pressure is unlikely to change by as much as 25 kPa, but at an altitude of 5 km (for example in the Alps) pressures of 50 kPa would be experienced by alpine and migrating insects.

There is no simple explanation for the reduction of WBF at low pressure that we measured. It is unlikely that a shortage of oxygen reduced the WBF of mosquitoes because a mosquito the size of *A. aegypti* probably uses no more than 20 μl h⁻¹ (Hocking 1953), and Sotavolta (1952) saw no difference in other insects when he used pure oxygen rather than air as a medium. Further investigation of the amplitude of the wing beat and the resonance of the flight mechanism are needed. The effect of pressure on frequency is in any case small and unlikely to be of practical significance to man or insect.

Laird (1948) found that mosquitoes in an unpressurised aircraft would not fly at altitudes above 10 km (30 kPa) and at this pressure were reversibly anaesthetised at temperatures below about 8 °C.

4.4 Illumination

Light intensity has a marked effect on releasing and inhibiting flight behaviour (Haskell 1966), but no investigation seem to have been made on the effect of light on WBF. Buchner (1972) found that *Drosophila* and *Musca* doubled their thrust in the dark, but did not examine the wing beat. This could be important in insects that swarm at dusk and dawn and use flight sound to attract males. Costello (1974) was able to record flight sounds from 25 4- to 5-day-old unmated *A. aegypti* females which had been left in complete darkness for 15 min. He found that they did not differ significantly in frequency from the sounds of the same females after 15 min of illumination at 700 lx from a cool, white fluorescent lamp (Table 2). The slightly higher frequency in the light may well have been caused by an unmeasurable increase in temperature.

4.5 Wind Velocity

As with tethering, the effect of wind velocity would depend on the behaviour of the insect under study. Sense organs in several regions of the body, particularly antennae, head, and wings, can operate in feedback loops to adjust flight speed (Gewecke 1974).

Ignoring the effect of the angle of attack of the wings, the net effect of increased wind velocity on WBF and amplitude would be positive if an insect was maintaining a constant position or speed over the ground, but negative if it was maintaining a constant air velocity over the body. Experiments to test this would be complex particularly as there is evidence that, depending on conditions, amplitude and frequency in the locust can be negatively or positively correlated (Gewecke 1972, 1975).

My observations are restricted to some swarming male mosquitoes which maintained their position over a swarm marker in winds gusting to 6 km h^{-1}. Between 0 and 3 km h^{-1} there was no obvious change in WBF, but between 3 and 6 km h^{-1}, WBF was reduced by 15%, from 517–440 Hz and the amplitude of the wing beat must have increased, perhaps with some change in its angle of attack, to double the thrust (Table 2). This result appears to be in direct contradiction to Gewecke's (1974) generalisation that in locusts, bees, and flies air speed has a negative effect on flight speed. It is evidently not possible to use results from the laboratory to predict all types of behaviour in the field.

5 Conclusions

From the point of view of insects that attract mates with sound, only temperature, of the environmental factors discussed, would have a great enough effect to influence the success of mating. Costello (1974) found that a change of 15% (75 Hz in 500) which would be produced by a temperature change of about 10 °C was sufficient to reduce mating frequency in *A. aegypti*. Mosquitoes evidently compensate for this because the tuning of the male antenna changes with temperature in step with the female WBF and under natural conditions the frequency of successful mating is evidently not affected.

From the point of view of identifying insects by their WBF, Schaefer's (1976) chart of frequency and wing length can be extended to include mosquitoes (Fig. 3). Our regression line for mosquitoes appears to be almost identical in slope ($f = 822.1^{-0.73}$) with that compiled for acridids by Schaefer ($f = 400.1^{-0.78}$).

Despite their different flight mechanisms, it is interesting that on Schaefer's chart the sphingids and Coleoptera with synchronous and asynchronous flight muscles respectively differ in WBF by only 7% for a particular wing length. It also seems surprising that the Lepidoptera should be so much more diverse in their relationship between WBF and wing length than the Coleoptera and there is evidently scope for much more research in this area. If Pringle (1976) is correct, acridids, Lepidoptera and other orders of insects with synchronous flight muscles cannot flap their wings at frequencies much above 100 Hz and, if so, only Diptera, Coleoptera, Hymenoptera, and the asynchronous

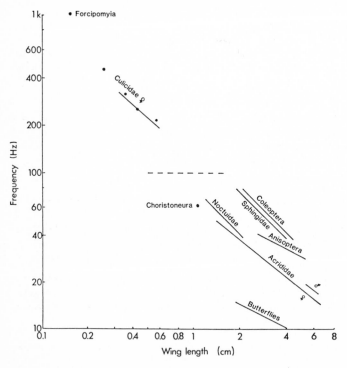

Fig. 3. Identification of insects by wing-beat frequencies. Measurements of ♀ culicids original, others redrawn from Schaefer

Hemiptera would be expected on the upper half of the chart. The detection by radar of such insects with wing lengths of 5 mm and WBFs of 250 Hz is technically possible (see Riley in this volume).

References

Belton P (1974) An analysis of direction finding in male mosquitoes. In: Barton Brown L (ed) Experimental analysis of insect behaviour. Springer, Berlin Heidelberg New York, pp 139–148

Belton P, Costello RA (1979) Flight sounds of the females of some mosquitoes of Western Canada. Entomol Exp Appl 26:105–114

Bienz-Isler G (1968) Electronenmikroscopische Untersuchungen über die imaginale Struktur der dorsolongitudalen Flugmuskeln von *Antherea pernyi* Guen (Lep). Acta Anat 70:416–433

Buchan PB, Satelle DB (1979) A radar-Doppler autocorrelation analysis of insect activity. Physiol Entomol 4:103–109

Buchner E (1972) Dark activation of stationary flight of the fruit fly *Drosophila*. In: Welmer R (ed) Information processing in the visual systems of Arthropods. Springer, Berlin Heidelberg New York, pp 141–146

Chadwick LE (1953) The motion of the wings. In: Roeder K (ed) Insect physiology. Wiley, New York, pp 577–614

Chadwick LE, Williams CM (1949) The effects of atmospheric pressure and composition on the flight of *Drosophila*. Biol Bull (Woods Hole) 97:115–137

Costello RA (1974) Effect of environmental and physiological factors on the acoustic behaviour of *Aedes aegypti* (L.) (Diptera, Culicidae). Ph.D. Thesis, Simon Fraser University, Burnaby

Farnworth EG (1972) Effects of ambient temperature, humidity and age on wing-beat frequency of *Periplaneta* species. J Insect Physiol 18:827–840

Gewecke M (1972) Antennen und Stirn-Scheitelhaare von *Locusta migratoria* L. als Luftströmungs-Sinnesorgane bei der Flugsteuerung. J Comp Physiol 80:57–94

Gewecke M (1974) The antennae of insects as air-current sense organs and their relationship to the control of flight. In: Barton Brown L (ed) Experimental analysis of insect behaviour. Springer, Berlin Heidelberg New York, pp 100–113

Gewecke M (1975) The influence of the air-current sense organs on the flight behaviour of *Locusta migratoria*. J Comp Physiol 103:79–95

Haskell PT (1966) Flight behaviour. In: Haskell PT (ed) Insect behaviour, 3rd symp R Entomol Soc, Lond. London, pp 29–45

Heinrich B (1971) Temperature regulation of the sphinx moth, *Manduca sexta* I. Flight energetics and body temperature during free and tethered flight. J Exp Biol 54:141–152

Heinrich B (1976) Heat exchange in relation to blood flow between thorax and abdomen in bumble bees. J Exp Biol 64:561–585

Hocking B (1953) The intrinsic range and speed of flight of insects. Trans R Entomol Soc Lond 104:223–345

Johnson CG (1976) Lability of the flight system: a context for functional adaptation. In: Rainey RC (ed) Insect flight. 7th symp R Entomol Soc Lond. Blackwell, Oxford, pp 217–234

Kammer AE, Heinrich B (1978) Insect flight metabolism. In: Advances in insect physiology, vol 13. Academic Press, New York, pp 133–228

Kennedy JS (1961) A turning point in the study of insect migration. Nature 189:785–791

Laird M (1948) Reactions of mosquitoes to the aircraft environment. Trans R Soc NZ 77:93–114

Nachtigall W (1976) Wing movement and the generation of aerodynamic forces by some medium-sized insects. In: Rainey RC (ed) Insect flight. 7th symp R Entomol Soc Lond. Blackwell, Oxford, pp 31–47

Pringle JWS (1976) The muscles and sense organs involved in insect flight. In: Rainey RC (ed) Insect flight. 7th symp R Entomol Soc Lond. Blackwell, Oxford, pp 3–15

Rainey RC (ed) (1976) Insect flight. 7th symp R Entomol Soc Lond. Blackwell, Oxford, pp 1–287

Schaefer GW (1976) Radar observations of insect flight. In: Rainey RC (ed) Insect flight. 7th Symp R Entomol Soc Lond. Blackwell, Oxford, pp 157–197

Sotavolta O (1947) The flight-tone (wing-stroke frequency) of insects. Acta Entomol Fenn 4:1–117
Sotavolta O (1952) The essential factor regulating the wing-stroke frequency of insects in wing
    mutilation and loading experiments and in experiments at subatmospheric pressure. Ann Zool
    Soc Zool Bot Fenn Vanamo 15:1–67
Unwin DM, Corbet SA (1984) Wingbeat frequency, temperature and body size in bees and flies.
    Physiol Entomol 9:115–121
Weis-Fogh T (1972) Energetics of hovering flight in hummingbirds and in *Drosophila*. J Exp Biol
    56:79–104

6 Orientation at Night by High-Flying Insects

J. R. RILEY and D. R. REYNOLDS[1]

1 Introduction

The substance of this paper deals with one striking feature of insect migration – common orientation at night by larger-sized (30–2000 mg), high-flying insects. This phenomenon was unsuspected before it was made dramatically obvious by the fortuitous sensitivity of simple scanning radars to non-random orientation in populations of airborne insects (Schaefer 1969). Since then, common orientation has been widely observed in many radar studies (J. Roffey, personal communication 1972; Riley 1975; Schaefer 1976; Reid et al. 1979; Riley and Reynolds 1979, 1983; W.W. Wolf, personal communication 1980; Greenbank et al. 1980; Drake et al. 1981; Drake 1983, 1984) and it appears to be a very common feature of nocturnal migratory flight. All authors agree that orientation occurs in both the presence and absence of moonlight.

There is clear evidence that the mean direction of collectively oriented night-flying insects is often related to wind direction (Schaefer 1976; Riley and Reynolds 1979, 1983; Drake 1983), but this relation appears to be a complex one. The only general rule to emerge is that in winds above the insects' flying speed, the angle between mean orientation and the downwind direction seems to be always less than $90°$ and is sometimes very small. In contrast to this wind-dependent behaviour, Schaefer (1976) has also reported common orientation by Sudan grasshoppers and noctuid moths in a fixed (SSW) direction, irrespective of the direction of the (light) wind.

In spite of the widespread nature of common orientation almost nothing is known about the factors which control either the mean direction or the degree of orientation of nocturnally migrating insect populations. There has been only one systematic study of the phenomenon published to date, and this has largely served to emphasise the lack of any consistent relation between orientation and obvious directional cues – at least in the case of the Australian plague locust, *Chortoicetes terminifera* (Walker) (Drake 1983). In this paper we compare earlier observations with new results obtained using improved radar techniques, and we describe some mechanisms which might be adopted by migrating insects to orientate themselves. The adaptive value of common orientation will be discussed in a later paper (D.R. Reynolds and J.R. Riley, in preparation).

[1] TDRI Radar Unit, RSRE Leigh Sinton Road, Malvern, Worcs. WR14 1LL, United Kingdom.

Insect Flight: Dispersal and Migration
Edited by W. Danthanarayana
© Springer-Verlag Berlin Heidelberg 1986

2 Methods

Three methods of using radar to study collective orientation are outlined below. Many of our results were obtained by the complementary use of all three.

2.1 Trajectory Measurements

When a flying insect is detected by a scanning radar it produces a string of dots or "footprints" on the radar screen. Measurement of the sequential position of the "footprints" allows one to compute the insect's ground velocity (Riley 1974). Similarly, the "footprints" of a freely drifting balloon carrying metal foil give wind velocity. In cases where these wind soundings are available in the locality of airborne insects, vector subtraction of wind velocity from an insect's ground velocity gives its air speed and orientation. A rough measure of general orientation direction may be obtained in this way, but detailed descriptions of heading distribution require many such measurements and the method becomes very tedious. It also becomes inaccurate when the wind speed substantially exceeds the insect air speed, and in addition is subject to complex bias effects which can be corrected only if the target's scattering properties are known (Riley 1979).

2.2 The Sector Patterning Method

A uniformly distributed aerial population of insects which has a degree of common orientation will produce a striking alternate quadrant (or dumb-bell shaped) pattern on the Plan Position Indicator (PPI) screen of a scanning radar. This is because the insects presenting side views to the radar reflect radio waves more effectively than those seen head or tail on, and they are consequently more readily detectable (Riley 1975). In the case of unimodal heading distributions, sector patterning is a very sensitive indicator of non-random headings, and the mean alignment direction of the insects can be readily determined. The method yields the average axis of *alignment*, but normally leaves a 180° ambiguity about the mean *heading*. This ambiguity may be resolved by comparing the mean direction of target displacement with that of a drifting balloon at the same time and altitude, and thus determining the general direction of flight. In instances where the target insects show differences in front/rear radar reflectivity, as well as the (much larger) side/end differences, it is also possible in principle to resolve the 180° ambiguity from the fine detail of the sector patterning (J.R. Riley, unpublished).

PPI sector patterns are not generally suited to quantitative measurements of the *degree* of alignment because interpretation of the patterns depends on a detailed knowledge of the angular dependence of the targets' radar reflectivity and on assumptions about the form of their heading distribution. These patterns have nevertheless been used to estimate heading distributions on several occasions when the identity of the targets was known, and where experimental measurements of cross-section were available (Schaefer 1976; Drake 1983; Riley et al. 1981, 1983). The method will, however, lead to misleading results if the heading distribution being observed is not unimodal.

2.3 Vertical-Looking Radar

This method uses a stationary, vertically-pointing radar beam in which the plane of (linear) polarisation is rotated. Insects displacing individually through the beam normally produce maxima in the radar signal when the rotating plane of polarisation of the electric field becomes aligned with their longitudinal body axes. Measurements of the position of these maxima in the rotation cycle allow estimation of the geographical alignment of the body axis with an accuracy of $\pm 1.25°$ (J.R. Riley, unpublished). The technique also allows the acquisition of wing-beat frequency, displacement speed and parameters related to the body shape of each target (Riley and Reynolds 1979).

In the case of the largest insects (mass ~ 1 g), the longitudinal maxima may be replaced by circumferential ones which occur when the body axis is at right angles to the electric field. Fortunately these cases are readily detectable because the rotational modulation is then relatively shallow (Riley 1985). The few signals displaying this characteristic have been excluded from our analysis.

3 Results

We present here a small, but representative selection of the results we have acquired during the past 12 years in both East and West Africa. The selection has been chosen to illustrate different aspects of the phenomenon and to emphasize some new facets made particularly evident by the simultaneous use of vertical-looking (VLR) and scanning radars.

The alignment distributions obtained from the vertical-looking radar data are plotted as equi-areal polar diagrams (Batschelet 1965) with arbitrarily selected 10° bin widths, and their mean values were computed using the method of doubling the angles (Batschelet 1981). The parameters, however, used as a measure of dispersion, was calculated using the expression in Mardia (1972) for double angles:

$$s = [- 2 \ln (r)]^{1/2}/2 \text{ radians}$$

because this expression produces values which correspond closely to the standard deviation in linear statistics, up to s values of 45° [r is the root of the mean value of the sum of squares of the sines and cosines of twice the individual orientation angles (Batschelet 1981)].

The term used by Batschelet for double angles:

$$s = [2 (1 - r)]^{1/2}/2 \text{ radians}$$

is equivalent to standard deviation only up to 25° (r > 0.9) and thereafter becomes a progressively less sensitive indicator of dispersion.

The precision of alignment measurement ($\pm 1.25°$) meant that no correction for grouping was necessary. The "forward" or heading halves of the alignment distributions were usually identified by comparison of balloon and mean insect trajectories as observed simultaneously with a scanning radar, and are shown hatched. In cases where ambiguity remained (for example, instances of small groups aligned at large angles to the bulk of the distribution) no attempt was made to determine heading. The direction of mean insect displacement (D) and of balloon movement (W) (when available at the appropriate altitude) are shown on the histograms, as well as the moon's azimuth (M).

3.1 Downwind Unimodal Distributions

The distributions shown in Fig. 1a were obtained in southwestern Kenya at a site near Mara River ($1°03'$ S, $35°15'$ E) and are typical of many observed. They are symmetrical about the mean and are not significantly different from the circular normal by the Chi^2 test (Batschelet 1981). The aerial density was of the order of one insect per 10^4 m^3. Overall displacement was nearly parallel to the alignment axis and observed to be at 5.1 ± 1 m s^{-1} at 250 m above ground level (agl), and at 6.1 ± 1 m s^{-1} at 460 m agl. We deduce from this that the insects were oriented close to the downwind direction, and as they were flying with an average air speed of 2.5 m s^{-1} (J.R. Riley and D.R. Reynolds, unpublished) the wind would have been $\cong$ 3 m s^{-1}. Balloon soundings were not made during this period, but the alternative possibility that the insects were heading into an 8 m s^{-1} wind and displacing backwards was ruled out because of the presence of a small number of very slowly moving ($<$ 1 m s^{-1}) insects (an 8 m s^{-1} headwind would give minimum ground velocities of $\sim$5 m s^{-1}). In the many simultaneous observations we have made of balloon and insect trajectories, overall backwards displacement has never been detected for oriented groups.

It may be noted that orientation was not directed towards the full moon which was clearly visible at an elevation of $26°$. The identity of the insect targets was not definitely established, but it seems certain from the wide range of wing-beat frequencies (20–42 Hz) extracted from their radar signatures that the common orientation was shared by a mixture of different species, perhaps noctuid and sphingid moths. The wing-beat frequency histograms in Fig. 1a show that the species with the higher frequencies were dominant at low altitudes, but were outnumbered in the 540–555 m range by larger insects with lower wing-beat frequencies. At the intermediate level of 390–405 m both groups seem equally represented. It is to be noted that although some decrease in frequency with altitude would be expected because of decreasing air temperature, the decrease over 300 m would be much smaller than that observed. For example, in the case of the noctuid moth, *Spodoptera exempta* (Walker) with a temperature coefficient of 1.2 Hz °C^{-1} (Farmery 1982), even an extreme lapse rate of 1 °C per 100 m would produce only a 3–4 Hz decrease in frequency.

3.2 Crosswind Unimodal Distributions

In contrast to the example shown in Fig. 1a, the orientation distributions in Fig. 1b are aligned at $63°$ to the direction of average insect movement, which in this case was

---

Fig. 1a,b. Alignment distributions, at three altitudes, for insects showing orientation in: (a) the downwind, and (b) a crosswind direction, together with corresponding wing-beat frequency distributions for each altitude. D = direction of displacement; M = moon's azimuth. *Shaded* section indicates deduced heading distribution

(a)

Mara River 9 March 04·22 – 04·47 Hrs

Altitude range = 240 to 255m
Mean angle = 347·87°
S = 21·91°

Altitude range = 390 to 405m
Mean angle = 350·33°
S = 14·73°

Altitude range = 540 to 555m
Mean angle = 344·43°
S = 14·86°

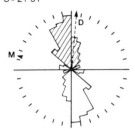

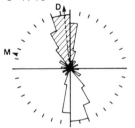

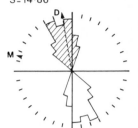

No. plotted 145

No. plotted 140

No. plotted 72

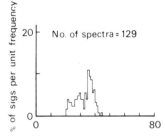

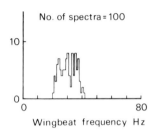

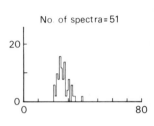

No. of spectra = 129

No. of spectra = 100

No. of spectra = 51

Wingbeat frequency Hz

(b)

Mara River   9 March 1982 20·41 – 21·02 Hrs

Altitude range = 300 to 360m
Mean angle = 336°
S = 32·84°

Altitude range = 540 to 600m
Mean angle = 338°
S = 24·78°

Altitude range = 780 to 840m
Mean angle = 338°
S = 16·70°

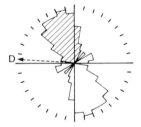

No. plotted = 135

No. plotted = 86

No. plotted = 12

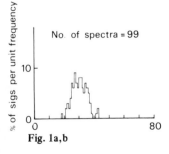

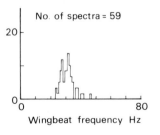

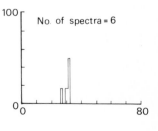

No. of spectra = 99

No. of spectra = 59

No. of spectra = 6

Fig. 1a,b

Wingbeat frequency Hz

seen on the PPI to be towards 270°–280° at 11 ± 2 m s^{-1}. The insects were thus aligned largely across the wind. If it is assumed that a representative individual flying speed is 2.5 m s^{-1} directed to the northwest quadrant, then the average air speed for a population with a standard deviation, s, of 25° would be 2.3 m s^{-1}. Vector subtraction of this figure from the observed displacement velocity gives an average wind of 10 m s^{-1} directed *towards* 264°, i.e. the downwind direction is some 74° away from the mean insect orientation direction. As in Fig. 1a, the broad spread of wing-beat frequencies shows that a mixture of species was present, especially at the lowest altitude (300–360 m agl).

It was noted from the PPI display that in the altitude range 230–570 m agl the average insect displacement direction had changed from 310° to 275° at some time between 20.34 and 20.48 h, and that the mean orientation had also moved from approximately 005° to 335° during the same period. Vector subtraction of an assumed flying speed of 2.3 m s^{-1} from the observed displacement vectors indicated that the wind had backed from towards 300° to towards 264° and slackened slightly from 10.5 to 10 m s^{-1}. *Thus the average orientation appeared to be at rather similar angles to the downwind (65° and 71°) before and after the wind shift.* The difference between these figures and the value of 74° found from the VLR results probably occurred because the PPI results are spot measurements, whereas the VLR produced a value averaged over 25 min.

3.3 Bimodal Distributions

As well as the more common unimodal distribution to the type illustrated in Fig. 1 we have occasionally observed clear instances of bimodal or split distributions. One example observed in Mali in West Africa has been described previously (Riley and Reynolds 1979) and another more recent observation, also from Mali (at Daoga, 15° 53′ N, 0° 14′ E), is shown in Fig. 2. In this example, a fairly tight distribution centred on 41° between 225 and 270 m agl is largely replaced by a northwards orienting group at 675–720 m, but the groups clearly overlap at the intermediate heights. This pattern is reflected in the wing-beat frequency histograms which show a dominance of higher frequencies (33–42 Hz) at the lowest altitude, and of lower frequencies (26–33 Hz) above 675 m. It would thus appear that several different species were present, with overlapping vertical density profiles and with two different mean orientation directions. PPI observations of a balloon trajectory at 20.40 h showed that the wind was towards 348° at 7 m s^{-1} up to 600 m, the highest altitude observed, thus the northwards group was heading close to the downwind direction. The mean direction of insect displacement at all altitudes seen on the PPI was very close to northwards, the lowest fliers having a mean ground direction towards only 002° in spite of their northeast heading. This result is consistent with the vector addition of a mean air speed of 3 m s^{-1} (found to be typical for the insects we were observing in the area) at 41° to the wind vector. The mean orientation direction deduced from PPI sector patterning backed steadily with altitude from 45° to 360° but as would be expected, gave no indication of the heading split made evident from the VLR records.

The identity of the targets contributing to both the orienting groups was not certain. However, comparing the wing-beat frequencies of the Acridid species taken at our light

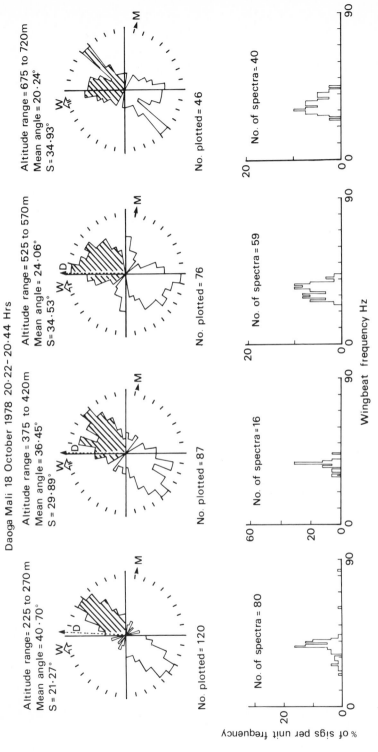

Fig. 2. Alignment distributions at four altitudes, for insects (probably Acridids) showing collective orientation about two different mean headings, together with corresponding wing-beat frequency distributions for each altitude. D = direction of displacement; W = direction towards which wind was blowing; M = moon's azimuth. *Shaded* section indicates deduced heading distribution

traps with those measured from the overflying insects, we consider it likely that *Oedaleus senegalensis* Krauss was predominant in the low frequency group, and that smaller grasshoppers such as *Acrotylus* spp. and *Pyrgomorpha conica* (Ol.) might have been the dominant species in the higher frequency, northeast-heading population.

3.4 Accumulated Results

We found that in four particularly clear examples of bimodal distribution, the mean orientation of one of the groups was within ±5° of the downwind direction. It therefore seemed likely that these groups were perceiving wind direction and responding to

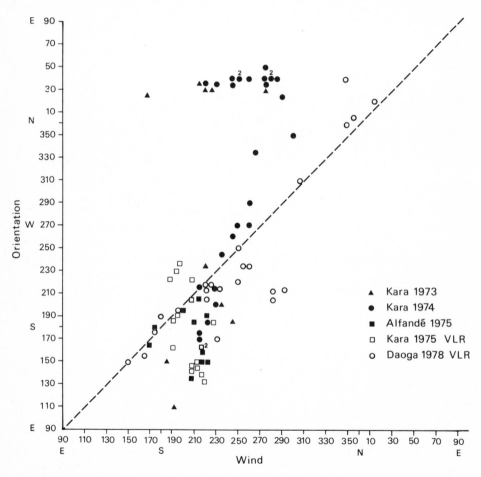

Fig. 3. Mean heading of orienting groups of insects (believed to be mainly Acridids) plotted against the direction towards which the wind was blowing. Wind direction was established from the movement of freely drifting balloons at the same altitude as the insects. The radar data was recorded in Mali, West Africa from three sites: Kara (14° 10′N, 5° 1′W); Alfande (15° 48′N, 3° 4′W), and Daoga (15° 53′N, 0° 14′E). The Daoga (1978) and Kara (1975) data are from a vertical-looking (VLR) radar, the rest are from PPI recordings from a scanning radar

it, and that the other groups were aligned to some other cues. In none of the cases was there an obvious cue like the moon's azimuth, the twilight arch or a significant feature on the horizon.

To determine the relative incidence of downwind orientation, we pooled both VLR and PPI results for several years and from three sites, and plotted orientation against wind direction. The data from Mali produced a total of 87 points (Fig. 3) of which only 24% were within $\pm 10°$ of the downwind line. There was strong evidence of a preferred orientation towards $25°-50°$ in light winds blowing towards a wide variety of directions ($165°-350°$), and in both the presence and absence of moonlight. There was also some concentration of off-wind headings within the much wider range of $130°-210°$ in winds towards the southwest.

4 Discussion

In most of the examples of nocturnal collective orientation which we have seen, the aerial density of the insects has been well below one insect per 10^4 m^3 and the average separation between individuals has ranged from 30 to 150 m, so it seems most unlikely that the insects maintained common orientation by visual reference to each other. This conclusion is supported by the observation that degree of orientation is not affected by moonlight nor, in our experience, by variations in aerial density. In any event, mutual references of any type would not provide a mechanism by which a non-random mean heading could be selected and sustained over large areas and for long periods without the accumulation of drift due to systematic errors or biases in the reference process. It is therefore necessary to invoke an environmental directional cue (or hierarchy of cues) to which the insects individually respond, and in the sections below we discuss a number of candidate cues. Throughout the discussion we assume that different species may on occasions simultaneously respond to different cues in the same environment [as implied by Fig. 2 and by our earlier results (Riley and Reynolds 1979)].

4.1 Orientation to the Wind

There is now a weight of evidence to suggest that at least on some occasions, orientation direction is related to wind direction. For example, we have described an observation in which the orientation of Sahel grasshoppers veered with altitude and time in a similar way to wind direction, so that orientation was maintained close to the downwind (Riley and Reynolds 1983). Schaefer (1976) has reported similar behaviour in spruce budworm moths, although in this case the maintainance of downwind direction became less accurate above 400 m agl. Wind-related orientation is not always, however, close to downwind, as we demonstrated in Sect. 3.2. In this observation mean orientation was maintained at $65°$ to $71°$ to the downwind direction after the wind had backed by $36°$. Drake (1983) noted a similar effect when the orientation of migrating *Chortoicetes* apparently remained at some $40°$ to the wind after a wind shift of more than $90°$.

These observations prompt enquiry into the mechanisms by which the wind might provide cues to which insects could orientate themselves. As we have pointed out previously (Riley and Reynolds 1983), any ability on the part of the airborne insects to establish geographical direction (using for example, celestial cues or the earth's magnetic field) would not have enabled them to deduce the direction of the wind. To do this they must either have been able to detect the direction of apparent ground movement relative to their own alignment, or else have been able to sense some accelerative anisotropy of air movement associated with wind direction.

4.1.1 Visual Perception of Relative Ground Movement

Kennedy (1951) has hypothesised that visual perception of wind-induced movement relative to the ground is a factor which governs the orientation of locusts flying at low altitude during the day. The conditions for which this optomotor response was postulated were, however, very different from those in which we detected wind-related orientation. The illumination levels were very much higher (ca. 10^4 lx compared to ca. 10^{-2} lx) and the angular rates of wind-induced movement were also higher. For example Kennedy (op.cit. p. 218) gives a rough estimate of preferred retinal velocity as that produced by a displacement speed of 3 m s^{-1} at 6 m agl, i.e. 27° s^{-1}, whereas our observations typically relate to insects at several hundred metres agl and experiencing longitudinal angular rates of ca. 0.2° to 3° s^{-1}. If insects maintain their heading with respect to the wind by observing relative ground movement, it might be hypothesised that they do so by orienting themselves so that the transverse angular velocity of ground features remains within preferred limits. An estimate of required limits of this type may be obtained from the downwind distribution described in Sect. 3.1. Thus the estimated wind speed at 545 m agl was 3 m s^{-1} along the mean orientation direction, and the component of this speed transverse to the longitudinal axes of those insects near the edges of the distribution ($\pm15^\circ$) would be $\pm3 \sin(15)$ m s^{-1}. Insects at 545 m agl would therefore have to maintain their transverse angular velocities within the range of $0 \pm 0.08^\circ$ s^{-1} to achieve the observed heading distribution.

It is perhaps more instructive to examine the case of wind-related, off-wind headings. Consideration of the example described in Sect. 3.2 shows that the observed distribution with a standard deviation of $\pm25^\circ$, at some 74° to a wind of 10 m s^{-1} at 570 m agl, would produce lateral angular velocities within the range 0.759° to 0.993° s^{-1}, with a mean of 0.966° s^{-1}, i.e. *the angular velocity distribution would be highly skewed.* Trigonometric considerations show that this is generally the case for symmetrical off-wind distributions when the mean heading is greater than 45° away from the direction towards which the wind is blowing. Conversely, if the insects oriented themselves so as to produce a symmetric distribution of transverse angular rates about a preferred mean (as one might expect), a skewed heading distribution would result. Our observations show, however, that both downwind and crosswind headings are usually symmetric about the mean. Furthermore, the maintainance of off-wind headings by the adoption of a 'preferred' lateral angular velocity would introduce an altitude dependence into the resultant heading, i.e. the higher the insects were flying, the more they would have to head off-wind to maintain the same lateral angular velocity. This is clearly *not* the case in the example shown in Fig. 1b where the same mean heading was maintained at all altitudes, even though the lateral angular velocity experienced by the lowest group was more than twice that seen by the highest group. (The displacement speed and direction were virtually unchanged over this altitude range.)

Overall it seems clear that the transverse angular velocity hypothesis does not account for observed off-wind orientation. Nor is it consistent with the independence of standard deviation on altitude which is often observed in both crosswind and downwind heading distributions (Fig. 1a and b; and Riley and Reynolds 1983).

An alternative hypothesis for the maintainance of off-wind headings is that the *ratio* of lateral to forward angular movement might be used as an orientation cue (Baker et al. 1984). This would not predict symmetrical distributions for off-wind headings, but provided that angular rates remained high enough to be quantitatively perceived, it *would* explain the altitude independence of both standard deviation and the direction of off-wind headings.

4.1.2 Laboratory Measurements of Response to Angular Movements

It is of interest to enquire whether there is any evidence from the laboratory that insects can visually perceive and respond to the angular rates required to explain wind-related orientation. Thorson (1966) has shown that rigidly mounted desert locusts *(Schistocerca gregaria)* respond to very small (5×10^{-3} deg) angular movements of boldly striped patterns. These movements corresponded in his experiments to maximum angular rates of 4×10^{-3} deg s^{-1}, and the response was achieved at an average pattern brightness of 1 L. Decreasing the brightness lowered sensitivity to movement, and at 10^{-4} L a response was just perceptible to a pattern movement of $0.1°$ ($0.08°$ s^{-1}). This is comparable to the rates that we have seen are necessary to maintain wind-related headings and it is therefore pertinent to compare the laboratory visual stimuli with those available to insects in the field.

The pattern used by Thorson was of higher contrast (86%) and regularity than that likely to be found in the natural landscapes in which we have made our observations, but it is possible to make a comparison of the brightness of the visual environment in the two cases. If one assumes that a reflectivity of 0.2 in the wavelength range of 400–700 nm is representative of the terrain in which we obtained our results (M.B. Allenson, personal communication), then a full moon at zenith and a clear starlit sky, giving illuminations respectively of 3×10^{-1} and 10^{-3} lx, would produce surface brightnesses of 6×10^{-6} and 2×10^{-8} L. It would appear from these figures that to achieve by visual means the wind-related orientation that we have observed when the moon was at low elevation or below the horizon, the insects concerned would have had to respond to ground patterns whose brightness levels were at least two orders of magnitude below those at which a response could be elicited from locusts in the laboratory.

A further point of relevance is that Thorson found responses only when the angular period λ, of the moving pattern exceeded $3°$ (most of his data were obtained with λ = $8°$) and he noted that this result was compatible with a light- (or dim-)adapted visual acuity of ca. $3°$. The implication of this is that dark-adapted locusts, with visual acuity of ca. $6°$ (Horridge 1965, quoted in Thorson op.cit.) and flying at 400 m agl, could respond only to movement relative to those features on the ground which showed substantial contrast over scales of $\geqslant$ 40 m. Individual trees and bushes would thus not provide a usable pattern. In the case of the moth *Manduca sexta* (Joh.), the minimum angular period required to induce a response in motion-detecting neurons in the dark-adapted eye is reported to be $16°$ (Rind 1983), so even if this moth could perceive ground contrast, it could presumably orient itself to the wind from an altitude of 400 m agl only when flying over contrasting features on a scale of $\geqslant 100$ m.

The angular movements response experiments of Kien (1974) with single-edge patterns showed a very much higher threshold level ($0.4°$ s^{-1}) for the Australian plague locust, *Chortoicetes terminifera* than Thorson found for the desert locust, and this was at quite high brightness levels (6.8×10^{-3} L). In Kien's experiments the locusts were able to move their heads and her results may therefore be more applicable to the conditions of free flight. If this is the case, then her work would indicate even more strongly that visual perception of the apparent direction of ground movement provides an inadequate explanation of wind-related common orientation.

Quite apart from constraints due to the limits of visual perception, the detection of low angular rates normally requires that the viewing platform be stabilised within much smaller limits than the rates to be detected. In the cases we have considered, this would seem to require the insects to control their head attitudes within, say 10^{-2} to 10^{-3} degrees about their roll axes – an improbable feat for an insect in free flight (Kennedy 1975, p. 112).

The presence of irregular, flight-induced movements may not, however, completely exclude the possibility that airborne insects can detect low angular rates of motion. Horridge (1966a) has shown that the eye of the crab *Carcinus* responds to pattern movements as low as 7×10^{-4} deg s^{-1}, in spite of the fact that the eye is subject to continual tremor movements of $0.04°-0.2°$, at a frequency of 1 Hz. He also found evidence for a similar opto-kinetic memory in *Locusta migratoria* (L.) (Horrdige 1966b). It thus seems conceivable that some insects may be able to make very precise assessments of the average angular position of visible features, even when subject to flight perturbation. The average angular positions of the horizon and any visible ground features underneath the insect flight path might then presumably be compared to reveal small, relative angular rates associated with displacement.

We note in conclusion that the general absence of 'backwards' displacement would seem to imply that nocturnally flying insects can usually maintain enough visual contact with the ground to detect and avoid gross front-to-back movement of ground features. The threshold at which reverse angular rates become unacceptable has not been established, but careful examination of some of our balloon and insect trajectories suggests that it is at least $0.2°$ s^{-1} under starlight illumination (J.R. Riley, unpublished).

4.1.3 Anisotropic Air Movement

The use of anisotropic accelerations or gusts in the wind as a cue to fix wind direction was postulated many years ago by Williams et al. (1942) and Nisbet (1955) to explain wind-related orientation. Since then it has been invoked by a number of authors including most recently Larkin (1980) and Able et al. (1982) for birds, and ourselves in relation to insects (Riley and Reynolds 1983), but there has been a singular lack of experimental evidence to show that small-scale anisotropy is a widespread phenomenon.

The turbulence induced by wind shear at altitude is believed to be substantially iso-tropic below length scales of the order of a kilometre, except in cases where the shear produces Kelvin-Helmholtz (KH) waves (E.A. Gossard, personal communication).

Atlas et al. (1970) have described an example of KH instability in which insect horizontal displacement velocities apparently oscillated by ± 2 to 3 m s^{-1} in the vicin-ity of the wave, and it was deduced that the maximum velocity perturbation was in the direction of wave motion. It seems very probable that the accelerations associated with velocity changes of 2-3 m s^{-1} occurring in a period of ca. 20 s (the average time for an insect to fly between a trough and a crest of the observed wave) would be per-ceptible to airborne insects, either directly, or because their inertia within the accelerat-ing/decelerating air mass would produce cyclic changes in their air speed. Atlas et al. (1971) concluded that the insects were travelling (and therefore presumably orienting) in a direction opposite to that of the wave propagation, which in this case was down-wind. The wind speed at the altitude of the waves (365 m) was 1 m s^{-1}, so upwind dis-placement would indeed have been possible for insects with an air speed of 3 m s^{-1}.

It would appear from the observations of Atlas et al., that wave-induced anisotropy provides a directional cue which could be used by insects to maintain wind-related orientation, and the fact that the direction of wave propagation (and therefore of maximum anisotropy) is often not downwind (E.A. Gossard, personal communication) could explain the occurrence of wind-related, but off-wind headings.

The lack of experimental evidence for the common occurrence of anisotropies may in part reflect the scarcity of instruments able to detect fine-scale structure in the wind at altitudes of several hundred metres. For example, the results of Atlas et al. were made possible only by the advent of a specially constructed, ultra-high resolution radar. On the other hand, subsequent use of this radar by Noonkester (1973) for a total of 849 h detected breaking KH waves for only 0.6% of the operational period, and Noon-kester considered that meteorological conditions in the area in which he was working would generally favour wave phenomena. If this result is representative, then KH waves producing refractive index gradients severe enough to register on even the most sensitive radars must be considered rate. It remains *conceivable*, however, that "mild" wave events, which generate sufficient anisotropy for orientation but do not break, and so do not create the refractive index gradients required for radar detection, occur quite regularly (K.A. Browning, personal communication). We have therefore begun to re-examine our VLR data to see if there is any evidence from individual insect displace-ment speeds for anisotropies in the wind.

4.2 Orientation to Compass Directions

Our West African observations of persistent northeast headings in a wide range of wind directions provides convincing evidence that the insects concerned (probably Acridids), were orienting to a preferred geographical direction. Because these north-eastwards results were all obtained (with one exception) at one site (Kara), we cannot rigorously exclude the possibility that the effect was merely a localised, site-specific response to some distinctive topographical feature, but we consider this to be unlikely. By far the most striking topographical feature in the area was the River Diaka, and while this undoubtedly influenced the behaviour of some insects, the influence was highly localised and obviously associated with the outline of the river (Reynolds and Riley 1979). The northeast headings were on the other hand uniformly maintained over an area of 7 km^2 and showed no evidence of being related to any localised stimulus.

Overall, we consider that the data provide good evidence for the adoption of compass headings by some insects flying at high altitude in darkness.

In the following sections we briefly consider two of the mechanisms by which compass orientation might be achieved.

4.2.1 Use of Celestial Cues

Wehner (1984) has reviewed the evidence that nighttime skylight cues might be used for compass orientation, and notes that insects with apposition compound eyes can detect the moon as a point source, and that those with superposition eyes may in addition be able to detect the brightest stars. He draws attention to the fact that use of the moon as a compass presents formidable problems because of the complexity of its azimuth – time relationship.

Moon Compass. Sotthibandhu and Baker (1979) report observations of orientation at a fixed (non-time-compensated) angle to the moon's azimuth for tethered large yellow underwing moths (*Noctua pronuba* L.), but there is general agreement amongst radar observers that collective orientation in high-flying insects is unaffected by the presence of the moon. The northeast heading groups we described in Sect. 3.3 mainly occurred when the moon was well below the horizon so the apparent compass orientation exhibited by these groups was clearly not a moon-related phenomenon.

Stellar Compass. In the absence of moonlight, Sotthibandhu and Baker (1979) detected a response of *Noctua pronuba* to rotation of the celestial sphere and deduced that the moths were using stellar cues to maintain non-time-compensated orientation. Our observations of geographical heading could conceivably be explained if the insects concerned had been capable of *time-compensated stellar* orientation. However, stellar orientation of any type would presumably not have been available to the northeast heading Acridids we observed, because their apposition eyes are unlikely to resolve individual stars (Wehner 1984).

Polarised Light. A non-localised skylight cue to direction which is available to day-flying insects is the pattern of polarisation in the sunlight scattered from the atmosphere [see Wehner (1984) for a review]. Once the sun is more than 7° below the horizon (30 min after sunset), however, the atmosphere overhead is no longer effectively illuminated by the sun and polarisation information is lost. Insects which rely on polarisation patterns then become disoriented (R. Wehner, personal communication). Scattering of moonlight by the atmosphere generates similar patterns to those caused by sunlight, but the scattered light is of very low intensity (sky brightness $\sim 10^{-6}$ L). No experiments on the response of insects to low light polarisation patterns have been reported to date, so orientation to night-sky polarisation must for the time being be considered speculative. We note, however, that we have observed common orientation on occasions when the moon has been more than $7°$ below the horizon, and on moonless nights, so it is certain that night-sky polarisation cannot in itself account for the common orientation phenomenon.

4.2.2 Terrestrial Magnetic Cues

There is no doubt that a number of biological organisms exhibit great sensitivity to various features of the earth's magnetic field [see for example a summary by Gould (1980)]. Perhaps the most convincing evidence that this sensitivity is shared by at least one species of flying insect is provided by the behavioural experiments of Lindauer and Martin (1972). They showed that the error in the 'waggle' dance direction of bees on a vertical comb was apparently related to the rate of change of the magnitude of the earth's magnetic field. What is particularly surprising about this result is the tiny size of the field changes (ca. 200 gammas in a normal field of 50,000 gammas). Even greater sensitivity has been suggested ($2-5$ gammas or 0.3 gammas min^{-1}) to explain the maintainance of circadian rhythm in bees deprived of all temporal cues except the diurnal variations in terrestrial field strength (Martin and Lindauer 1977; Gould 1980). Sensitivity to changes as small as this suggests that simple orientation to the normal field would be easy to accomplish. Nevertheless, evidence that insect species orientate to magnetic cues appears to be limited to the work of Baker and Mather (1982) on *Noctua pronuba*.

In summary, the question of whether insects at large are able to use terrestrial magnetic fields for orientation remains unanswered. Given, however, the apparently extreme magnetic sensitivity of bees, it would not be surprising if at least some other insect species were able to perceive the gross features of the terrestrial field well enough to orientate to it.

5 Conclusions

The common orientation which insects regularly exhibit when flying at high altitude at night, has been shown to be sometimes related to wind direction, and sometimes to geographical direction. Occasionally, one species will adopt one of these strategies, whilst another, in the same aerial environment, adopts the other. Evidence has been given which demonstrates that wind-related common orientation is almost certainly not maintained by visual references to the ground, but ground reference may well be used to detect and avoid gross backwards displacement.

It has been argued that accelerative anisotropies of air movement associated with shear-induced (Kelvin-Helmholtz) waves in the atmosphere provide insects with directional cues related to wind direction, but it remains to be seen whether anisotropies of this type are widespread enough to account for wind-related orientation.

Of the candidate cues examined to explain compass orientation, perception of the earth's magnetic field, and time-compensated stellar orientation would seem to be the most plausible, although neither has been demonstrated for insects in free flight.

We would expect nocturnally migrating insects to adopt wind-related cues or compass orientation in an hierarchical manner which would depend on wind conditions and on their individual flight strategies.

Acknowledgements. The computer programs used in the orientation analysis were written by one of us (JRR) whilst he was being supported by a NASA-NRC Research Associateship at Wallops Island, Virginia. Particular thanks are due to Mrs. G. Barfield for her painstaking assistance with the data analysis, and especially to Mr. A.D. Smith for his comprehensive technical support.

References

Able KP, Bingman VP, Kerlinger P, Gergits W (1982) Field studies of avian nocturnal migratory orientation II. Experimental manipulation of orientation in white-throated sparrows *(Zonstrichia albicollis)* released aloft. Anim Behav 30:768–773

Atlas D, Metcalf JI, Richter JH, Gossard EE (1970) The birth of "CAT" and microscale turbulence. J Atmos Sci 27:903–913

Atlas D, Harris FI, Richter JH (1971) The measurement of point target speeds with incoherent non-tracking radar: insect speeds in atmospheric waves. Proc 14th Radar Meteor Conf, Nov 17–20, 1970, Tucson, Arizona. Am Met Soc, Boston, pp 73–78

Baker RR, Mather JG (1982) Magnetic compass sense in the large yellow underwing moth, *Noctua pronuba* L. Anim Behav 30:543–548

Baker PS, Gewecke M, Cooter RJ (1984) Flight orientation of swarming *Locusta migratoria*. Physiol Entomol 9:247–252

Batschelet E (1965) Statistical methods for the analysis of problems in animal orientation and certain biological rhythms. Am Inst Biol Sci, Wash DC

Batschelet E (1981) Circular statistics in biology. Academic Press, London

Drake VA (1983) Collective orientation by nocturnally migrating Australian plague locusts, *Chortoicetes terminifera* (Walker) (Orthoptera: Acrididae): a radar study. Bull Entomol Res 73: 679–692

Drake VA (1984) The vertical distribution of macro-insects migrating in the nocturnal boundary layer: a radar study. Boundary-Layer Meteorol 28:353–374

Drake VA, Helm KF, Readshaw JL, Reid DG (1981) Insect migration across Bass Strait during spring: a radar study. Bull Entomol Res 71:449–466

Farmery MJ (1982) The effect of air temperature on the wingbeat frequency of naturally flying armyworm moths *(Spodoptera exempta)*. Entomol Exp Appl 32:193–194

Gould JL (1980) The case for magnetic sensitivity in birds and bees (such as it is). Am Sci 68:256–267

Greenbank DO, Schaefer GW, Rainey RC (1980) Spruce budworm (Lepidoptera: Tortricidae) moth flight and dispersal: new understanding from canopy observations, radar and aircraft. Mem Entomol Soc Can 110:1–49

Horridge GA (1966a) Optokinetic memory in the crab *Carcinus*. J Exp Biol 44:233–245

Horridge GA (1966b) Optokinetic memory in the Locust. J Exp Biol 44:255–261

Kennedy JS (1951) The migration of the desert locust (*Schistocerca gregaria* Forsk.) I. The behaviour of swarms. II. A theory of long-range migrations. Philos Trans Roy Soc B Biol Sci 235:163–290

Kennedy JS (1975) Insect dispersal. In: Pimental D (ed) Insects, science and society. Academic Press, New York, pp 103–119

Kien J (1974) Sensory integration in the locust optomotor system I. Behavioural analysis. Vision Res 14:1245–1254

Larkin RP (1980) Transoceanic bird migration: evidence for detection of wind direction. Behav Ecol Sociobiol 6:229–232

Lindauer M, Martin H (1972) Magnetic effects on dancing bees. In: Galler SR, Schmidt-Koenig K, Jacobs GJ, Belleville RE (eds) Animal orientation and navigation. US Gov Print Office, Wash DC, pp 559–567

Mardia KV (1972) Statistics of directional data. Academic Press, London

Martin H, Lindauer M (1977) Der Einfluß des Erdmagnetfeldes auf die Schwereorientierung der Honigbiene. J Comp Physiol 122:145–187

Nisbet ICT (1955) Atmospheric turbulence and bird flight. Br Birds 48:557–559

Noonkester VR (1973) Breaking wave characteristics determined from FM-CW radar observations. Bull Am Meteorol Soc 54:937–941

Reid DG, Wardaugh KG, Roffey J (1979) Radar studies of insect flight at Benalla, Victoria in February 1974. CSIRO Aust Div Entomol Tech paper no 16

Reynolds DR, Riley JR (1979) Radar observations of concentrations of insects above a river in Mali, West Africa. Ecol Entomol 4:161–174

Riley JR (1974) Radar observations of individual desert locusts. Bull Entomol Res 64:19–32

Riley JR (1975) Collective orientation in night-flying insects. Nature 253:113–114

Riley JR (1979) Quantitative analysis of radar returns from insects. In: Vaughn CR, Wolf W, Klassen W (eds) Radar, insect population, ecology, and pest management. NASA Conf Pub 2070, Wallops Island, Virginia, pp 131–158

Riley JR (1985) Radar cross-section of insects. Proc IEEE 73:228–232

Riley JR, Reynolds DR (1979) Radar-based studies of the migratory flight of grasshoppers in the middle Niger area of Mali. Proc R Soc Lond B Biol Sci 204:67–82

Riley JR, Reynolds DR (1983) A long-range migration of grasshoppers observed in the Sahalian Zone of Mali by two radars. J Anim Ecol 52:167–183

Riley JR, Reynolds DR, Farmery MJ (1981) Radar observations of *Spodoptera exempta*. Kenya, March-April 1979. Misc Rep no 54 Centre of Overseas Pest Research, London

Riley JR, Reynolds DR, Farmery MJ (1983) Observations of the flight behaviour of the armyworm moth *Spodoptera exempta*, at an emergence site using radar and infra-red optical techniques. Ecol Entomol 8:395–418

Rind FC (1983) A directionally sensitive motion detecting neuron in the brain of a moth. J Exp Biol 102:253–271

Schaefer (1969) Radar studies of locust, moth and butterfly migration in the Sahara. Proc R Entomol Soc Lond Ser A Gen Entomol 34:33, 39, 40

Schaefer (1976) Radar observations of insect flight. In: Rainey RC (ed) Insect flight. Symp Roy Entomol Soc no 7. Blackwell, Oxford, pp 157–197

Sotthibandhu S, Baker RR (1979) Celestial orientation by the large yellow underwing moth, *Noctua pronuba* L. Anim Behav 27:786–800

Thorson J (1966) Small-signal analysis of a visual reflex in the locust I. Input parameters. Kybernetik 3:41–53

Wehner R (1984) Astronavigation in insects. Annu Rev Entomol 29:277–298

Williams CB, Cockbill GF, Gibbs ME, Downes JA (1942) Studies in the migration of lepidoptera. Trans R Entomol Soc Lond 92:101–283

7 Lunar Periodicity of Insect Flight and Migration

W. DANTHANARAYANA[1]

1 Introduction

The desirability of studying lunar periodicity of insect flight was first pointed out by Hora (1927). Hartland-Row (1955) and Corbet (1958) provided experimental evidence in support of this phenomenon. They found that some Ephemeroptera, Trichoptera, and Diptera (Chironomidae) showed periodic fluctuations in numbers caught in light traps, and that this correlated closely with the age of the moon. The species differed in the position of their peak abundance in the lunar cycle and in the amplitude of the fluctuations. Peaks occurred 2–5 or 23–26 days before or after new moon or within 5 days of full moon (moon age 9–19 days) (Fig. 1a,b). The occurrences of peaks and troughs relative to full moon and new moon were attributed to a lunar rhythm of emergence.

Subsequent work on lunar periodicity caused much controversy and confusion, with the various workers suggesting a full moon periodicity (e.g. Bidlingmayer 1964; Bowden and Morris 1975; Duviard 1974; El-Ziady 1957), others a new moon periodicity (e.g. Miller et al. 1970; Nemec 1971; Provost 1957) and still others no periodicity (e.g. Brown and Taylor 1971; Williams 1936; Williams and Singh 1951; Williams et al. 1956). Thus, lunar periodicity of insect flight has been a perplexing problem ever since it was proposed in 1927. Also, the evolutionary significance or the adaptive value of this phenomenon, if one exists, has not been seriously considered until 1976 when it was suggested that lunar periodicity may be associated with migration (Danthanarayana 1976).

It is now well-established that the greatest potential for migration of insects occurs when they become airborne and enter air streams that may move faster than their intrinsic flight velocity (Johnson 1969; Rainey 1976; Taylor 1960, 1974). This may be the only means of achieving long-range displacement by small insects. Evolution of behavioural patterns and life-cycle strategies have enabled insects to make use of the inexhaustible source of energy provided by the winds (Dingle 1972; Johnson 1960, 1969; Kennedy 1975; Taylor 1960, 1974).

A mechanism for the initiation of migration in diurnal insects is to use sunlight (skylight) as a visual cue, to launch themselves into the air and fly upwards, often aided by thermals (Johnson 1969). These insects leave the boundary layer (Taylor 1958, 1960, 1974) of relatively still air close to the ground and enter the horizontally moving air

---

[1] Department of Zoology, The University of New England, Armidale, NSW 2351, Australia.

Insect Flight: Dispersal and Migration
Edited by W. Danthanarayana
© Springer-Verlag Berlin Heidelberg 1986

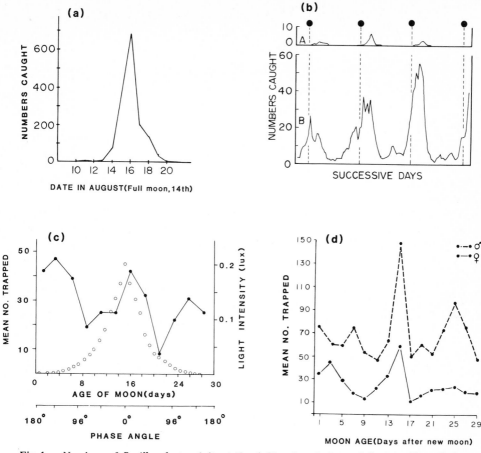

Fig. 1. a Numbers of *Povilla adusta* adults at Kaazl, Uganda, relative to full moon (from Corbet 1958). b Numbers caught nightly of (*A*) *Clinotanypus claripennis* and (*B*) *Tanytarsus balteatus* in relation to new moon (from Corbet 1958). c The trimodal-lunar periodicity curve of *Epiphyas postvittana* (●─●) based on suction-trap data over seven lunar cycles (Danthanarayana 1976a). d Mean numbers of *Plutella xylostella* in suction-trap catches over five lunar cycles (Goodwin and Danthanarayana 1984)

higher up, in order to be transported by wind to which insects orientate for downwind movement (Greenbank et al. 1980; Rainey 1976; Taylor 1974). In contrast to that of diurnal fliers, mechanisms of initiation of migratory flights of night-flying insects are not well understood (see Kennedy 1975). In a previous study on the flight activity of the tortricid moth *Epiphyas postvittana* (Walker), a lunar periodicity was demonstrated with peaks of activity occurring shortly after new moon, around full moon and shortly before new moon and this phenomenon was also found to occur in the cabbage moth, *Plutella xylostella* (L.) (Goodwin and Danthanarayana 1984) (Fig. 1c,d). These pre- and post-new moon peaks and the full moon peak produced a trimodal-flight periodicity curve within the lunar month. It was hypothesized that if insects are able to fly to upper levels of the atmosphere during these peak activity periods at night, then migration is feasible, and that the moon may be associated with this process (Danthanarayana

Fig. 2. Experimental site at La Trobe University. The position of the 0 m and 2.5 m suction traps are shown with *arrows*. The 5, 10, and 20 m traps are seen on the towers

1976), and recent radar studies show that nocturnal species do migrate in low-level wind jet at 100–300 m (Drake 1985). Experiments in support of this hypothesis were carried out between 1978 and 1980 using a vertical series of suction traps, in an experimental set-up similar to that used by Taylor (1974) (Fig. 2). Much of the material collected in these traps is still being worked out, but a complete set of data on the mosquito *Culex pipiens australicus* (Dobrotworsky and Drummond) and *Plutella xylostella* have been analysed. Hourly data obtained during the seasons 1978/79 and 1979/80 provided information over ten lunar cycles. Some data were also obtained during the 1984–1985 summer season.

2 Methods

The sampling site was located at La Trobe University, about 15 km from the centre of Melbourne. Samples were taken with 30.5 cm (12″) diameter Vent-Axia type Johnson-Taylor suction traps with automatic hourly catch-segregation mechanisms (Johnson 1950; Johnson and Taylor 1955a,b). Five traps were set up in a vertical series to sample at 0, 2.5, 5, 10, and 20 m above ground level. The three higher traps are visible in Fig. 2. The 2.5 m trap on the stand and the position of the 0 m trap operating from a pit in the ground are shown by arrows. The traps were located 10 m apart in a circle 20 m in diameter within a 40 × 50 m closely mown grass patch. At the northern edge of the grass patch there are four artificial freshwater ponds, each 9 m in area and 0.5 m deep at the centre. Behind this area, towards the north, is a small newly planted 0.5 ha apple orchard, a 40 m deep strip of grassland and a *Eucalyptus* woodland as seen in the background in Fig. 2. Bordering on the west is pasture land and on the east natural grassland, a barn and an *Acacia* plantation is extending beyond 200 m from the towers. To the south, the site is bordered by a narrow road, a treed area, a large, open car park, parkland and university buildings respectively. Wind speeds at six levels (0.3, 1.25, 2.5, 5, 10, and 20 m) were measured with Gill 3-cup micro-anemometers (ex. R.M. Young, Michigan, USA). Temperature and humidity measurements were made with thermocouple probes at 0.3, 7, 13, and 20 m placed within naturally ventilated radiation shields (ex. R.M. Young, Michigan) fixed along the 20 m tower. Light intensity measurements were made continuously at 20 m using a wide-spectrum silicon photo diode (type Photops 500, United Detector Technology Inc, California) connected to a logarithmic amplifier. All measurements were recorded by analogue recorders housed within a small meteorological station (wooden hut in Fig. 2). Rainfall measurements were obtained from the adjoining La Trobe University School of Agriculture farm. The aerial insect population at the five levels were sampled simultaneously and the traps were emptied at 0900-0920 h daily. Further sampling was carried out between October 28, 1984 and January 31, 1985 using only three traps placed at 1.25, 2.5, and 5 m from the ground.

Threshold levels of environmental components for the onset and cessation of flight were determined at 50% flight occurrence for the recorded measurements of temperature, wind speed, % relative humidity and light (in relation to time of sunset and sunrise) from the hourly catches at 2.5 m for the period October 22, 1978 – November 30, 1978 following the method of Taylor (1963). The flight speed was determined by the formula $y = 25\ x^{0.49}$ (Lewis and Taylor 1967) where y is the flight speed in cms^{-1} and x the geometric mean size (wing span x body length in mm^2). Sixty specimens of each species were measured for this purpose.

Mosquitoes trapped were dissected to determine their physiological age using the method of Detinova and Bertram (1962). The ovaries were teased out on a drop of normal saline and their condition recorded as to whether they were nulliparous (yet to lay the first batch of eggs), parous (gravid females with 1–3 follicular relics) or multiparous (females with more than four follicular relics).

3 Results

3.1 Diel and Lunar Flight Periodicity

In all, 26 439 female and 1289 male *Cx. p. australicus* and 1781 female and 1326 male *P. xylostella* were captured in the five traps. Pooled data on *Plutella* and only the data on female *Culex* were used for analysis. *Cx. p. australicus* and *P. xylostella* are nocturnal fliers with peak activities occurring between 1900 h and 0500 h (Fig. 3). The daily catches in the five traps show lunar periodicity as well as a buildup and decline in population size for each season (Figs. 4 and 5) with the population fluctuations tending to mask the lunar effect. The trimodal-lunar periodicity curve becomes clear when the daily means from nine lunar cycles (based on data for periods October 3, 1978 – Febru-

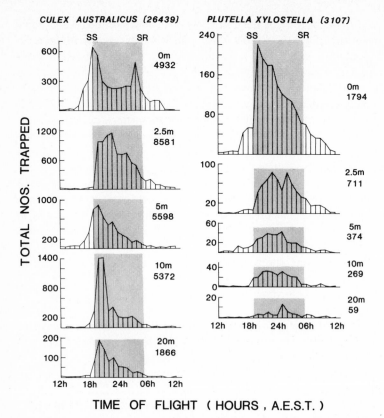

TIME OF FLIGHT (HOURS , A.E.S.T.)

Fig. 3. Diel flight of 26 439 female *Culex pipiens australicus* and 3107 *Plutella xylostella* (males and females) caught in the five suction traps. The numbers caught at each height, the times of sunset (*SS*) and sunrise (*SR*) and the nocturnal period (*shaded area*) are also shown

ary 1979 and October 22, 1979–February 16, 1980), when the insects were most abundant each season, are plotted (Fig. 6). In view of the large within and between seasonal variations in population size and structure (Figs. 4 and 5), and because this data strongly follow Taylor's power law (Taylor 1961, 1965, 1971) (unpublished information), the calculation of standard deviations (hence error bars) for untransformed data was considered inappropriate. Evidence for the existence of these peaks, therefore, comes from their consistency at different trapping heights in both species, repeatability in various seasons (1978/79, 1979/80, 1984/85), previous work on *E. postvittana* (Danthanarayana 1976) and *P. xylostella* (Goodwin and Danthanarayana 1984) and from other examples cited in Sect. 4.

This seemingly characteristic (see below) trimodal curve could not be explained in terms of a number of typical environmental variables, viz. temperature, % relative

Fig. 4a,b. Logarithmic plot of the total numbers of (a) *Plutella xylostella* and (b) *Culex pipiens australicus* caught in the five suction traps during the first season (1978/79) shows the buildup and decline of the populations. Data for the second season (1979/80) are similar and not shown. The times of full moon and new moon are indicated by *open* and *closed circles* respectively

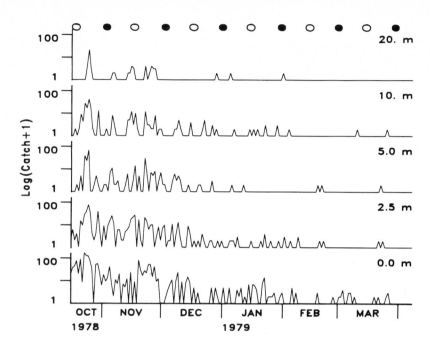

(a)   *Plutella xylostella* (♂♀)

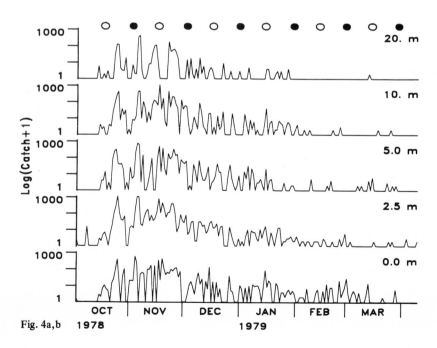

(b)   *Culex p. australicus* (♀♀)

Fig. 4a,b

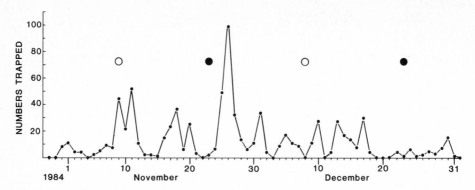

Fig. 5. Plot of the total numbers of *Culex pipiens australicus* caught in three suction traps during the season 1984/85. Data for January-March 1985 and for *P. xylostella* are not shown because of very low numbers. *Open* and *closed circles* show the times of full moon and new moon respectively

humidity, time of sunset and sunrise, rainfall and wind speed (Appendix 1), but a correlation was found to exist between the numbers trapped and the amount of polarization of moonlight. The flight speeds of the two species and the upper and lower flight threshold values for a number of physical environmental variables are given in Table 1.

The total brightness of the moon and the amount of polarization of moonlight in relation to the phase angle, α, and the age of the moon are shown in Fig. 7 and described below. For statistical analysis of the data, the lunar month was divided into three periods based on the transition points of the polarization and brightness curves which cross at two points (Fig. 7c). This creates two periods of dim-polarized light between days 1 and 11 ($\alpha = 180°-48°$) on the waxing side and between days 19 and 30 ($\alpha = 48°-180°$) on the waning side, and a single period of bright moonlight between days 11 and 19 ($\alpha = 48°$ on either side of full moon). There were significant positive correlations between the mean numbers of insects caught and the amount of polarization of moonlight. This relationship was statistically significant for the pooled-trap data (Fig. 7d,e) as well as for the data from each of the four traps above the ground level with *Cx. p. australicus* (not shown). With *P. xylostella* the correlation was significant for the pooled-trap data only, probably because of the low numbers trapped in individual traps. There was a lag of about 2 days in peak activity in relation to the day of full moon (Figs. 6 and 7a,b).

Table 1. Flight speeds and upper and lower threshold levels of environmental components for onset and cessation of flight

	Cx. p. australicus	*P. xylostella*
Flight speed (air speed)	1.51 m s^{-1}	1.87 m s^{-1}
Temperature thresholds (lower; upper)	$4 °C; 25.5 °C$	$7 °C; > 26 °C$
Wind-speed thresholds (lower; upper)	$< 0.4 \text{ m s}^{-1}; 5.2 \text{ m s}^{-1}$	$< 0.4 \text{ m s}^{-1}; 2.2 \text{ m s}^{-1}$
Humidity thresholds (R.H.) (lower; upper)	$< 75\%; 100\%$	$< 74\%; 100\%$
Light threshold (time after sunset)	1 h	1 h
Light threshold (time before sunrise)	2 h	1 h

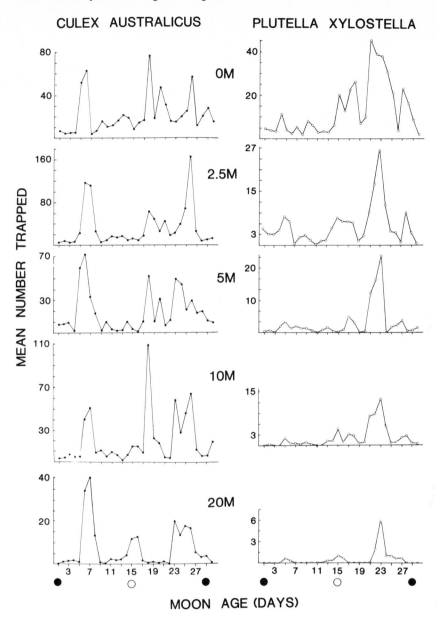

Fig. 6. Mean numbers of *Culex pipiens australicus* females and *Plutella xylostella* (males and females) caught in suction traps per lunar month at each trapping height based on data from nine lunar cycles

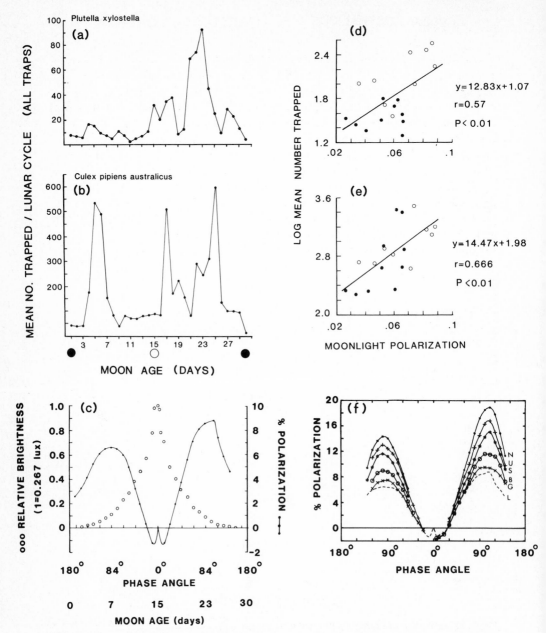

Fig. 7a–f. Mean numbers of (a) *Plutella xylostella* and (b) *Culex pipiens australicus* caught in the five suction traps per lunar day. Data from nine lunar cycles. **c** The illumination (○○○○) and polarization (●–●–●) curves of moonlight (after Lyot 1929). Relationship between the log mean catch and the degree of moonlight polarization of (d) *P. xylostella* and (e) *Cx. p. australicus*, using data from moon-age days 1–10 (waxin moon; ○) and days 20–28 (waning moon, ●). **f** Polarization vs phase angle at five wavelengths for integrated disc of the moon: *N*, 336 nm; *U* 367 nm; *S*, 383 nm; *B*, 449 nm; *G*, 519 nm (after Pellicori 1971). *Dashed curve* is the polarization curve of Lyot (*L*) (1929) for integrated moonlight

3.2 The Aerial Density Profiles

Whether or not *Cx. p. australicus* and *P. xylostella* enter the upper levels of air for wind-assisted migration, in relation to the lunar phases described above, was tested by determining the hourly vertical-density profiles (Johnson et al. 1962) for the three periods of peak-flight activity as well as for the two periods of minimal-flight activity (i.e. at new moon and a trough period between new moon and full moon). The numbers of insects trapped were converted to aerial density (mean No./283 × 10^3 m^3/h) by allowing for the amount of air sampled by the traps and their efficiency for various wind speeds (Taylor 1962; Service 1976). To minimize irregularity which the small insect catches give to the profiles for separate hours, densities were pooled for the same hours on the 3 days of each flight peak and of each trough period to give mean hourly density profiles (Johnson 1957; Johnson et al. 1962). Vertical-density profiles (height × density) were plotted on a double-log scale using data from the four heights from 2.5 m–20 m (Fig. 8). Captures at 0 m were mostly lower than those at 2.5 m and are not included

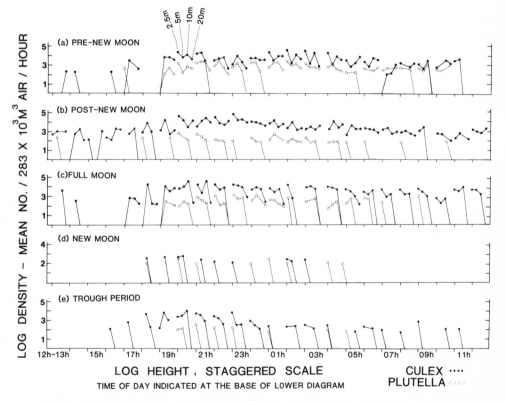

Fig. 8a–e. Observed mean hourly density profiles from 2.5 m to 20 m during five lunar phases. The profiles are on a double-log scale, each being slightly displaced to the right to avoid overlap (method of Johnson et al. 1962). *Culex p. australicus* data (•–•). *Plutella* data (o--o). a Pre-new moon peak period: October 24–26, 1978. b Post-new moon period: November 6–8, 1978. c Full moon period: November 16–18, 1978. d New moon period: October 31–November 2, 1978. e Trough period between post-new moon and full moon: November 10–12, 1978. Results of the second lunar cycle (not shown) are similar

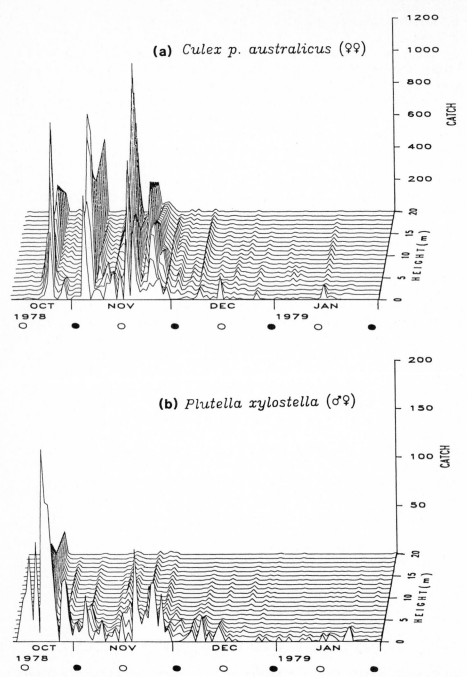

Fig. 9a,b. Plots of daily catches (noon-noon) as a surface in time and space . The surface was calculated from actual measurements at 0, 2.5, 5.0, 10, and 20 m with linear interpolation at 0.5 m intervals. Data are for the period October 16, 1978–January 31, 1979 from the 1978/79 season (Fig. 4). The calculations and the plotting were done by a FORTRAN program using PLOT 79 plotting package on a VAX 11/780 computer. The full and new moons are indicated by *open* and *closed circles* respectively in the graphs and the regressions. Surface views of the daily catches plotted in time and space for the period October 16, 1978 – January 31, 1979 are shown in Fig. 9.

The graphs in Fig. 8 are displaced to the right to avoid overlap. The extent to which insects use wind as a vehicle for dispersal can be gauged by the slope of the regression line of log density on log height which indicates the relative proportions of the airborne populations at different heights in the air (Johnson 1957, 1969; Johnson et al. 1962). The low negative values and positive values of the regression coefficient, b, and the shape of the profiles, tending to become shallow and parallel to the height axis, clearly show the times when the populations move to the upper levels of air. A large negative value of the regression coefficient denotes that relatively more of the total aerial population is in the lower layers of air, while a small negative value shows that a greater proportion of the population is in the upper layers of air. A few positive values of b obtained in the present study indicate infinitely great densities in the air at infinitely great altitudes. These unrealistic positive values of b result from larger catches at 20 m than at lower levels and indicate that the expected inverse relationship between the density and altitude (Johnson 1957, 1969) begins above the maximum trapping height of 20 m on these occasions.

For the vertical-density profile data presented in Fig. 8, regression coefficients and their standard errors were evaluated whenever there were four points in a density profile (Table 2). Since both species were trapped only in one or two traps (at 2.5 m and/or 5 m) at new moon and during the entire trough period with *Plutella*, b values could not be calculated for these respective sets of data. As seen from the results shown in Fig. 8 and from the b values presented in Table 2, densities at the upper levels of air were significantly greater during the pre- and post-new moon periods than at full moon, new moon and the trough periods at most times between 1800 and 0700 h. At full moon high densities occurred at the upper levels generally between 1900 and 0600 h, whereas at new moon and the trough period upward movements of insects did not occur. A summarised computation of this upward movement was obtained by calculating the common b value based on data from 13 h of the maximum flight-activity times, between 1800 and 0700 h. For *Cx. p. australicus* the b values and their standard errors during the pre-new moon peak, the post-new moon peak, the full moon peak and the trough period were found to be $-0.165 \pm 0.722, -0.243 \pm 0.287, -2.438 \pm 1.013$, and -3.072 ± 0.902 respectively. The respective values for *P. xylostella* were -1.23 ± 0.768, -1.215 ± 0.107, and -1.746 ± 0.848 (the new moon and/or the trough period data not being suitable for analysis for reasons given above). Similar results were obtained when data from the second lunar cycle were analysed. Estimates of the total number of insects in the air between two heights were obtained by the graphical integration method of determining the average total number of insects in a column of air with a base area 9.20×10^5 m^2 for each hour (Johnson et al. 1962). It must be noted that these authors trapped insects up to a height of 305 m and extrapolated, on good theoretical grounds, the density on height profiles to 914 m (approximate cloud base) to determine the total number of insects in the air column up to the cloud base. In the present study, extrapolation was made up to 80 m, since 70 m is regarded as the height above which there is usually sufficient wind for migration (see Danthanarayana 1976; Geiger 1966). The number of insects in air columns, for each hour, up to the maximum trapping height (20 m) and up to 80 m are presented in Table 3. The graphical data in Fig. 8, the regres-

Table 2. Regression coefficients (*b* values) and their standard errors of hourly density profiles during the lunar phases indicated, based on results shown in Fig. 8, depending on the availability of 4 data points per profile. Hours in Australia Eastern Standard Time. New moon data on both species and trough period data on *P. xylostella* were not available for analysis (see text)

Hours	17–18	18–19	19–20	20–21	21–22	22–23	23–24	24–01	01–02	02–03	03–04	04–05	05–06	06–07	07–08
Cx. p. australicus															
Pre-new moon (24–26 Oct. 1978)	—	3.561 ±2.319	-0.624 ±0.403	-4.469 ±1.839	-0.876 ±0.771	-0.515 ±0.933	-0.229 ±0.121	-0.273 ±0.658	-0.578 ±0.924	-0.951 ±1.913	-0.455 ±0.94	0.46 ±0.698	0.567 ±0.668	2.997 ±0.928	0.835 ±2.744
Post-new moon (6–8 Nov. 1978)	-0.777 ±1.045	-0.472 ±1.134	-0.776 ±0.683	-0.567 ±0.62	-0.473 ±0.528	-0.866 ±0.305	-0.433 ±0.106	-0.312 ±0.459	-0.352 ±0.674	-0.176 ±0.339	-0.184 ±0.658	-0.517 ±0.723	-0.484 ±0.159	-0.476 ±0.406	-0.416 ±0.237
Trough period (10–12 Nov. 1978)	—	-2.478 ±2.525	-3.168 ±2.534	-3.944 ±1.542	-3.621 ±1.456	-3.873 ±1.136	-3.134 ±0.991	—	—	—	—	—	—	—	—
Full moon (16–18 Nov. 1978)	—	—	-1.297 ±1.47	-1.64 ±1.485	-4.095 ±2.142	-1.574 ±0.657	-1.413 ±0.91	-1.557 ±0.851	—	-4.014 ±2.174	-3.954 ±2.016	-0.611 ±2.876	-3.257 ±2.055	-3.567 ±1.587	—
P. xylostella															
Pre-new moon (24–26 Oct. 1978)	—	—	-0.306 ±0.521	-3.226 ±1.932	-0.918 ±0.815	-2.981 ±1.479	-3.568 ±1.249	-0.195 ±0.658	-0.68 ±0.231	-0.527 ±0.182	-0.546 ±0.273	-1.417 ±2.086	-0.92 ±2.31	-0.6 ±2.357	-1.847 ±2.588
Post-new moon (6–8 Nov. 1978)	—	—	—	-2.923 ±1.025	—	-2.35 ±0.97	—	-0.964 ±2.038	—	-2.317 ±0.971	-1.83 ±1.154	—	—	—	—

Table 3. Integrated hourly totals, between 1700 h and 0800 h (AEST), representing the number of insects in an air mass 283 × 10³ m³ up to the maximum trapping height (20 m) and to 80 m (extrapolated) during 5 lunar phases[a]

Hours	17–18	18–19	19–20	20–21	21–22	22–23	23–24	24–01	01–02	02–03	03–04	04–05	05–06	06–07	07–08	Total
Cx. p. australicus (in 1000 s)																
20 m																
Pre-new moon	0	113.4	132.2	64.8	29.3	24.9	61.6	121.6	108.4	37.5	64.7	47.8	46.7	7.2	4.7	864.8
Post-new moon	15.7	28.6	177.5	235.6	161.6	289.1	104.2	96.3	40.1	39.8	73.6	33.1	21.3	17.1	9.0	1342.6
Full moon	0	0	70.0	61.7	50.2	85.5	41.4	29.0	6.5	46.3	34.7	2.3	10.8	9.8	0	448.2
Trough period	0	4.1	16.8	19.1	9.9	8.6	2.5	0	0	0	0	0	0	0	0	61.0
New moon	0	0	0	0	0	0	0	0	0	0	0	0	0	0	0	0

80 m

Pre-new moon	0	631.8	298.8	64.8	53.5	62.7	209.2	393.0	256.1	64.6	172.3	377.7	424.1	184.0	61.6	3254.2
Post-new moon	31.2	75.0	351.2	2142.3	422.8	523.6	284.2	298.2	120.5	219.0	408.5	281.9	55.1	44.9	25.0	5283.4
Full moon	0	0	97.6	74.7	50.2	105.6	54.7	36.1	6.5	46.3	34.7	5.1	10.8	9.8	0	532.1
Trough period	0	4.3	17.0	19.1	9.9	8.6	2.5	0	0	0	0	0	0	0	0	61.4
New moon	0	0	0	0	0	0	0	0	0	0	0	0	0	0	0	0

P. xylostella (in 100 s)

20 m

Pre-new moon	0	0	88.0	90.9	102.0	31.3	65.7	124.6	142.3	89.5	64.7	1.7	24.1	13.6	3.4	84.3
Post-new moon	0	0	0	18.4	1.8	0.3	7.0	7.4	1.7	6.5	4.2	0	0	0	0	4.7
Full moon	0	0	7.5	23.4	19.2	43.6	12.1	8.2	0	0	0	0	0	0	0	11.2
Trough period	0	0	0	0	0	0	0	0	0	0	0	0	0	0	0	0
New moon	0	0	0	0	0	0	0	0	0	0	0	0	0	0	0	0

80 m

Pre-new moon	0	0	274.3	91.2	179.9	31.3	65.7	439.1	305.4	221.5	157.3	1.7	352.2	31.2	0	2150.8
Post-new moon	0	0	0	18.4	6.2	0.3	7.0	12.2	6.5	6.5	4.3	0	0	0	0	61.4
Full moon	0	0	7.5	23.5	19.2	43.7	14.9	37.3	0	0	0	0	0	0	0	146.1
Trough period	0	0	0	0	0	0	0	0	0	0	0	0	0	0	0	0
New moon	0	0	0	0	0	0	0	0	0	0	0	0	0	0	0	0

Mean wind speed ($m\,s^{-1}$) at 0.3–20 m from ground level

Pre-new moon	1.6	1.7	1.7	1.9	1.8	1.8	0.7	0.9	1.0	1.4	1.9	2.2	2.9	3.8	3.8
Post-new moon	2.5	2.3	2.1	2.0	1.5	1.8	1.7	1.3	1.5	1.4	1.8	1.8	1.7	1.5	1.7
Full moon	1.6	1.5	1.8	3.1	3.2	2.7	3.1	3.6	4.0	4.3	4.5	4.3	4.4	4.8	5.9

[a] The mean wind speed between 0.3 and 20 m, i.e., the estimated average distance travelled is given for each hour. Dates for lunar phases are as given in the legend of Fig. 8.

sion coefficients (Table 2), as well as the integrated population totals (Table 3) show that during the lunar cycle, upward movements of both *Cx. p. australicus* and *P. xylostella* occurred during the three peak periods of flight (pre- and post-new moon and full moon) rather than at new moon or during the other trough period. The density profiles (Fig. 8) for almost all hours of the day, indicate a wide spectrum of activity, particularly in mosquitoes, at pre- and post-new moon periods and greater numbers were engaged in upward flights at these times than at full moon (Tables 2 and 3). Comparison of the wind profiles (Fig. 10) and the flight profiles (Fig. 8) show that many of the flights during the three lunar flight peaks occurred above the boundary layer.

In *Cx. p. australicus* flight activity was intense from 1800–0700 h during the pre- and post-new moon periods and 1900–0700 h during the full moon period; in *P. xylostella* intense flight occurred from 1900–0700 h at pre-new moon, 2000–0400 h at

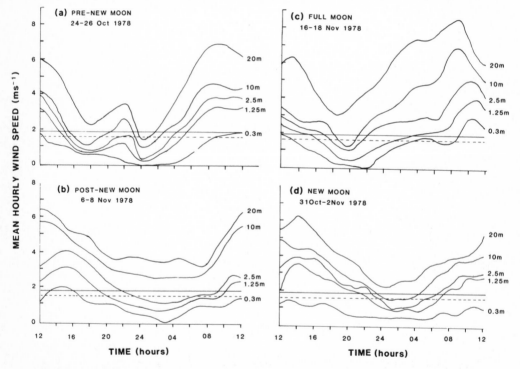

Fig. 10a–d. Wind-profile data at the trapping site. Diel variation of mean wind speed at five heights during (a) pre-new moon period; (b) post-new moon period; (c) full moon period; and (d) new moon period. Dates of these periods are same as in Fig. 8. Boundary layer of flight is indicated by a *dashed-horizontal line (Culex p. australicus)* and a *continuous-horizontal line (Plutella xylostella)*

Fig. 11. Twenty-four-hour population actograms of *Plutella xylostella* (males and females) and *Culex p. australicus* (females) based on captures in the four upper traps (2.5–20 m) during two lunar cycles from October 16–December 23, 1978. Total daily captures (noon-noon) are shown alongside the hourly catch histograms (log scale). The times of sunrise and sunset are given by the two continuous *vertical lines*, and the times of moonrise and moonset by *upward* and *downward arrowheads* respectively. The moon phases are indicated by the symbols: ○ full moon; ● new moon; ◑ first quarter; ◐ last quarter

POPULATION ACTOGRAMS

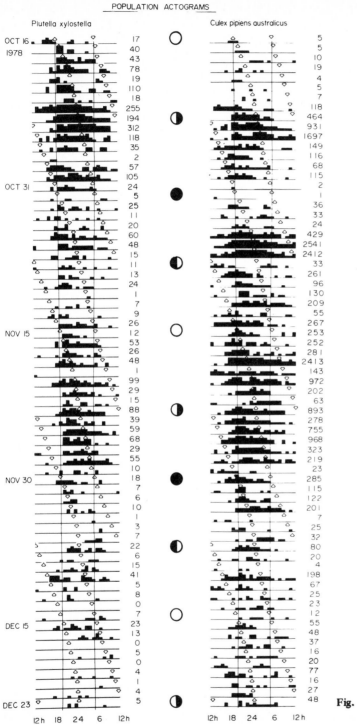

Fig. 11

post-new moon and 1900–0100 h at full moon (Table 3, Fig. 8). From the average wind-speed data at the site (Table 3), the above boundary-layer displacement within the 20 m height was estimated as 71 km, 55 km, and 158 km for *Cx. p. australicus* and 46 km, 7 km, and 56 km for *P. xylostella* during the pre-new moon, post-new moon and full moon nights respectively; the distances traversed at and above 20 m would be even greater (see Fig. 10).

A detailed examination of the hourly and daily flight activity was made by preparing population actograms for the period October 16–December 23, 1978, covering two lunar cycles (Fig. 11). In addition to information on times of peak activity, the data show that flights may begin and continue before moonrise and after moonset. A rhythmic-activity pattern (pre- and post-new moon peaks and a lagged full moon peak) associated with the moon's age is suggested by the actograms despite the large decline of population size with time.

Age distribution of mosquitoes during the five periods show that most of the females flying at the higher levels were nulliparous (Table 4). Of more significance is the fact that at full moon 48% of females flying near the ground level were parous compared with 18.4%–39.0% at the other four periods (Table 4). The proportions of parous females flying were generally higher during the full moon and the new moon periods.

Table 4. Total number trapped and the proportion of females from each age group during the three peak and two trough periods[a]

Trap height	No. and % trapped at different heights					
	0 m	2.5 m	5 m	10 m	20 m	All traps
Post-new moon Nos.	506	2218	1458	890	779	5851
% Nulliparous	81.5	99.7	99.6	99.4	99.1	97.6
% Parous	18.4	0.2	0.4	0.3	0.7	2.4
% Multiparous	0.1	0.1	0.0	0.3	0.2	0.03
Pre-new moon Nos.	777	2401	941	1142	424	5685
% Nulliparous	76.7	99.2	99.7	99.8	100	97.1
% Parous	23.3	0.7	0.3	0.1	0.1	2.8
% Multiparous	0.0	0.1	0.3	0.1	0.0	0.0
Full moon Nos.	742	752	597	1189	129	3409
% Nulliparous	50.7	97.9	99.5	99.7	100	88.6
% Parous	48.0	2.0	0.2	0.3	0.0	11.0
% Multiparous	1.3	0.1	0.0	0.0	0.0	0.4
Trough period Nos.	336	427	124	160	74	1121
% Nulliparous	70.5	96.7	97.6	95.0	51.3	85.7
% Parous	24.7	2.1	0.0	3.7	20.3	10.1
% Multiparous	4.8	1.2	2.4	1.3	28.3	4.2
New moon Nos.	182	164	156	189	7	698
% Nulliparous	60.4	89.0	98.1	98.4	85.7	86.1
% Parous	39.0	8.5	0.6	0.5	14.3	12.6
% Multiparous	0.5	2.4	1.3	1.0	0.0	1.3

[a] Moon age days 5–7 post-new moon, days 10–12 trough period, days 15–17 full moon, days 23–25 pre-new moon, days 29–1 new moon trough period. Data from both seasons.

3.3 The Moon and Moonlight

The moon moves around the earth in an eccentric orbit at a mean distance of 384492 ± 1 km (range: 364400−406730 km) each month (Kopal 1960, 1962). The moon rotates slowly making one complete turn on its axis in 27.32 days. This is the same period as its revolution around the earth (the siderial month) and thus it always presents the same hemisphere to the earth. By the time the moon has completed one revolution of the earth, the earth-moon system itself has moved a fractional distance along its orbit around the sun. Therefore, the time between successive full moons or successive new moons (called syzygies) is a little longer than the siderial period. The period between one full moon and the next full moon, called the synodic month or lunar month is 29.53 days and has a variation of 29.25−29.83 days arising out of the moon's eccentric motion. The approximate value of 30 days used for the synodic month in the present investigation introduces only a small error of 0.0157.

Moonlight is partially reflected sunlight and is spectroscopically similar, though distinctly redder, than the illuminating sunlight (Kopal 1962, 1963). The colour index of the integrated moonlight varies somewhat with the phase of the moon so that it is bluest at full moon and growing slightly redder towards both quarters. The maximum intensity of moonlight, under clear atmospheric conditions at the earth's surface, is 0.267 lx when the full moon is at the zenith (Anonymous 1974). The illumination of the moon (Fig. 7c) is a function of: (1) the angle of vision or phase angle, defined as the selenocentric angle or the angle of vision between the direction to the sun and the earth (Fig. 7c); (2) the differences in the reflectance (albedo) of the different parts of the lunar surface; (3) the variation in the earth-moon distance during the cycle; (4) the altitude angle of the moon above the earth's horizon; and (5) the atmospheric effects (Kopal 1960, 1962; Pellicori 1971). The albedo effect makes the waxing moon about 20% brighter than the waning moon (Anonymous 1974). The most characteristic photometric feature of the moon is the steepness of its light curve − in particular, a rapid rise of light towards full moon, and equally rapid diminution of it after the full phase, so that at quarter phase, when the size of the illuminated disc is cut by half, the brightness is one-ninth of that at full moon (Kopal 1960, 1962; Pellicori 1971) (Fig. 7c). At new moon ±2 days illumination is very low, measuring about 0.0005 lx.

Moonlight is partially plane-polarized at the lunar surface (Kopal 1960, 1962; Lyot 1929). The amount of polarization is a function of the phase angle and, of the polarizing properties of the particles on the lunar surface (Kopal 1960, 1962; Lyot 1929; Pellicori 1971). The direction of polarization is always exactly perpendicular or exactly parallel to the plane of scattering. The plane of vibration of the electric vector is the same at every point on the moon; it is perpendicular to the plane of scattering (called positive polarization by convention) for phase angles greater than $23° 30'$ and is parallel to the plane (called negative polarization) for angles $0°-23° 30'$. At $23° 30'$ (about 2 days from full moon) and at $0°$ (full moon) the polarization goes through zero. The properties of the moon's polarized light is described by means of a single 'curve of polarization' (Fig. 7c) (Lyot 1929). Moonlight polarization is greatest near the quadrature when $\alpha = 84°-110°$ (6- to 8-day-old and 22- to 24-day-old moon). For integrated moonlight the maximum is 6.6% near the first quarter and 8.8% near the last quarter because more maria are illuminated during the last quarter. The maximum negative polarization of -1.2% is reached at $11°$ (1 day from full moon). Polarization of moonlight for short wavelengths, which are more attractive to insects (Mazokhin-Porshnyakov 1969) is greater than that for longer wavelengths. The respective polarization peaks for the wavelengths 336 nm, 367 nm, 383 nm, 449 nm, and 519 nm are 14.2%, 12.9%, 11.4%, 8.9%, and 7.3% in waxing moonlight and 18.9%, 17.0%, 15.2%, 10.5%, and 9.3% in waning moonlight (Pellicori 1971). Polarization of moonlight is thus a continuously changing phenomenon that varies roughly inversely with the albedo as well as with the wavelength and peaks near the first and the last quarters (Fig. 7c,f). It has been shown that moonlight can extend or modify the crepuscular flight activity of some insects (see below). In this context it is of interest to note that in ants and bees only those visual cells in the eye that are responsive to ultraviolet wavelengths are involved in the perception of polarized light (Halverson and Edrich 1974; Wehner 1976, 1984) and this probably applies to insects in general (Brines and Gould 1982). The adaptive value of sensitivity to polarized ultraviolet light, is that it is in the ultraviolet range of wavelengths that the polarization of skylight is least affected by atmospheric disturbances and is therefore the most stable (von Frisch 1967; Wehner 1976). Also insects may use ultraviolet light in discriminating between the sky and ground, e.g. in detecting the sky when taking off the ground (Kennedy et al. 1961; Mazokhin-Porshnyakov 1969; Menzel 1979), detecting air-water interphases and waxy surfaces of plant leaves (Chen and Rao 1968; Waterman 1974), or in any kind of course control in which skylight is involved (Wehner 1984). The degree of polarization required for satisfactory operation of biological polarimeters has not been studied extensively. In the honey bee (von Frisch 1967) and in *Daphnia* (Watermann 1966a), the behavioural threshold for E-vector discrimination is somewhere about 10% linear polarization. More recent studies tend to show that ants and bees ignore the degree of polarization (Brines and Gould 1979; Wehner 1984). Whichever view is correct it would seem that the amount of polarization generated at the lunar surface is sufficient to stimulate insect photoreceptors.

Since much of the natural irradiation is plane polarized, it is not surprising that many animals can perceive the E-vector of linearly-polarized light (Watermann 1966b). Animal polarimeters, including those of insects were evolved in nature because of the adaptive advantage of the organisms being able to "read" just one more potentially informative parameter in the visual environment (Watermann 1974). So far it has been recognized that there are three particular sources of polarized light that compose a substantial, if not ubiquitous, part of an animal's visual environment: (1) Rayleigh-scattered sunlight in the atmosphere; (2) Rayleigh-scattered sunlight of the hydrosphere; and (3) differentially reflected and refracted light from natural air-water interfaces and terrestrial surfaces [e.g. wet surfaces in the air and waxy surfaces of plant leaves (Waterman 1974, 1981)]. Polarized moonlight is a fourth source of polarized light in the environment that has not been considered before. All of the biological significance so far known for natural polarized light is restricted to sunlight. Since moonlight is reflected sunlight and is partially plane-polarized at the lunar surface itself, polarized moonlight should occur from moonset to moonrise at any point in the sky (this aspect has not been studied) unlike sunlight where the polarization is dependent on the sun's position in the sky. Despite the widely occurring lunar periodicity of activity in insects and other animals (Anderson 1983; Neumann 1981), hardly any research has been done on the possible influence of moonlight itself on these activities.

Now turning to the question of whether insects use the moon for orientation or navigation, there are no clear examples from among insects where the moon is used for compass navigation during long-distance travels. There are several difficulties in using the moon as a compass. First, the moon may not be up when the animal starts out or may set before the night is over as it is visible for only a part of the night and, on successive nights, for different parts of the night (Fig. 11). Second, though the moon travels at about the same speed as the earth, its timing (LD cycle) shifts 50 min or $12°$ each day and is, therefore, inconsistent. Third, its image size and the position in the sky (the moon-azimuth/time) change drastically from night to night. Fourth, an insect would have to possess an internal clock in phase with the lunar time, operating independently of the circadian clock and synchronized by a signal other than the visible moon or moonlight (see Sect. 4). Such lunar clocks do exist (Neumann 1981; Palmer 1976). There is strong evidence that beach amphipods *(Talitrus, Orchestoides, Talorchestia)* use a time-compensated moon compass to orient their nocturnal movements (Enright 1972; Papi 1960; Papi and Pardi 1963). The possibility that insects may use the moon, when it is visible, for orientation and navigation, at least during short-distance travels should not, therefore, be ruled out. It would seem that skepticism on research efforts relating to the influence of the moon, and the belief that those who work on the influence of the moon on living beings are anthroposophists may have retarded progress in this field (see Korringa 1957).

4 Discussion

These results provide indisputable evidence on densities of *Culex p. australicus* and *Plutella xylostella* at various heights and the change of profiles during the day and night in relation to phases of the moon. Density estimates at five heights over nine lunar cycles show that there is a trimodal lunar periodicity with peak flight activity occurring at pre- and post-new moon periods (about 4–8 days before and after new moon, i.e. near the first and the last quarters) and at full moon. This is in agreement with the previous findings on *Epiphyas postvittana* (Danthanarayana 1976) and *P. xylostella* (Goodwin and Danthanarayana 1984).

Both species (*Cx. p. australicus* and *P. xylostella*) are nocturnal fliers with maximum flight activity extending from 1 h after sunset to 1 or 2 h before sunrise. The lunar periodicity of flight is thus superimposed on that of the circadian cycle. Since the average flight speeds of *Cx. p. australicus* and *P. xylostella* were determined as $1.51 \ \mathrm{m\,s}^{-1}$ and $1.87 \ \mathrm{m\,s}^{-1}$ respectively, an individual travelling in wind speeds in excess of these values would have crossed its boundary layer (Taylor 1958, 1960, 1974) and its track will mainly be downwind. Such movements generally result in wind-assisted migration (Johnson 1969; Johnson et al. 1962; Taylor 1958, 1974). Evidence presented here show quite well that *Cx. p. australicus* as well as *P. xylostella* migrate during the pre- and post-new moon and full moon periods.

The only significant correlation between lunar periodicity of flight activity and the physical environmental factors considered (temperature, humidity, rainfall, sunlight, moonlight, wind speed) was between the numbers of insects trapped and the amount of polarization of moonlight. This is unlikely to be a cause-and-effect relationship as pointed out below. The independence of peak-activity patterns with the times of moon-rise and moonset suggests that the observed lunar periodicity is rhythmic, but it is not known whether there is an endogenous lunar-flight rhythm as in the chironomid *Clunio marinus* Haliday and the mayfly *Povilla adusta* Navás (Hartland-Rowe 1958; Neumann 1975). Extensive studies of entrainment by moonlight on *Clunio marinus* showed that cultures of the midge kept in LD 12:12 exhibit only 24 h rhythmicity in their emergence for mating flights, but cultures exposed to simulated moonlight (0.3 lx) during a run of four nights every month, or even one sequence of three "moonlit" nights re-estab-lishes the circasemilunar rhythm, thus confirming it is truly endogenous (Neumann 1978). A rhythmic-activity pattern associated with the moon's age is faintly suggested by the actograms. In nature, endogenous rhythms do not normally free run, because conditions are never constant and any daily phase-shift in the endogenous clock is reset each cycle and this may bring about slight variations (see Brady 1982; Saunders 1982). Corbet (1960) has pointed out, those cyclical activities which are poorly synchronised, and which occur when environmental changes are small, must be regarded as likely to have a strong endogenous component.

A lunar periodicity independent of the timing of the moon's appearance in the sky precludes the possibility that the observed flight behaviour results from a *direct* visual response to the moon or moonlight, unless there is a delayed response to moonlight as in some other organisms (see Bünning 1973). But it is known that moonlight is used by some insects flying during moonlit nights, as with the temporal extension of swarming in the mosquito *Psorophora confinnis* (Lynch Arrib) when moonlight postponed the usual fall of darkness (Provost 1958) and the continuation of the crepuscular foraging flights of the nocturnal bee *Sphecodogastra texana* (Cresson) into the night when the moon is visible (Kerfoot 1967). It would seem that the trimodal periodicity is a popula-tion phenomenon, probably endogenous, with three unimodal periodicities, but with different phase relationships to a single Zeitgeber.

The possibility that dim polarized light available at night and twilight may be used for vision, orientation, short-distance navigation or merely as a cue for flight initiation should not be ignored, however, for in addition to the above examples it is known that mosquitoes stop travelling as soon as overhead polarization is interrupted (Wellington 1974, 1976). A number of ant species are known to show moon-compass reaction (photomenotaxis). The workers of the ant *Monomorium salomonis* (L.) marching in a straight line at night, are suddenly disoriented when the moon becomes obscured by clouds, and in *Formica rufa* L. assumption of a determined angle of orientation to the moon has been demonstrated by means of experiments with mirrors (reviewed by Cloudsley-Thompson 1980; Jander 1957; Papi 1960). The African desert ant *Catagly-phis bicolor* Fab. can employ the moon for orientation, for a positive or negative tropo-taxis, but not for photomenotaxis (Wehner and Duelli 1971). In the beetle *Calandra granaria* L., there are cycles of phototactic responses that correspond to the lunar phases (Birukow 1964). Also the nocturnal migratory moth *Noctua pronuba* L. uses the moon's azimuth as an orientation cue, but does not compensate for the movement of the moon across the sky (Sotthibandhu and Baker 1976). The periodicity of foraging activities of *S. texana* is based upon the lunar cycle of 29 days, as an extension of its regular crepuscular foraging activity (Kerfoot 1967), and certain night-flying insect species and groups show a preferential response to dim polarized light over unpolarized light of the same intensity (Danthanarayana and Dashper, this volume; Kovrov and Monchadskiy 1963). Although it can be argued that the upward flights at dusk and dawn when polarization is greatest (Rozenberg 1966) is associated with sunlight, such movements did not occur to any great extent during periods other than the pre- and post-new moon peaks (Fig. 8). This implies that polarized sunlight is unlikely to be associated with the upward flight phenomenon.

A notable lag, of about 2 days, occurred in the full moon flight peak of *Cx. p. australicus* as well as that of *P. xylostella*. The cause of this lag, which also occurs in *E. postvittana* (Danthanarayana 1976), is not known from the present study. A similar lag caused by the timing of emergence of adults occurs in *Povilla adusta* (Hartland-Rowe 1955) and *Clunio marinus* (Neumann 1976) (Fig. 12a,b). Emergence of adults as well as reproduction of *P. adusta* and other Ephemeroptera peaked shortly after the day of full moon, and this was rhythmic, persisting in the laboratory when nymphs were reared in the dark (Hartland-Rowe 1958). In species of Chironomidae, Culicidae, Ephemeroptera and Plecoptera, periodic emergence of a generation is associated with lunar periodicity, the emergence peak coming 2–5 days after full moon (Johnson 1969). This synchronization may also be accompanied by a diapause in the later larval stages and is then associated with the typical skew curve of emergence and flight near full moon (Johnson 1969), a pattern similar to that of *Cx. p. australicus, P. xylostella,* and *E. postvittana*. In *C. marinus* the midges emerge shortly after full moon and shortly after new moon (Fig. 12a) and the timing of this semilunar periodicity of emergence from pupation is controlled by an endogenous rhythm (Neumann 1967, 1975, 1976).

With respect to *Cx. p. australicus* it was demonstrated that at full moon, there is a very large proportion of parous females flying near ground level than at other times. Full moon flights are, therefore, in part, of the appetitive type. It is noteworthy that in a similar way the feeding activity of *Anopheles albimanus* (Widemann) peaks at full moon (Pratt 1948) (Fig. 12f).

The question arises as to whether the trimodal lunar periodicity is unique to *Cx. p. australicus, P. xylostella,* and *E. postvittana* (Danthanarayana 1976) or is common place among other insects and insect groups. The occurrence of the trimodal curve in other insect species and groups is implicit in an array of published work, provided the shortcomings of the sampling methods are appreciated (Appendix 2). In light-trap catches the full moon peak is obliterated as a consequence of either the moon competing with the trap light and/or trap inefficiency during brightly moonlit nights (Bowden 1973; Bowden and Church 1973; Bowden and Morris 1975; see Fig. 12c,d,f). Nowinszky et al. (1979) found that the relative abundance of seven species of insects caught in light traps in Hungary peaked at the first and the last quarters of the moon and that this correlated with moonlight polarization (Appendix 2). Brown and Taylor (1971) using suction traps found that the vertical distribution of total flying insects in East Africa had a significant lunar cycle, out of phase with the moon, insects flying higher in the first quarter of the moon than the third. A trimodal lunar periodicity curve is also implicit in their data. In the simultaneous light and animal-bait trapping of *Anopheles albimanus* carried out by Pratt (1948) only the pre- and post-new moon peaks occur in the light-trap data, whereas the full moon peak appears to be clear and distinct in the bait-trap data. The suppression of the full moon peak is a common feature in light-trap catches as pointed out above (Appendix 2). Radar observations of Schaefer (1976) show

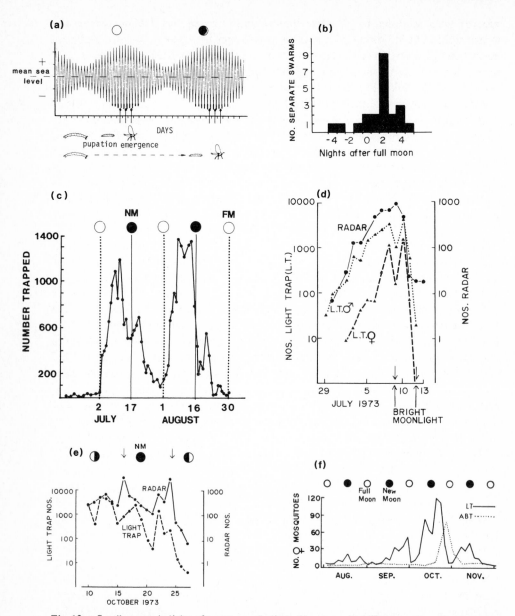

Fig. 12. a Semilunar periodicity of emergence (indicated by *arrows*) of *Clunio marinus* showing post-new moon and lagged full moon peaks (from Neumann 1975). b Frequency of 22 *Povilla adusta* swarms shows a lagged full moon peak (from Hartland-Rowe 1955). c Numbers of *Heliothis zea* moths caught in two light traps show the pre- and post-new moon peaks (from Nemec 1971; Bowden 1973). d Numbers of *Choristoneura fumiferana* during take-off period peaking at full moon in radar observations compared with the light-trap catches (*L.T.*) (from Schaefer 1976). e Radar observations and light-trap catches of the Sudan grasshopper, *Aiolopus simulatrix*, showing the pre- and post-new moon peaks (from Schaefer 1976). f Comparison of *Anopheles albimanus* in light-trap and animal-bait trap collections demonstrate the pre- and post-new moon peaks in the former and the full moon peak in the latter traps (after Pratt 1948)

the pre- and post-new moon peaks for the Sudan grasshopper *Aiolopus simulatrix* (Walker) and the full moon peak for the spruce budworm, *Choristoneura fumiferana* (Clemens) (Fig. 12d,e). Both these species are long-distance migrants flying high at mean altitudes of 600 m (maximum: 1500 m) and 200–250 m (maximum: 600 m) respectively (Greenbank et al. 1980; Schaefer 1976; Riley and Reynolds 1983); unfortunately, data for an entire lunar month are not available in both cases. These and other examples referred to in Appendix 2 illustrate that the detection of the occurrence of one, two or all three of the characteristic flight peaks may depend on the method and frequency of sampling adopted and/or the timing of the emergence of the adults. Wide use of light traps (e.g. Biddingmayer 1964; Haddow et al. 1961; Nemec 1971; Williams 1940), expression of data in terms of the four quarters of the moon (e.g. El-Ziady 1957; Miller et al. 1970; Williams 1940) and lack of sufficient numbers of insects in trap catches (e.g. in suction traps) (Brown and Taylor 1971) appear to be the reasons for the non-detection of the trimodal, lunar-periodicity curve.

Previous accounts (Johnson 1969, p. 276; Williams et al. 1956) may suggest that the most convincing lunar periodicities in insects are found in species with aquatic life histories. Information documented here (Appendix 2) shows that lunar periodicity of insect flight is widespread and, in fact, extends to wholly terrestrial forms (cf. Bowden 1973). In this context it is of interest and perhaps intriguing to note that there is a semilunar periodicity in the cardio-acceleratory activity of extracts of cockroach nervous tissue and haemolymph on the heart of *Periplaneta americana* L. (Rounds 1981). The influences of the moon that trigger this physiological activity (Fig. 13) are not yet understood. A close examination of the data of Rounds (Figs. 1–6 of Rounds 1981) does show indeed that the pre- and post-new moon peaks and the lagged full moon peak occur in the cardio-acceleratory activity in extracts of suboesophageal, supraoesophageal, first thoracic and abdominal ganglia of *P. americana*. The question that arises is whether such a lunar periodicity also exists in the production and/or release of hormones associated with insect migratory activity.

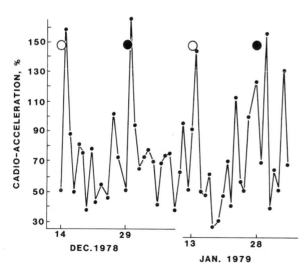

Fig. 13. Cardio-acceleratory activity of the suboesophageal ganglion extracts (ethanol extracted, dried and re-constituted in insect saline) of male *Periplaneta americana* on the cockroach heart. *Large open* and *closed circles* represent full moon and new moon respectively (after Rounds 1981)

Results reported here have shown that there is a lunar rhythm in the flight and migratory activities of *P. xylostella* and *Cx. p. australicus* and this probably applies to many other night-flying insect species. The nature of the timing mechanism involved is, however, not known. Although the lunar rhythms in *Clunio marinus* and *Povilla adusta* have been shown to be of endogenous origin, the usual constant conditions under which the experiments were carried out did not ordinarily exclude the gravitational effects of the moon. Since the lunar periodicity of flight activity of *Apis mellifera* does not depend on photoperiod, Oehmke (1973) considered the potential of gravitational variations caused by the moon as a Zeitgeber. In the same way that lunar periodical changes in gravity results in the phenomenon of tides of the sea, there are such tides in the atmosphere as well, the existence of which has been verified by Bartels and Chapman (see Oehmke 1973) and others (Lieber 1978). The lunar periodicity in *A. mellifera* is attributed to the perception of the lunar gravitational Zeitgeber, for there is a correlation between flight activity and gravitational oscillations in each season (Oehmke 1973). Similarly, the possibility that pulses of gravitational maxima or associated atmospheric tides may act to trigger the semilunar cyclicity of the neurophysiological process in the cockroach has been suggested by Rounds (1981).

Adaptive advantages of lunar periodicity appear to be that it enables survival by migration, promotes the synchronization of emergence and flight for mating and reproductive purposes, the moon or moonlight may be used for vision, orientation or short-distance navigation, and the occurrence of the flight peaks at times when the moon is not at its brightest should facilitate escape from predators.

These findings, discussed in relation to 61 insect species from 9 orders and 22 families (Appendix 2), show that the moon is more closely implicated in insect flight and migration than was previously known. They also provide an entirely new perspective on the understanding of the phenomenon of lunar periodicity of flight. It is imperative, therefore, that careful consideration be given to the timing of observations, and the methods of sampling adopted as well as to the interpretation of data in studies dealing with flight behaviour and migration of night-flying insects.

Acknowledgements. I thank Anthony Wohlers for his help with the data analysis, Chutamas Satasook for collecting the 1984/85 data, Nina Bate, Patrick Vaughan, and Stephen Pearson for the careful technical assistance. An earlier version of this paper was reviewed by four anonymous referees through Mary Ann Rankin of the University of Texas and L.R. Taylor of the Rothamsted Experimental Station. I thank them for their constructive criticism and Tim New and Ian Thornton for reading the manuscript. This work was supported by grants from the Australian Research Grants Scheme (1978 Committee) and the Sunshine Foundation of Victoria.

Appendix 1. Daily numbers of *Culex p. australicus* and *P. xylostella* trapped and the measurements of physical environmental factors between 1800 h and 0600 h each day for the period October 16 to December 16, 1978

Day	Moon age	Numbers trapped		Mean	Precipitation	Humidity	Wind run
		Culex	Plutella	Temp. (°C)	(mm)	(% R.H.)	(km)
Oct. 16	13.8	5	17	9.6	0.0	98	33
17	14.8	5	40	13.1	2.9	98	45
18	15.8	10	43	8.6	5.8	94	80
19	16.8	19	78	10.7	0	91	32
20	17.8	4	19	9.0	0	90	17

Appendix 1 (continued)

Day	Moon age	Numbers trapped		Mean temp. (°C)	Precipitation (mm)	Humidity (% R.H.)	Wind run (km)
		Culex	*Plutella*				
Oct. 21	18.8	5	110	13.1	0	89	13
22	19.8	7	18	13.0	0	91	27
23	20.8	118	255	16.1	0	94	45
24	21.8	464	194	17.3	0	95	42
25	22.8	931	312	23.3	0	92	81
26	23.8	1697	118	18.0	0.3	93	94
27	24.8	149	35	7.6	1.0	93	35
28	25.8	116	2	8.7	11.8	94	40
29	26.8	68	57	9.5	1.0	98	30
30	27.8	115	105	14.0	0	97	26
31	28.8	2	24	10.5	0	94	82
Nov. 1	29.8	1	5	8.2	0	92	70
2	0.7	36	25	13.3	0	91	23
3	1.7	33	11	11.6	0	94	36
4	2.7	24	20	12.1	0	98	22
5	3.7	429	60	13.5	0.2	99	43
6	4.7	2541	48	16.8	3.3	100	39
7	5.7	2412	15	16.6	1.0	99	38
8	6.7	33	11	13.3	6.3	94	70
9	7.7	261	13	11.5	0	96	25
10	8.7	96	24	8.7	0	94	78
11	9.7	130	1	11.1	0	92	59
12	10.7	209	7	11.0	4.9	96	48
13	11.7	55	9	12.8	9.5	95	58
14	12.7	267	26	13.4	0.2	93	24
15	13.7	253	12	13.8	0	97	22
16	14.7	252	53	18.9	0	95	30
17	15.7	281	26	22.7	0	91	102
18	16.7	2413	48	14.9	0	95	63
19	17.7	143	1	12.1	36.6	100	70
20	18.7	972	99	15.9	4.0	100	56
21	19.7	202	29	18.0	0	91	87
22	20.7	63	15	14.8	0	93	83
23	21.7	893	88	17.3	0	95	26
24	22.7	278	39	17.3	0	98	27
25	23.7	755	59	22.8	0	78	64
26	24.7	968	68	22.5	0	74	60
27	25.7	323	29	17.4	0	86	37
28	26.7	219	55	22.6	0	84	44
29	27.7	23	10	13.5	0.5	82	83
30	28.7	285	18	11.7	7.0	96	32
Dec. 1	0.2	115	7	15.5	0	84	45
2	1.2	122	6	21.9	0	78	107
3	2.2	201	10	18.8	0	71	78
4	3.2	7	1	11.3	9.4	99	73
5	4.2	25	3	13.4	0.2	83	71
6	5.2	32	7	13.0	1.2	93	69
7	6.2	80	22	11.9	0.2	97	50
8	7.2	20	6	13.5	0	95	24

Appendix 1 (continued)

Day	Moon age	Numbers trapped		Mean temp. (°C)	Precipitation (mm)	Humidity (% R.H.)	Wind run (km)
		Culex	*Plutella*				
Dec. 9	8.2	4	15	13.3	0	89	70
10	9.2	198	41	19.1	1.8	95	29
11	10.2	67	5	15.9	0	71	62
12	11.2	25	8	12.0	0.3	88	40
13	12.2	23	0	13.3	5.8	99	89
14	13.2	12	7	13.2	2.0	89	61
15	14.2	55	23	15.3	0	90	30
16	15.2	48	13	17.4	0	90	59

Appendix 2. Records of presence (+) or absence (–) of peak flight activity implicit in published work (0 denotes no observation made)

Insect group/species	Type of study[a]	Post-new moon	Full moon	Pre-new moon	Reference
All insects	LT	+	–	+	Williams (1940)
Coleoptera					
Bostrychidae	LT	–	+	–	Bowden and Morris (1975)
Scarabaeidae					
Melolontha melolontha L.	LT	+	–	+	Nowinszky et al. (1979)
Staphylinidae	LT	+	–	+	Bowden and Morris (1975)
Diptera	ST	+	+	+	El-Ziady (1957)
Diptera	LT	+	–	+	Bowden and Morris (1975)
Chaoboridae					
Chaoborus spp.	VSC	+	–	+	MacDonald (1956)
Chaoborus edulis Edwards	LT	–	–	+	Corbet (1958)
Chironomidae					
Clinotanypus claripennis Kieffer	LT	+	–	–	Corbet (1958)
Clunio marinus Haliday	SE	+	+	–	Neumann (1975, 1976)
Tanytarsus balteatus Freeman	LT	+	–	–	Corbet (1958)
Culicidae					
Aedes leneatopennis Ludlow	LT	+	–	+	Miller et al. (1970)
Aedes taeniorhynchus (Wiedemann)	LT	+	–	+	Pratt (1948), Provost (1959)
Aedes taeniorhynchus (Wiedemann)	TFT	+	–	+	Provost (1959)
Anopheles albimanus (Wiedemann)	LT	+	–	+	Pratt (1948)
Anopheles albimanus (Wiedemann)	ABT	–	+	–	Pratt (1948)
Anopheles crucians (Wiedemann)	LT	+	–	+	Provost (1948)
Anopheles philippinensis Ludlow	LT	+	–	+	Miller et al. (1970)
Anopheles quadrimaculatus Say.	LT	+	–	+	Pratt (1948)
Anopheles subspictus Grassi	LT	+	–	+	Miller et al. (1970)
Anopheles vagus Doritz	LT	+	–	+	Miller et al. (1970)
Culex spp.	LT	+	–	+	Pratt 1948
Culex annulus Theobald	LT	+	–	+	Miller et al. (1970)
Culex gelidus Theobald	LT	+	–	+	Miller et al. (1970)
Culex fuscucephalus Theobald	LT	+	–	+	Miller et al. (1970)
Culex pipiens australicus (Dob. and Drum.)	ST	+	+	+	Present Study
Culex tritaeniorhynchus Giles	LT	+	–	+	Miller et al. (1970)

Appendix 2 (continued)

Insect group species	Type of study[a]	Post- new moon	Full moon	Pre- new moon	Reference
Mansonia annulifera Theobald	LT	+	−	+	Miller et al. (1970)
Mansonia indiana Edwards	LT	+	−	+	Miller et al. (1970)
Mansonia indubitans Dyar	LT	+	−	+	Pratt (1948)
Mansonia perturbans (Walker)	LT	+	−	+	Miller et al. (1970)
Ephemeroptera	SW	0	+	0	Hora (1927)
Baetidae					
Oligoneura rhenanna Imhoff	SW	−	+	−	Hora (1927)
Ephemeridae					
Hexagenia bilineata Imhoff	SW	0	+	+	Hora (1927)
Palingenia robusta Eaton	SW	+	0	+	Hora (1927)
Polymitarcys virgo Imhoff	SW	+	0	+	Hora (1927)
Povilla adusta Navas	SW	−	+	−	Hartland-Rowe (1955)
Povilla adusta Navas	LT	−	+	−	Hartland-Rowe (1955); Corbet (1958)
Heptagenidae					
Rhithrogena mimus Imhoff	SW	0	+	0	Hora (1927)
Heteroptera, Belastomatidae					
Sphaerodema severinii Melicher	LT	−	+	−	Bowden (1964)
Hymenoptera, Formicidae					
Dorylus atriceps Shuckard	LT	+	−	+	Bowden and Morris (1975)
Apidae					
Apis mellifera L.	VSC	+	+	+	Oehmke (1973)
Apis mellifera carnica Pollm.	VSC	−	+	−	Oehmke (1973)
Sphecodogastra texana (Cresson.)	LT	+	+	+	Kerfoot (1967)
Isoptera	LT	−	+	−	Bowden and Morris (1975)
Lepidoptera, Arctiidae	ST(5)	+	+	+	Brown and Taylor (1971)
← Arctiidae	ST(30)	+	+	+	Brown and Taylor (1971)
Rhodogastria luteibarba Hmps.	LT	+	−	+	Bowden and Morris (1975)
Lymantriidae ↗					
Hyphantria cunea (Drury)	LT	+	−	+	Nowinszky et al. (1979)
Scotia segetum Schiff	LT	+	−	+	Nowinszky et al. (1979)
Geometridae					
Operophtera brumata L.	LT	+	−	+	Nowinszky et al. (1979)
Pyralidae					
Diatraea saccharalis (F.)	LT	+	−	+	Bowden (1964)
Marasmia trapezalis (Gn.)	LT	+	−	+	Bowden and Morris (1975)
Stemmorrhages sericea (Dru.)	LT	+	−	+	Bowden and Morris (1975)
Noctuidae	LT	+	−	+	Williams (1936)
Heliothis armiger (Hubner)	LT	+	−	+	Morton et al. (1981)
Heliothis punctiger Wallengren	LT	+	−	+	Morton et al. (1981)
Heliothis zea Boddie	LT	+	−	+	Nemec (1971)
Orthosia cruda Den et Schiff	LT	+	−	+	Nowinszky et al. (1979)
Orthosia golica L.	LT	+	−	+	Nowinszky et al. (1979)
Schalidomitra variegata (Holland)	LT	−	−	+	Bowden (1973)
Scotia segetum Schiff	LT	+	−	+	Nowinszky et al. (1979)
Spodoptera triturata (Wlk.)	LT	−	+	−	Bowden and Morris (1975)
Tortricidae					
Choristoneura fumiferana (Clem.)	RAD	0	+	0	Schaefer (1976)
Epiphyas postvittana (Walk.)	ST	+	+	+	Danthanarayana (1976)

Appendix 2 (continued)

Insect group/species	Type of study[a]	Post-new moon	Full moon	Pre-new moon	Reference
Yponomeutidae					
Plutella xylostella L.	ST	+	+	+	Goodwin and
Orthoptera, Acrididae					Danthanarayana (1984)
Aiolopus simulatrix (Walk.)	RAD	+	0	+	Schaefer (1976)
Aiolepus simulatrix (Walk.)	LT	+	0	+	Schaefer (1976)
Trichoptera, Leptoceridae					
Atripsodes stigma Kimmins	LT	–	–	+	Corbet (1958)
Atripsodes ugandanus Kimmins	LT	+	–	–	Corbet (1958)

[a] Abbreviations: *LT* = Light trapping; *ST* = suction trapping (height in feet); *VSC* = visual swarm counts; *SE* = swarm emergence; *TFT* = truck-mounted funnel trapping; *ABT* = animal-bait trapping; *SW* = swarm observation; *RAD* = radar observation.

References

Anderson EW (1983) Animals as navigators. Van Nostrand Reinhold, New York

Anonymous (1974) Electro-optics handbook. Technical Ser EOH-11. RCA Corporation, Pennsylvania

Bidlingmayer WL (1964) The effect of moonlight on the flight activity of mosquitoes. Ecology 45L:87–94

Birukow G (1964) Aktivitäts- und Orientierungsrhythmik beim Kornkäfer (*Calandra granaria* L.). Z Tierpsychol 21:279–301

Bowden J (1964) The relation of activity of two species of Belastomatidae to rainfall and moonlight in Ghana (Hemiptera:Heteroptera). J Entomol Soc South Afr 26:293–301

Bowden J (1973) The significance of moonlight in photoperiodic responses of insects. Bull Entomol Res 62:605–612

Bowden J, Church BM (1973) The influence of moonlight on catches of insects in light-traps in Africa II. The effect of moon phase on light-trap catches. Bull Entomol Res 63:129–142

Bowden J, Morris MG (1975) The influence of moonlight on catches of insects in light-traps in Africa III. The effective radius of a mercury-vapour light trap and the analysis of catches using effective radius. Bull Entomol Res 65:303–348

Brines ML, Gould JL (1979) Bees have rules. Science 206:571–573

Brines ML, Gould JL (1982) Skylight polarization patterns and animal orientation. J Exp Biol 96:69–91

Brown ES, Taylor LR (1971) Lunar cycles in the distribution and abundance of airborne insects in the equatorial highlands of East Africa. J Anim Ecol 40:767–779

Brady J (ed) (1982) Biological time keeping. Cambridge University Press, Cambridge

Bünning W (1973) The physiological clock. Circadian rhythms and biological chronometry, 3rd edn. The English Universities Press, London

Chen HS, Rao CRN (1968) Polarization of light reflection by some natural surfaces. Br J Appl Phys Ser 2 1:1191–1200

Cloudsley-Thompson JL (1980) Biological clocks. Their functions and nature. Weidenfeld and Nicholson, London

Corbet PS (1958) Lunar periodicity of aquatic insects in Lake Victoria. Nature 182:303–331

Corbet PS (1960) Patterns of circadian rhythms in insects. Cold Spring Harbor Symp Quant Biol 25:357–360

Danthanarayana W (1976) Diel and lunar flight periodicities in the light brown apple moth, *Epiphyas postvittana* (Walker) (Tortricidae) and their possible adaptive significance. Aust J Zool 24:65–73

Detinova TS, Bertram DS (1962) Age-grouping methods in Diptera of medical importance. World Health Organization, Geneva

Dingle H (1972) Migration strategies of insects. Science 175:1327--1335

Drake VA (1985) Radar observations of moths migrating in a nocturnal low-level jet. Ecol Entomol 10:259--265

Duviard D (1974) Flight activity of Belastomatidae in central Ivory Coast. Oecologia (Berl) 15: 321--328

El-Ziady S (1957) A probable effect of the moonlight on the vertical distribution of Diptera. Bull Soc Entomol Egypté 41:655--662

Enright JT (1972) When the beachhopper looks at the moon: the moon-compass hypothesis. In: Galler SR, Schimdt-Koenig K, Jacobs GJ, Belville RE (eds) Animal orientation and navigation. NASA Spec Publ 262:523--555, Wash DC

Geiger R (1966) The climate near the ground. Harvard University Press

Goodwin S, Danthanarayana W (1984) Flight activity of *Plutella xylostella* (L.) (Lepidoptera: Yponomeutidae). J Aust Entomol Soc 23:235--240

Greenbank DO, Schaefer GW, Rainey RC (1980) Sprucebudworm (Lepidoptera:Tortricidae) moth flight and dispersal: new understanding from canopy observations, radar, and aircraft. Mem Entomol Soc Can no 110

Haddow AJ, Brown KW, Corbet PS, Dirmhirn I, Gillett JD, Jackson THE (1961) Entomological studies from a high tower in Mpanga forest, Uganda I--XII. Trans R Entomol Soc Lond 113: 249--368

Halverson O von, Edrich W (1974) Der Polarisationsempfänger im Bienenauge. ein Ultraviolettrezeptor. J Comp Physiol 94:33--47

Hartland-Rowe R (1955) Lunar rhythm in the emergence of an ephemeropteran. Nature 176:657

Hartland-Rowe R (1958) The biology of a tropical mayfly, *Povilla adusta* Navas with special reference to lunar rhythm. Rev Zool Bot Afr 58:185--202

Hora SL (1927) Lunar periodicity in the reproduction of insects. J Asiatic Soc Bengal 23:339--341

Jander R (1957) Die optische Richtungsorientierung der roten Waldameise *(Formica rufa)*. Z Vgl Physiol 40:162--238

Johnson CG (1950) A suction trap for small airborne insects which automatically segregates the catch into successive hourly samples. Ann App Biol 37:80--91

Johnson CG (1957) The distribution of insects in the air and the empirical relation of density to height. J Anim Ecol 26:479--494

Johnson CG (1960) A basis for a general system of insect migration and dispersal by flight. Nature 186:348--350

Johnson CG (1969) Migration and dispersal of insects by flight. Methuen, London

Johnson CG, Taylor LR (1955a) The development of large suction traps for airborne insects. Ann App Biol 43:51--61

Johnson CG, Taylor LR (1955b) The measurement of insect density in the air. Lab Pract 4:187--192, 235--239

Johnson CG, Taylor LR, Southwood TRE (1962) High altitude migration of *Oscinella frit* L. (Diptera:Chloropidae). J Anim Ecol 31:373--383

Kennedy JS (1975) Insect dispersal. In: Pimental D (ed) Insects science & society. Academic Press, New York, pp 103--119

Kennedy JS, Booth S, Kershaw WJS (1961) Host finding by aphids in the field III. Visual attraction. Ann Appl Biol 49:1--21

Kerfoot WB (1967) The lunar periodicity of *Specodogastra texana*, a nocturnal bee (Hymenoptera, Halictidae). Anim Behav 15:479--486

Kopal Z (1960) An introduction to the study of the moon. Reidel, Dordrecht, Holland

Kopal Z (1962) Physics and astronomy of the moon. Academic Press, New York

Kopal Z (1963) The moon: Our nearest celestial neighbour. Chapman and Hall, London

Korringa P (1957) Lunar periodicity. Geol Soc Am Mem 67:917--934

Kovrov BG, Monchadskiy AS (1963) The possibility of using polarized light to attract insects. Entomol Rev 42:25--28

Lewis T, Taylor LR (1967) Introduction to experimental ecology. Academic Press, London

Lieber AL (1978) The lunar effect. Anchor/Doubleday, New York

Lyot B (1929) Reserches sur la polarisation de la lumiere des planetes et de quelques substances terrestres. Ann Obs Paris (Meudon) 8 (Trans: Research on the polarization of light from planets and from some terrestrial substances. TTF-187, Wash DC. National Aeronautics and Space Administration, 1964)

MacDonald WW (1956) Observations on the ecology of chaoborids and chironomids in Lake Victoria and on the feeding habits of the elephantsnout fish (*Mormyrus kannume* Forsk.). J Anim Ecol 25:36–53

Mazokhin-Porshnyakov GA (1969) Insect vision. Plenum, New York

Menzel R (1979) Spectral sensitivity and colour vision in invertebrates. In: Autrum H (ed) Handbook of sensory physiology. Springer, Berlin Heidelberg New York

Miller TA, Stryker RG, Wilkinson RN, Esah S (1970) The influence of moonlight and other environmental factors on the abundance of certain mosquito species in light-trap collections in Thailand. J Med Entomol 7:755–761

Morton R, Tuart LD, Wardhaugh KG (1981) The analysis and standardisation of light-trap catches of *Heliothis armiger* (Hubner) and *H. punctiger* Wallengren (Lepidoptera:Noctuidae). Bull Entomol Res 71:207–222

Nemec SJ (1971) Effects of lunar phases on light-trap collections and populations of bollworm moths. J Econ Entomol 64:860–864

Neumann D (1967) Genetic adaptation in the emergence time of *Clunio* populations to different tidal conditions. Helgol Wiss Meeresunters 15:163–171

Neumann D (1975) Lunar and tidal rhythms in the development of reproduction of an intertidal organism. In: Vernberg FJ (ed) Physiological adaptation to the environment. Intext, New York

Neumann D (1976) Adaptations of chironomids to intertidal environments. Annu Rev Entomol 21:387–414

Neumann D (1978) Entrainment of a semi-lunar rhythm by simulated tidal cycles of mechanical disturbance. J Exp Ecol 35:73–83

Neumann D (1981) Tidal and lunar rhythms. In: Aschoff JA (ed) Handbook of behavioural neurobiology. Plenum, New York

Nowinszky L, Szobo S, Toth G, Ekk I, Kiss M (1979) The effect of the moon phases and of the intensity of polarized moonlight on the light-trap catches. Z Angew Entomol 88:337–353

Oehmke MG (1973) Lunar periodicity in flight activity of honey bees. J Interdiscip Cycle Res 4:319–335

Palmer JD (1976) An introduction to biological rhythms. Academic Press, New York

Papi F (1960) Orientation by night: the moon. Cold Spring Harbor Symp Quant Biol 25:475–480

Papi F, Pardi L (1963) On the lunar orientation of sandhoppers (Amphipoda Talitridae). Biol Bull (Woods Hole) 124:97–105

Pellicori SF (1971) Polarizing properties of pulverized materials with special reference to the lunar surface. Appl Optics 10:270–285

Pratt HD (1948) Influence of the moon on light trap collections of *Anopheles albimanus* in Puerto Rico. J Nat Malaria Soc 7:212–220

Provost MW (1957) The dispersal of *Aedes taeniorhynchus* II. The second experiment. Mosq News 17:235–247

Provost MW (1958) Mating and male swarming in *Psorophora* mosquitoes. Proc Xth Int Congr Entomol 2:553–561

Provost MW (1959) The influence of moonlight on light-trap catches of mosquitoes. Ann Entomol Soc Am 52:261–272

Rainey RC (1976) Flight behaviour and features of the atmospheric environment. In: Rainey RC (ed) Insect flight. Blackwell, Oxford, pp 75–112

Rainey RC (1979) Dispersal and redistribution of some Orthoptera and Lepidoptera. Mitt Bull Soc Entomol Suisse 52:125–132

Riley JR, Reynolds DR (1983) A long-range migration of grasshoppers observed in the Sahelian zone of Mali by two radars. J Anim Ecol 52:167–183

Rounds HD (1981) Semi-lunar cyclicity of neurotransmitter-like substances in the CNS of *Periplaneta americana* (L.). Comp Biochem Physiol Comp Pharmacol 69:293–299

Rozenberg GV (1966) Twilight. Plenum, New York

Saunders DS (1982) Insect clocks. Pergamon, Oxford

Service M (1976) Mosquito ecology. Applied Science, London

Schaefer GW (1976) Radar observations of insect flight. In: Rainey RC (ed) Insect flight. Blackwell, Oxford, pp 157–197

Sotthibandhu S, Baker RR (1979) Celestial orientation by the large yellow underwing moth, *Noctua pronuba* L. Anim Behav 27:786–800

Southwood TRE (1962) Migration of terrestrial arthropods in relation to habitat. Biol Rev 37: 171–214

Taylor LR (1958) Aphid dispersal and diurnal periodicity. Proc Linn Soc Lond 169:67–73

Taylor LR (1960) The distributions of insects at low levels in the air. J Anim Ecol 29:45–63

Taylor LR (1961) Aggregation, variance and mean. Nature 189:68–77

Taylor LR (1962) The absolute efficiency of suction traps. Ann Appl Biol 50:405–421

Taylor LR (1963) Analysis of the effect of temperature on insects in flight. J Anim Ecol 32: 99–117

Taylor LR (1965) A natural law for the spacial deposition of insects. Proc XIIth Int Congr Entomol 396–397

Taylor LR (1971) Aggregation as a species characteristic. In: Patil GP, Pielou EC, Waters EW (eds) Statistical ecology, vol 1. Pennsylvania State University Press, Pennsylvania

Taylor LR (1974) Insect migration, flight periodicity and the boundary layer. J Anim Ecol 43: 225–238

von Frisch K (1967) The dance language and orientation of bees. Harvard University Press, Cambridge, Mass

Watermann TH (1966a) Polarotaxis and primary photoreceptor events in Crustacea. In: Bernhard CG (ed) The functional organization of the compound eye. Pergamon, Oxford

Waterman TH (1966b) The specific effects of polarized light in organisms. In: Setman PK, Dittmer DS (eds) Environmental biology. Fed Am Soc Exp Biol, 2nd edn. Bethesda, Maryland

Waterman TH (1974) Polarimeters in animals. In: Gehrels T (ed) Planets, stars and nebulae studied with polarometry. University of Arizona Press, Tuscon, Arizona

Waterman TH (1981) Polarization sensitivity. In: Autrum H (ed) Handbook of sensory physiology, vol 7/6B. Springer, Berlin Heidelberg New York

Wehner R (1976) Polarized light navigation by insects. Sci Am 235:106–115

Wehner R (1984) Astronavigation in insects. Annu Rev Entomol 29:277–298

Wehner R, Duelli P (1971) The spacial orientation of desert ants, *Cataglyphis biolor*. Experientia (Basel) 27:1364–1366

Wellington WG (1974) Changes in mosquito flight associated with natural changes in polarized light. Can Entomol 106:941–948

Wellington WG (1976) Applying behavioural studies in entomological problems. In: Anderson JF, Kaya HK (eds) Perspectives in forest entomology. Academic Press, New York, pp 87–97

Williams CB (1936) The influence of moonlight on the activity of certain nocturnal insects, particularly of the family Noctuidae, as indicated by the light trap. Philos Trans R Soc Lond B Biol Sci 226:357–389

Williams CB (1940) An analysis of four years captures of insects in a light trap II. The effect of weather conditions on insect activity; and the estimation and forecasting of changes in the insect population. Trans R Entomol Soc Lond 90:227–306

Williams CB, Singh BP (1951) Effect of moonlight on insect activity. Nature 167:853–854

Williams CB, Singh BP, El-Ziady S (1956) An investigation into the possible effects of moonlight on the activity of insects in the field. Proc R Entomol Soc Lond Ser A Gen Entomol 31:135–144

8 Response of Some Night-Flying Insects to Polarized Light

W. DANTHANARAYANA[1] and S. DASHPER[2]

1 Introduction

The possibility that night-flying insects may use naturally polarized light as a cue for flight, dispersal, migratory activities and for orientation during movements (at night) has rarely been considered. The reason for this neglect may well be the belief that the amounts of polarized light that are available to insects at night are too small to influence them. During recent investigations on the flight activity of the mosquito *Culex pipiens australicus* (Dobrotworsky and Drummond) and the moths *Plutella xylostella* (L.) and *Epiphyas postvittana* (Walk.), it was found that there are three peaks of flight activity during the lunar cycle (Danthanarayana 1976, this volume; Goodwin and Danthanarayana 1984). Two of these peaks correlated with the degree of polarization of moon light (which increases at the first and last quarters) and the third occurred at the time of full moon when moonlight is more intense, but less polarized. It is not known, however, whether the correlation between the degree of polarization of moonlight and the amount of flight activity is a cause-and-effect relationship. As a preliminary step towards resolving this problem, the response of night-flying insects to polarized light was examined with the aid of light traps, activity-meter studies and by histological determination of the pigment position of superposition eyes of moths. Moths were selected for laboratory experiments because there is much published work in the eye-pigment movements of this group (see below); in previous studies more Lepidoptera have been trapped under polarized light than under nonpolarized light (Kovrov and Monchadskiy 1963) and observations on the light-brown apple moth, *Epiphyas postvittana*, suggested the existence of profound restlessness (Zugunruhe) during the above phases of the moon (unpublished information). Results obtained during these studies, as reported below, confirm that some night-flying insects respond to dim-polarized light to a greater extent than to nonpolarized light of the same intensity.

2 Methods

Observations on the flight behaviour of insects in the field with respect to polarized light were made with the aid of three light traps of the Pennsylvanian design (Southwood 1966) simultaneously operated and emitting equal amounts of light (5500 lx,

---

[1] Department of Zoology, The University of New England, N.S.W. 2351, Australia.
[2] Department of Zoology, La Trobe University, Victoria 3083, Australia.

Insect Flight: Dispersal and Migration
Edited by W. Danthanarayana
© Springer-Verlag Berlin Heidelberg 1986

measured at a distance of 1 m in a darkroom). The lamp used was a 12 V Toshiba LF 6W cool-white fluorescent tube. In two traps the tubes were covered on the four sides with a polaroid sheet (HN38 of the Polarized Corporation, a neutral-colour linear polarizer having a total luminous transmittance of 38% within the 375–750 nm light spectrum). In one polarized light trap the arrangement produced vertically-plane polarized light, and in the other horizontally-plane polarized light. The third trap, which emitted nonpolarized light, was covered with partially exposed photographic celluloid (film) that transmitted approximately the same amount of light (5500 lx) as the polaroid sheet, i.e. with the same amount (38%) of luminous transmittance as the polaroid sheet. The traps were arranged in an equilateral triangle 10 m apart, and the positions were rotated daily. The traps were operated from 1900 h to 1600 h and the catch collected every morning. Insects were classified to the family level. The trapping periods extended from February 25–April 28, and August 6–October 13, 1982. These field observations were made at the La Trobe University entomology experiment site (Danthanarayana, this volume).

The ability to respond to polarized light was investigated by measuring the activity of the tortricid moths *E. postvittana* and *Laspeyresia pomonella* (L.) in an activity meter (Fig. 1). The activity chamber is a glass cylinder 8.5 cm long and 3.5 cm wide, closed at one end. Wrapped around its inner walls are two alternating tinned-copper wires which form the input to an amplifier with an input impedance of 100 mega ohms. Currents of 0.5 μA (which have no effect on insects) were passed so that a minimum resistance of 100 mega ohms is detected by a one-shot pulse producer which creates a 300 ns pulse recorded in a four-digit counter. The activity meter was used within a flight chamber (Kennedy and Booth 1963; Laughlin 1974) having a vertical beam of light from the ceiling generated by 1000 W quartz halogen lamp operated via a variable intensity switch. Polarized light was produced by sliding a polaroid sheet (type HNP'B of Polaroid Corporation, a linear polarizer for photometric instruments effective in the UV range, from 280 nm wavelength, and visible light). The light intensity with and without polarization was standardized using a Gossen Lunar 6 photometer. All observations were carried out at a temperature of 20.5° ± 1.5 °C within the flight chamber. This temperature is within the optimal temperature range for flight activities of *E. postvittana* (Danthanarayana 1976) and *L. pomonella* (Pristayko and Chernii 1974). Observations were restricted to times between 1930 h and 2300 h for *E. postvittana* (Danthanarayana 1976) and 1745 h and 1945 h for *L. pomonella* (Batiste et al. 1973; Mani et al. 1974) which are the peak flying times of the two species. Prior to each observation moths were totally dark-adapted for 45 min, a period greater than the maximum dark-adaption times of 27–37 min (Bernhard and Ottoson 1964; Collins 1934; Day 1941). For each species a dark-adapted moth was placed in the activity-meter chambers and exposed for 5 min each to polarized light having an intensity of 175 lx, to total darkness and nonpolarized light, also of 175 lx intensity. The experiment was replicated by using different numbers of insects (Table 2). The order of exposure to polarized and nonpolarized light was reversed for alternative moths though there was no difference in the results obtained from the two methods.

For histological studies, 7-day-old female moths were exposed to one of four light regimes (total darkness, 40000 lx nonpolarized, 16 lx dim-polarized; 16 lx dim-nonpolarized) using ten insects per treatment. At the conclusion of 1 h of exposure the

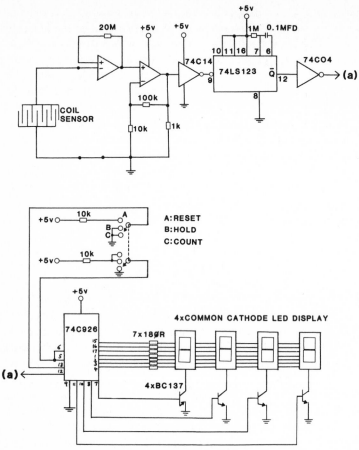

Fig. 1. The circuit diagram of activity meter

heads were excised, bisected, and fixed in alcoholic Bouin's fluid. The latency period for pigment migration to resume after a change in light intensity is 7 min (Bernhard and Ottoson 1964; Collins 1934; Day 1941). The eyes were subsequently sectioned at 6 μ and permanently mounted on glass slides without staining or depigmentation. Pigment position was determined by measuring, under the microscope (Table 3), (a) the distance between the basement membrane and the proximal secondary iris pigment level; (b) the distance from the proximal iris-pigment level to the base of the corneal lens; and (c) the distance between the basement membrane and the distal margin of the corneal lens. A measure of the pigment position, unaffected by variations in eye size, was obtained by taking the ratios a/c and b/c.

3 Results

Results of the light-trapping experiments show that some night-flying insect groups show a greater response to polarized light, confirming the findings of Kovrov and

Table 1. Data on insects caught in polarized and unpolarized light traps compared with those of Kovrov and Monchadskiy (1963)[a]

Insect order/ family	New data (6 W fluorescent light)				Kovrov and Monchadskiy data (1000 W mercury-quartz light)		
	Vertically-plane polarized trap	Horizontally-plane polarized trap	Unpolarized trap	Total catch	Polarized trap	Unpolarized trap	Total catch
(a) *All insects*							
Coleoptera	104(26)	104(26)	185(47)	393	23(62)	14(38)	37
Dermaptera	39(48)	20(24)	23(28)	82	—	—	—
Diptera	257(7)	3066(84)	318(9)	3641	68832(77)	20715(23)	89547
Hemiptera	24(28)	18(21)	43(51)	85	3(100)	(0)	3
Hymenoptera	17(24)	20(29)	33(47)	70	40(68)	19(32)	59
Lepidoptera	493(17)	558(19)	1821(63)	2872	759(68)	352(32)	1111
Neuroptera	1(20)	0(0)	4(80)	5	—	—	—
Trichoptera	11(26)	23(55)	8(19)	42	4(40)	6(60)	10
Total	946(13)	3810(53)	2435(34)	7191	69661(77)	21106(23)	90767
(b) *Diptera*							
Agromyzidae	11(18)	29(48)	21(34)	61	—	—	—
Bibionidae	0(0)	29(58)	21(42)	50	—	—	—
Chironomidae	176(6)	2897(89)	166(5)	3239	59543(79)	15594(21)	71537
Culicidae	2(67)	0(0)	1(33)	3	670(51)	642(49)	1312
Sciaridae	55(32)	52(30)	65(38)	172	—	—	—
Simuliidae	1(50)	0(0)	1(50)	2	6(50)	6(50)	12
Tipulidae	7(7)	50(54)	36(39)	93	—	—	—
Others	5(24)	9(43)	7(33)	21	8613(68)	4113(32)	12726
Total	257(7)	3066(84)	321(9)	3641	68832(77)	20355(23)	89187

[a] Percentages in parentheses.

Table 2. Mean activity-meter readings under different light regimes ± SE

	n	Polarized light (175 lx)	Normal light (175 lx)	Darkness
L. pomenella				
Males	23	157.6 ± 5.6	113.3 ± 5.9	36.8 ± 2.7
Females	69	178.0 ± 2.1	124.6 ± 2.1	26.6 ± 0.8
Total	92	172.9 ± 1.5	121.8 ± 1.5	29.2 ± 0.6
E. postvittana (culture 1)				
Males	29	97.4 ± 2.9	54.5 ± 1.6	16.1 ± 1.8
Females	62	87.8 ± 1.3	40.4 ± 0.8	14.9 ± 0.7
Total	81	90.8 ± 0.9	44.9 ± 0.5	15.3 ± 0.5
E. postvittana (culture 2)				
Males	33	44.0 ± 0.9	12.2 ± 0.4	9.3 ± 0.4
Females	49	56.1 ± 1.0	19.7 ± 0.4	12.0 ± 0.4
Total	82	51.2 ± 0.5	16.7 ± 0.2	10.0 ± 0.2

Table 3. Measurement of eye sections of *L. pomonella* exposed to four light regimes[a]

Treatment	Distance from basement membrane to lower pigment level (a)	Distance from lower pigment level to base of lens (b)	Eye depth from basement membrane to lens (c)	Ratio a/c	Ratio b/c
1. Fully dark-adapted	97 ± 5.9	52 ± 2.0	148 ± 6.1	0.647 ± 0.016	0.354 ± 0.016
2. Fully (40000 lx) normal-light adapted	34 ± 3.4	159 ± 6.0	193 ± 8.5	0.173 ± 0.013	0.819 ± 0.013
3. Dim (16 lx) polarized-light adapted	107 ± 7.7	58 ± 3.6	165 ± 7.4	0.643 ± 0.025	0.357 ± 0.024
4. Dim (16 lx) normal-light adapted	91 ± 5.2	82 ± 8.8	173 ± 8.4	0.534 ± 0.037	0.466 ± 0.037

[a] Values are the means from ten insects in microns ± SE.

Monchadskiy (1963) (Table 1). Diptera gave the largest catch during the short period of investigation, allowing further classification to family level (Table 1). More Diptera, particularly Chironomidae, and Trichoptera were attracted to horizontally-plane polarized light than to nonpolarized and vertically-plane polarized light. More Dermaptera were attracted to vertically-plane polarized light than to either horizontally-plane polarized or nonpolarized light. Nearly equal, but low numbers of Coleoptera, Hemiptera, Hymenoptera, and Neuroptera were captured in each type of trap. More Lepidoptera were caught in nonpolarized light trap, a contrasting result to that obtained by Kovrov and Monchadskiy (1963). This discrepancy is likely to be associated with high light intensity and/or the quality of light provided by the 1000 W mercury-quartz lamp used by these workers in comparison to the 6 W fluorescent tube used in the present study. Nevertheless, both investigations demonstrate that some night-flying insects, particularly Diptera (Chironomidae) respond to polarized light in preference to nonpolarized light.

Results of the activity meter studies with *E. postvittana* and *L. pomonella* are presented in Table 2. Both species are nocturnal fliers, and lunar periodicity has been demonstrated for the former (Danthanarayana 1976). Activity of both species was found to be higher in polarized light compared to that under nonpolarized light of the same intensity (128 lx) and in total darkness (Table 2). These differences were highly significant when tested by two-way analysis of variance ($P < 0.001$) and by Friedmann two-way analysis by ranks ($P < 0.01$) (Siegel 1956). Also both sexes showed the same types of responses. The activity-meter studies support the light-trap results of Kovrov and Monchadskiy (1963) on Lepidoptera.

With regard to the third series of experiments, measurements of the pigment position of the superposition eye of *L. pomonella* showed that eyes exposed to dim-polarized light (16 lx in these experiments) have a similar pigment position to that of the completely dark-adapted eye (Fig. 2b,d). This is in contrast to the results obtained with dim-nonpolarized light of the same intensity (16 lx) which showed partial light adaptation (Fig. 2c) in which the pigments occupy an intermediate position between complete dark adaptation (Fig. 2b) and complete light adaptation (Fig. 2a). There was no significant difference in the pigment position of the dark-adapted eye and that of the dim-

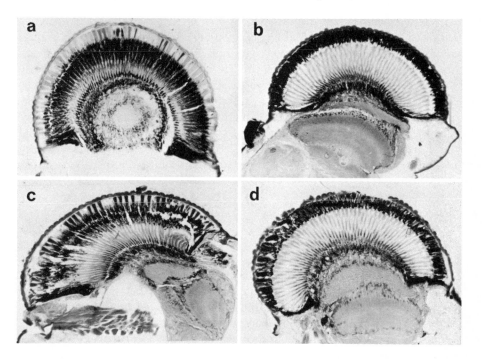

Fig. 2a–d. *L. pomonella* (see legend to Table 3). **a** Fully light-(40,000 lx)-adapted eye. **b** Fully dark-adapted eye. **c** Dim-(16 lx)-nonpolarized light-adapted eye. **d** Dim-(16 lx)-polarized light-adapted eye

polarized light-adapted eye, whereas there was a significant difference ($P = 0.05$) in the pigment position of the dark-adapted eye and that of the dim nonpolarized light adapted eye (Table 3). The dim-polarized light-adapted eye is, thus, as sensitive as the completely dark-adapted eye.

4 Discussion

The possibility of using polarized light to attract insects to light traps was first investigated by Kovrov and Monchadskiy (1963). This was done with the object of increasing the efficiency of light traps, for light traps have long held an important place in studies of the phenology, flight intensity and comparative studies on flight of many insect species and groups, particularly those of economic importance. These workers concluded that a polarized-light trap is at least twice as effective in trapping insects as a nonpolarized-light trap. Results obtained during the present investigations support their findings mainly with respect to Chironomidae. The attraction of Chironomidae to polarized light was clearly significant; the response was to horizontally- and not to vertically-plane polarized light. In the present study Dermaptera and Trichoptera also showed a greater attraction to vertically- and horizontally-plane polarized light respectively. It is interesting to note that the responses of the two groups were different, the former being more abundant in the vertically-plane polarized-light trap and the latter in the horizontally-plane polarized-light trap.

It is well-established that there is an increase in sensitivity of the eye of nocturnal insects with a decrease in the available light source, i.e. the dark-adapted eye is much more sensitive than the light-adapted eye (Goldsmith and Bernard 1974; Chapman 1969; Wigglesworth 1972). Adaptation to light of different intensities in insects may involve a variety of factors, including the availability of visual pigment, cytological changes in the eye, biochemical events, adaptation within the nervous system and pigment movement in the iris cells of the compound eye. The last factor, migration of pigment, is important in the superposition eyes of the nocturnal forms (Goldsmith and Bernard 1974; Chapman 1969; Wigglesworth 1972). That the position of the secondary iris pigment varies according to the intensity of light has been demonstrated in the tortricid moths *Ephestia kuhniella* Zeller, *L. pomonella* and several species of Neuroptera and Trichoptera (Day 1941; Collins 1934). In the dark-adapted moths, the secondary iris pigment is found as a dense layer surrounding the crystalline cones, extending from the distal end of the retinular cells to the base of the corneal lens; the retinular cells are not surrounded by any pigment (Fig. 2b). In this position the amount of light that reaches the basal cells is maximized, making the eye more sensitive to low intensity light. In the light-adapted position (Fig. 2a), the pigment granules migrate proximally, surrounding the retinular cells also, forming a 'curtain' that extends beyond the crystalline cones and thereby preventing the light from adjacent facets reaching on-axis rhabdoms (Goldsmith and Bernard 1974; Chapman 1969).

In nocturnal insects the maximum response of the light-adapted eye is about 20% of that of the dark-adapted eye; in diurnal insects about 60% (Wigglesworth 1972). It is known that the dark-adapted eye of the noctuid moth *Cerapteryx gramminis* (L.) is more sensitive to light than the light-adapted eye by a factor of 1000 (Bernhard and

Ottoson 1964). Polarized light itself, in comparison to nonpolarized light, increases the sensitivity of the insect eye by 15–30% (Mazokhin-Porshnyakov 1969). The eyes of night-flying insects should accordingly exhibit their greatest amount of sensitivity in the presence of dim-polarized light.

In conclusion, our results show that nocturnal insects are able to perceive polarized light and that their behaviour may be modified by polarized light. Exposure to dim-polarized light, but not to nonpolarized light of the same intensity, causes the eye to take on the histological appearance of a dark-adapted eye. This enhances sensitivity of the eye to light of even low intensities and would account for the increase activity observed in the presence of polarized light. An increase in the perception of and reaction to low levels of light would be of value to species which depend on and/or are activated by naturally occurring dim light, such as twilight and moonlight (see Danthanarayana, this volume).

Acknowledgements. We acknowledge the financial support given by the Sunshine Foundation of Victoria and the Australian Research Grants Committee of 1978. We thank Roger de Valle of the Zoology Department for the design of the activity meter and Prof. J.S. Maritz of the Department of Mathematical Statistics, La Trobe University for suggesting Friedmann's method of analysis. We are much obliged to Ian Thornton and Tim New for reviewing the manuscript.

References

Batiste WC, Olson WH, Berlowitz A (1973) Codling moth: influence of temperature and daylight intensity on periodicity of daily flight in the field. J Econ Entomol 66:883–888

Bernhard CG, Ottoson D (1964) Quantitative studies on pigment migration and light sensitivity in the compound eye at different light intensities. J Gen Physiol 47:465–478

Chapman RF (1969) The insects, structure and function. The English Universities Press, London

Collins DL (1934) Iris pigment migration and its relation to behaviour in the codling moth. J Exp Zool 69:165–198

Danthanarayana W (1976) Diel and lunar flight periodicities in the light brown apple moth, *Epiphyas postvittana* (Walker) (Tortricidae) and their possible adaptive significance. Aust J Zool 24:65–73

Day MF (1941) Pigment migration in the eyes of the moth *Ephestia kuehniella* Zeller. Biol Bull (Woods Hole) 80:275–291

Goodwin S, Danthanarayana W (1984) Flight activity of *Plutella xylostella* (L.) (Lepidoptera: Yponomentidae). J Aust Entomol Soc 23:235–240

Godsmith TH, Bernard GD (1974) The visual system of insects. In: Rockstein M (ed) The physiology of insecta, 2nd edn. Academic Press, New York

Kennedy JS, Booth CO (1963) Free flight of aphids in the laboratory. J Exp Biol 40:67–85

Kovorov BG, Monchadskiy AS (1963) The possibility of using polarized light to attract insects. Entomol Rev 42:25–28

Laughlin R (1974) A modified Kennedy flight chamber. J Aust Entomol Soc 13:151–153

Mazohkin-Porshnyakov GA (1969) Insect vision. Plenum, New York

Mani E, Riggenbach W, Mendick M (1974) Tagesrhythmus des Falterfangs und Beobachtungen über die Flugaktivität beim Apfelwickler (*Laspeyresiea pomonella* L.). Mitt Schweiz Entomol 47: 39–49

Pristavko VP, Cherni AM (1974) Effect of air temperature on diurnal rhythm and activity of the lesser apple worm moth. Ekologiya 5:63–66

Siegel S (1956) Non-parametric statistics for behaviour sciences. McGraw-Hill, New York

Southwood TRE (1966) Ecological methods. Methuen, London

Wigglesworth VB (1972) The principles of insect physiology, 7th edn. Chapman and Hall, London

9 Migration in the African Armyworm *Spodoptera exempta*: Genetic Determination of Migratory Capacity and a New Synthesis

A. G. GATEHOUSE[1]

1 Introduction

Migration is a key component of insect life histories and recent interest in their evolution has focussed attention on the interaction of genetic and environmental factors in the regulation of migratory capacity (Dingle, this volume). Furthermore, many migratory species are serious agricultural pests and successful forecasting, monitoring and management of their attacks on crops depends on understanding their migratory strategies (Stinner et al. 1983, this volume; Raulston et al., this volume).

The incidence of migration is highest in species whose habitats are patchy in space and time (Southwood 1962, 1977). Some very temporary habitats may only support a single generation, and insects exploiting these must necessarily exhibit obligatory pre-reproductive migration. More generally, habitats are capable of supporting a variable number of generations before their favourability for insect growth and development deteriorates. These changes in habitat quality are associated with physical and biotic cues and many insects have evolved facultative migratory responses to them, enabling them to escape periods of habitat unfavourability.

Ziegler (1976) pointed out that the degree to which migration is regulated by phenotypic responses to environmental cues is likely to depend on their reliability as predictors of impending habitat deterioration. In the temperate zone, the major reversible changes in habitat favourability associated with the seasonal cycle, are heralded by consistently reliable cues provided by the annual variations in photoperiod and temperature and these predominate in the regulation of obligate migration. In summer generations of multivoltine species, when changes in habitat quality are less regular, facultative migration has been shown to occur in response to proximate factors associated with habitat deterioration, such as food availability and quality, water-stress or crowding (reviewed by Johnson 1969; Dingle 1980; Harrison 1980).

The decline in the amplitude of the annual variation in photoperiod with decreasing latitude means that at low latitudes in the tropics, the change in photoperiod may be below the threshold of detection for insects with short generation intervals (Beck 1980), although not for longer-lived species (Norris 1965). Furthermore, the reliability of photoperiod and temperature as cues of the less regular changes in habitats associated with seasonal rainfall in the tropics and subtropics, is frequently poor. Environmental factors more directly associated with impending habitat deterioration would therefore

---

[1] School of Animal Biology, University College of North Wales, Bangor, Gwynedd LL57 2UW, United Kingdom.

Insect Flight: Dispersal and Migration
Edited by W. Danthanarayana
© Springer-Verlag Berlin Heidelberg 1986

be expected to play the major part in regulating migration, and the few studies of tropical species confirm this expectation (Johnson 1969; Rose 1972; Dingle and Arora 1973; Fuseini and Kumar 1975; McCaffery and Page 1978; Padgham 1983; Smith 1983).

Most critical genetic analyses of migratory capacity have produced results consistent with multilocus or multiallelic models of inheritance, with environmental factors determining the phenotypic expression of the genotypes (Harrison 1980; Zera et al. 1983). Estimates of the heritability of thresholds of response to environmental cues have demonstrated a high additive genetic contribution to total phenotypic variance, indicating that these traits respond rapidly to directional selection. It has been suggested that this high genetic variance is maintained by seasonal and geographic shifts in the direction of selection and allows populations to adapt rapidly to the conditions they encounter (Dingle et al. 1977). Similar high additive genetic variance in flight duration (Dingle 1968; Caldwell and Hegmann 1969) is maintained by the irregular patchiness of habitats and achieves efficient scanning of available habitats during migration (Kennedy 1961).

Ziegler (1976) also proposed that stereotyped genotypic determination of migration can be expected to evolve when reliable environmental cues are absent. However, the frequent demonstration of phenotypic plasticity and the scarcity of examples of genetic determination of migratory potential (Harrison 1980) suggest that adequate cues of impending habitat change, to which insects can evolve responses, are generally present. Recent work in our laboratory, on the African armyworm *Spodoptera exempta*, has drawn attention to other circumstances in which genetic determination of migratory potential can be expected to evolve.

1.1 Distribution of Spodoptera exempta and the Occurrence of Outbreaks

S. exempta occurs over much of tropical Africa and extends seasonally into southern Africa and south-western Arabia. The larvae feed on a wide range of Graminae and some Cyperaceae and Palmae, and high density outbreak populations can cause extensive and severe damage to crops. The species also occurs in Asia and Oceania but, although occasional outbreaks have been reported, it is in Africa that it is of major agricultural importance (reviewed by Haggis 1984).

There is abundant field (e.g. Brown and Swaine 1966; Brown et al. 1969; Tucker et al. 1982; Haggis 1984) and laboratory (Aidley 1974; den Boer 1978; Gatehouse and Hackett 1980) evidence that the moths are capable of migratory flights covering hundreds of kilometres. Records of outbreaks in Eastern Africa (Haggis 1984) show two seasonal progressions apparently originating in regions of Mozambique, Malawi, and Tanzania where continuous generations are known (G.K. Nyirenda 1982, unpublished report), or have been inferred (Rose 1979; Blair et al. 1980) to occur. This pattern of outbreaks is associated with the passage of synoptic-scale wind-convergence systems which bring the seasonal rains, notably the northward passage of the Inter-Tropical Convergence Zone in the north-eastern region of the continent, and has led to the hypothesis that *S. exempta* in East Africa is distributed as a series of mobile, high density populations which are relocated by convergent-wind systems (Rainey and Betts 1979). However, while outbreaks, during seasons in which they are frequent, must often constitute sources of moths for further outbreaks in the subsequent generation, there are now compelling reasons for revising this interpretation. Firstly, it implies that there must be a return movement of high density populations following the southward and northward progressions of outbreaks, to the regions where they occur early in the rains in Eastern Africa, and there is little supporting (Brown et al. 1969; Haggis 1984), and some contrary (Brown and Swaine 1965; Tucker et al. 1982) evidence for such movements. Secondly, it fails to take adequate account of the complex larval biology of the species, or of the existence during both rainy and dry seasons of significant low density populations (Rose 1979).

1.2 Larval Biology and Phase Polyphenism

In common with several other species of Lepidoptera, *S. exempta* exhibits a density-dependent polyphenism, which has been termed a phase transformation by analogy with locusts (Faure 1943). Larvae reared in isolation, or from low-density field populations, are light-green or brown and highly procryptic, in contrast to the conspicuous, heavily-pigmented, velvety-black form typical of high larval densities. These forms, termed by Faure *solitaria* and *gregaria* phases, are extremes of a continuum, and intermediate *transient* forms occur in field populations.

The phases also behave very differently. *Solitaria* larvae are sluggish and feed low among the grasses which enhances their crypsis, while *gregaria* caterpillars are very active, feeding voraciously high on the host plant. It is this phase that is generally recognised as the armyworm, whereas the *solitaria* larvae are rarely seen and easily confused with other grass-feeding Lepidoptera (Rose 1979). However, they may occur at relatively high densities, provided the larvae are feeding in thick grass and do not encounter one another during development. Densities of $75-150$ m^{-2} were reported by Whellan (1954), and Rose (1979) has reported comparable densities in Zimbabwe.

The procrypsis and cryptic behaviour of *solitaria* larvae must reduce their vulnerability to predation and parasitism and must have evolved in response to selection by these mortality factors. In outbreaks, predators and parasites are presumably saturated and unlikely to cause significant overall mortality unless populations remain in the same place for several generations. However, pathogens, in particular viruses, are the dominant cause of mortality in high density populations and are widespread in armyworm outbreaks in East Africa, causing up to 90% larval mortality (Brown and Swaine 1965). Direct (Persson 1980, unpublished report) and circumstantial (Brown and Swaine 1965; Blair 1972) evidence suggests that *solitaria* populations are less susceptible to viral disease.

The phases also differ in their rates of larval development at constant temperatures, *solitaria* taking $2-3$ days longer from hatching to pupation (Khasimuddin 1981). This difference is probably enhanced by insolation to which the *gregaria* larvae are exposed as they feed at the top of the vegetation. Their black pigmentation must make them efficient in absorbing radiant energy, increasing body temperatures and accelerating development (Rose 1979). The adaptive significance of this effect is indicated by the fact that shaded plants in heavily attacked maize fields are often largely undamaged, implying that *gregaria* larvae avoid shade. The pupal period is not affected by larval phase (Matthee 1946).

Physiological and biochemical differences between the phases have been reported by Khasimuddin (1981). Earlier, Matthee (1945) had shown that *solitaria* contain significantly less fat in the final larval instar and pupal stage. The moths from *solitaria* and *gregaria* larvae are morphologically indistinguishable, but biochemical analyses of pharate adults within 24 h of eclosion showed those developing from *gregaria* larvae to contain significantly more fat, but no more protein than *solitaria* (U.R. Muhsin, unpublished results).

1.3 Behaviour of Newly Emerged Moths

Emergence occurs between 1900 und 2300 h and the moths are capable of flight within about 2 h. They then fly into nearby trees where, in outbreaks, they aggregate in very large numbers (Rose and Dewhurst 1979). Observations of the flight behaviour of the moths using optical and radar techniques (Riley et al. 1983) have shown that in some outbreaks, a substantial proportion of them take off from the trees almost immediately, giving a peak of emigration between 2100 and 2400 h, while others take off at intervals through the remainder of the night. In other outbreaks variable proportions, sometimes a majority, remain in the trees until about 1 h before sunrise when they take off en masse on short flights which, although they may reach heights of 200–300 m, terminate quickly as the moths descend to find daytime refuges in crevices in bark or in grass tussocks (Rose and Dewhurst 1979). On the night following emergence, these moths have been observed to return briefly to aggregate in the trees at dusk before taking off on emigratory flights discernible on the radar as a large discrete peak between 1900 and 2000 h (Riley et al. 1983). The reason for this difference in behaviour at different outbreaks is unclear. However, whether the moths leave the site on the night of emergence or at dusk on the following night, very few remain at outbreak sites for more than 24 h (Rose and Dewhurst 1979).

Migration occurs pre-reproductively in *S. exempta*. All females sampled from the trees at night and from daytime refuges were reproductively immature and unmated, and pheromone traps at the outbreak site did not attract males (Rose and Dewhurst 1979).

The radar observations showed that the emigrants climbed steeply out of their boundary layer to reach maximum altitudes of 300–600 m, the mean height of the concentrations of flying moths being 200 m. They were then displaced rapidly downwind and were observed overflying a second radar sited 13.6 km down-range. The transit times of the moths between the sites gave estimated ground speeds of 6–12 m s^{-1} and, as they showed no indications of descent while within range of the second radar, the minimum displacement achieved was 18–20 km (Riley et al. 1983). The moths are known to be capable of flights of several hours duration in the laboratory (Aidley 1974; Gatehouse and Hackett 1980), so that displacements of several hundred kilometres can be expected on persistent, strong winds. However, in the latter part of the night (Riley et al. 1983) and in some locations (Tucker 1983), winds are generally lighter and more variable which would limit the displacement of moths taking off into them.

Although moths emerging at outbreak sites often take off in dense concentrations, there is no evidence of any interactive behaviour between individuals acting to maintain a cohesive swarm. It is, therefore, self-evident that individual variations in flight speed, heading and duration, and the occurrence of wind shifts and turbulence, will normally result in the rapid and extensive scattering of moths downwind. This is confirmed by the average 10:1 reduction in aerial density between the concentrations leaving the emergence site and those detected 13.6 km away by the down-range radar. If this rate of dispersal is maintained, the concentrations would be reduced to 1/1000 of their original density within 40 km from the emergence site (Riley et al. 1983). There can, therefore, be little doubt that oviposition following emigration from even the densest larval populations will normally be widely scattered and will result in widespread but low density populations of *solitaria* larvae in the succeeding generation.

Since the occurrence of high density *gregaria* phase larval populations implies locally high densities of their ovipositing mothers, it follows that the moths must sometimes encounter conditions that result in their reconcentration. We now know that reconcentration occurs under the influence of mesoscale meteorological disturbances, particularly those associated with convective storms. These systems of localised and intense wind convergence have been shown to be capable of raising the densities of flying insects by at least an order of magnitude in less than 1 h (Pedgley et al. 1982) and their role in concentrating armyworm moths is indicated by the association between outbreaks and rainstorms at estimated times of oviposition (Brown et al. 1969; Blair et al. 1980; Tucker and Pedgley 1983). Although other small-scale disturbances, e.g. topographically-induced rotors, can also achieve the necessary degree of concentration (Pedgley et al. 1982), aggregations may disperse again when these transitory systems dissipate. The particular importance of concentration by rainstorms is that rain induces flying *S. exempta* to descend (Riley et al. 1983). Furthermore, rain provides the moths with a readily available source of free water, the uptake of relatively large amounts of which has been shown to be an essential prerequisite for the completion of reproductive maturation in females (Gunn and Gatehouse 1985).

2 Laboratory Studies on Migratory Flight

2.1 Flight Recording Technique

Factors determining flight potential in *S. exempta* have been investigated using the tethered-flight technique described by Gatehouse and Hackett (1980) with minor modifications (Parker and Gatehouse 1985a). The operational criterion used to distinguish long-fliers (by inference migrants) from short-fliers was a total flight duration of > 120 min on the night after emergence, made up of individual flights of not less than 30 min. In fact, approximately 95% of moths categorised as long-fliers on this criterion gave individual flights of > 120 min and continuous flights of 6–12 h occurred frequently. Although the large majority of short-fliers gave no flights longer than 30 min, flights of 10–20 min were common and comparable flight durations in the field could result in displacements of several kilometres (Riley et al. 1983). Thus, many of these individuals must be considered capable of flying significant distances.

A high degree of variability in flight duration, which is a feature of tethered-flight studies (Johnson 1976), was apparent in both short- and long-flying categories of

S. exempta (Parker and Gatehouse 1985a). Such variation has been considered adaptive in that it results in efficient scanning of patchily distributed habitats (Kennedy 1961).

2.2 Genetic and Environmental Factors Regulating Migration

Genetic Factors. Selection experiments on moths reared from *gregaria* phase larvae and estimation of the heritability of flight capacity have shown that there is a major genetic component in the determination of flight potential in *S. exempta* (Parker and Gatehouse 1985b). Selection over two generations increased the proportion of long-fliers among females from 38% to 75%, and continued selection maintained this proportion at 70–93%. The effect of selection for long flight by males was slightly less rapid, although 69–80% long-fliers was achieved. Selection for short flight was also successful but, although long-flying males were rapidly eliminated, the proportion of long-flying females did not fall below 20% during the experiment. The reasons for these apparent differences between the sexes in the effects of selection for flight performance are not clear. However, it is likely that some errors in the categorisation of moths as short-fliers were inevitable. The limited number of channels on the flight-recording equipment meant that it was only practicable to fly moths for one night (the night after emergence) although it is known that in females at least, some individuals which do not give prolonged flights on this night may do so on the night of emergence, or the second night after emergence (A.G. Gatehouse, unpublished results).

Estimates of the heritability of flight potential (Parker and Gatehouse 1985b; Table 1) confirm the substantial contribution of additive genetic variance to the observed phenotypic variance in flight performance. However, these estimates must be treated with caution because the errors in the categorisation of moths as short-fliers will have resulted in underestimation of the true heritabilities. It is also possible that a difference between the sexes in the probability of these errors (e.g. if the males are more consistent in the expression of their flight potential over several nights) could account for the generally lower estimates obtained in regressions involving female parents and offspring. If these assumptions are valid, the very high probable real values for the heritability of flight potential indicate a predominant genetic component in the determination of flight potential in *S. exempta*.

It has not yet been established whether flight capacity is a continuous variable or whether the variability is discontinuous. A second important question that remains to

Table 1. Heritability estimates for flight capacity in *Spodoptera exempta* moths calculated from regression of offspring (males, females, sexes pooled) against parent flight durations (male, female, mean parent value) recorded on the night after emergence (Parker and Gatehouse 1985b)

Offspring	Parents		
	Male	Female	Mean parent value
Males	0.88	0.54	–
Females	0.71	0.50	–
Sexes pooled	–	–	0.40

be resolved is the mode of inheritance of flight capacity in *S. exempta*, although the present evidence is consistent with polygenic inheritance (Parker and Gatehouse 1985b). Polygenic inheritance does not preclude the existence of discontinuous variation in flight potential if the morphs are determined by a factor or factors which must be above or below given thresholds (Harrison 1980; Roff 1984). In fact, Roff suggests that natural selection will favour the development of polygenic control of traits such as flight capacity during evolution.

Environmental Factors. Much of the range of *S. exempta* lies within $10°$ latitude of the equator. At these latitudes, the change in photoperiod over the short interval between egg deposition and adult emergence in this species (minimum 19 days in a study in Kenya by Persson 1981) is very small (9–17 min per month at $10°$ latitude; Beck 1980). Furthermore, neither photoperiod nor temperature provide reliable cues of impending rainfall. We therefore examined the effects of early signs of habitat deterioration as likely environmental cues for migratory flight (Parker and Gatehouse 1985a).

In experiments controlled as far as possible to ensure genetic compatibility between control and treatment samples of insects, we failed to demonstrate any effects of water-stress in the larval host plant (at levels ranging from moderate to severe), or of food deprivation on the flight performance of the resulting moths.

A third candidate cue, larval density, did influence flight. There was a significant increase in the proportion of long-fliers among moths reared from *gregaria* larvae at very high densities, but the major effect was associated with larval phase. When the offspring of pairs of long-flying parents were split into two groups reared as *solitaria* and *gregaria* respectively, the moths from the *gregaria* larvae included 70% long-fliers as opposed to 6% in the *solitaria* group (Woodrow and Gatehouse, in preparation). These data refer to females, but preliminary results with males indicate a comparable effect. The results confirm earlier work in which the genetic matching of the treatment groups had been less precise (A.G. Gatehouse 1979, unpublished report). The failure to demonstrate any effect of food quality or availability on flight, even when the treatments imposed were extreme, suggests that larval density, acting principally through phase variation, is the only environmental cue having a significant influence on flight potential in *S. exempta*.

During the course of this study, three apparently distinct phenotypes could be recognised. The existence of obligate long- and short-fliers is evident from the occurrence in all experiments, irrespective of larval treatment or environmental conditions, of some individuals in both categories. The effect of larval phase on flight performance confirms the existence of a facultative long-flier phenotype. On the present evidence, it is not possible to determine whether these phenotypes represent distinct groups of genotypes or a genetically-determined variability in the threshold of the flight response to larval density, such that obligate short-fliers have thresholds so high that they exclude any response, obligate long-fliers have thresholds so low that long flight is switched on even in *solitaria* individuals, and facultative long-fliers have intermediate thresholds. Roff (1975) has produced simulation models which show that polymorphisms for dispersal tendency remain stable under a wide range of conditions, both when the probability of dispersal is determined at a single locus with two alleles (provided one homozygote and the heterozygote disperse) and when the tendency to disperse is a quantitatively inherited trait. In both cases, stability is maintained when the probability of dispersal is density-dependent.

2.3 Post-Emergence Reproductive Development

There is evidence from the field and laboratory (Brown et al. 1969; Aidley 1974; Tucker et al. 1982; A.G. Gatehouse, unpublished results) that *S. exempta* moths are capable of migratory flights over more than one successive night. Parker (1983) showed that the capacity of females for prolonged flight is curtailed after they have started reproducing, suggesting that the number of nights over which moths can express their flight potential is determined by the rate of post-emergence reproductive development. Recent work by W.W. Page (1982, personal communication) indicates that females show significant variation in their rates of development but, contrary to expectation based on Johnson's (1969) oogenesis-flight syndrome, Parker's results suggest that delayed maturation is independent of the capacity for prolonged flight. This raises the interesting possibility that variability in the time taken to mature the oocytes may constitute an independent mechanism to maximise dispersion, and therefore scanning of the available habitats, by modulating the number of nights over which moths express their potential (high or low) for flight. The importance of the availability of water to females (essential for the completion of oocyte development, Gunn and Gatehouse 1985) in terminating dispersal is indicated by laboratory evidence of some acceleration of oocyte development in females provided with water from the time of emergence (W.W. Page 1984, personal communication).

2.4 Pre-Imaginal Diapause

Diapause in *S. exempta* has been reported (Khasimuddin 1977) but extensive subsequent work by Khasimuddin and in our laboratory provided no indication of arrested development in the pre-imaginal stages. The absence of reliable reports of its occurrence in any part of the species' range justifies the conclusion that diapause plays no part in its life-history strategy.

2.5 Migratory Strategy of S. exempta

The availability of habitats favourable for larval development is determined by the seasonal rains and, during the dry season in East Africa, the extensive grasslands dry out rapidly. Populations of *S. exempta* can only persist where there is enough moisture to support continuing growth of host plants and the temperature remains within the range tolerated by the insects; for example, in the highlands, in marshy areas and in river-flood plains (Rose 1979). Extensive coastal-flood plains which are watered by sporadic rain through most of the year and where temperatures remain high, may be particularly important for the maintenance of populations through the dry season (D.J.W. Rose 1984, personal communication). This conclusion is supported by backtracks from early season outbreaks in Tanzania and Kenya which suggest that the moths causing them came from the east and probably from coastal regions where no preceeding outbreaks had been reported (Tucker et al. 1982). At this time of year, meteorological disturbances capable of concentrating flying moths to cause high densities (outbreaks) of *gregaria* larvae occur very infrequently so that larval populations will remain at low densities and in the inconspicuous *solitaria* phase. Evidence is accumulating that populations do indeed persist in such areas. Pheromone traps deployed in the Kenya highlands (W.W. Page 1982, unpublished report) and in regions of Malawi (G.K. Nyirenda 1982, unpublished report) have continued to catch small numbers of moths throughout the dry season, and current surveys in the coastal regions of Kenya and Tanzania also indicate the presence of these hidden populations. Although these areas may be large, they are typically isolated and widely separated from one another so that selection can be expected to act against long-distance migrants which are unlikely to locate suitable habitats. However, as the populations are in the *solitaria* phase, migration will not be cued in the facultative migrants and only the obligate-migrant phenotypes will be lost. Furthermore, the relatively small number of generations over the dry season suggests that the depletion of long-flight alleles will be limited.

At the onset of the rains, concentration of airborne moths by convective storms will result in localised, high-density populations and facultative and obligate migrants from these, together with obligate migrants from low-density populations, will be widely dispersed downwind. If the rains are abundant and widespread, these emigrants will have a very high probability of reaching favourable habitats provided by the flush of growth over the ubiquitous grasslands. The selection pressure against migration will then be reversed. In fact, in river-flood plains, mortality due to flooding may intensify selection for emigration. The increased levels of migratory flights will result in widely dispersed low-density populations, with high-density outbreaks occurring whenever sufficient moths are airborne and encounter concentrating meteorological systems. The proportion of the total population at high densities is therefore likely to be very variable within and between seasons and, because of the factors acting to achieve dispersal of flying moths, must often be low.

In less extreme dry seasons following good rains, favourable habitats may persist over larger areas and in additional locations, and the widespread dispersal during the rains will increase the probability that they are occupied by low-density populations. Similarly, when the rains are scattered and sporadic, although there will be an extension of suitable habitat, it will retain a greater degree of patchiness so that selection for emigration may be less consistent.

It now seems clear that concentration of flying moths into high densities, an inevitable consequence of migratory flight during the rains in the tropics, exposes the resulting larvae to the risk of heavy mortality, particularly from viral disease and the exhaustion of available food plants. The evidence suggests that the phase phenomenon in this species functions as an adaptation to achieve rapid re-dispersal from high densities. Larval development is accelerated in the *gregaria* phase (as a result of selection pressure due to these mortality factors), and migratory flight is cued in the facultative migrants. The accumulation of larger fat reserves by *gregaria* phase larvae provides the fuel for prolonged flight in these phenotypes (Sect. 1.2).

This evidence led Parker and Gatehouse (1985b) to propose that variability in the extent and degree of patchiness of the grassland habitat, between dry and rainy seasons and between successive dry seasons and rains, has resulted in the evolution of a migratory strategy, the principal feature of which is the maximisation of dispersal within and between habitat patches subject to rains. This strategy will ensure that there is a high probability, that areas in which favourable conditions are going to persist through the ensuing dry season are occupied by low density *solitaria* populations. The costs of migration and of dispersal within the habitat will be low because of the ubiquity of suitable host plants and of the large excess of fat, especially in *gregaria* phase individuals (U.R. Muhsin, unpublished results) over that required for reproduction (Gunn and Gatehouse, in preparation). On the other hand, the costs of maintaining populations in the same location at all, but especially at high densities, are likely to be high because of increasing risks of mortality from predation, parasites and disease (Blair 1972). It must be remembered that many (perhaps most) moths categorised as short-fliers are capable of flights which achieve significant dispersal, so that low-density populations will be redistributed at each generation. However, many shorter flights occur in the latter part of the night (Parker and Gatehouse 1985b) when winds are generally light and variable, limiting displacement to distances which probably keep most of the moths within the relatively large habitat patches (Sect. 1.3). Our evidence suggests that this strategy is achieved by cyclic selection acting on genetically-determined variability in potential flight capacity.

3 Discussion

During evolution, selection acts to balance the risks of migration against those of remaining in the current habitat (Southwood 1977). For the majority of migratory insect species in which the causal aspects of migration have been studied, the costs of migration in terms of mortality and loss of reproductive potential are expected to be high. These costs are related to the expectancy of the migrant reaching a new favourable habitat, which depends on the patchiness of its habitat in space and time (Southwood 1977). For example, in species with highly patchy habitats each offering abundant resources, selection can be expected to favour the evolution of responses to proximate cues associated with habitat quality, such as to maintain populations in those habitats for as long as they remain favourable, e.g. by wing-muscle histolysis following feeding in *Dysdercus* spp. (Dingle and Arora 1973). However, in many leaf-feeding continental insects which are oligophagous with generally distributed host plants such as grasses, or highly polyphagous, the costs of migration may be very low during favourable seasons, because of the large size and widespread distribution of habitat patches. In some species, like *S. exempta*, populations staying in the same location over several generations face increased risks of mortality from predators, parasites and disease, because of the opportunity for functional and numerical responses of predator and parasite populations and the possibility of rapid population growth to levels at which disease can cause significant mortality [like other colonising species, *S. exempta* has very high fecundity, Gunn and Gatehouse (1985); Southwood (1978)]. Thus the costs of failing to disperse may exceed those of emigration. In these circumstances, selection cannot favour migratory responses to environmental cues associated with habitat quality because it is advantageous to disperse even when the immediate habitat remains favourable. Instead, it will act to maintain in the population the proportions of individuals with the appropriate range of genetically determined flight potential to achieve the optimal balance of dispersal within and between the large and generally distributed habitat patches (see also Hamilton and May 1977). Furthermore, these penalties of high larval density can be expected to result in the evolution of density-dependent modulation of flight potential, as in *S. exempta*.

Favourable habitats associated with rainfall in the arid tropics are highly unpredictable in space between seasons and years, and often widely separated. The length of unfavourable periods is also often very variable. These conditions are likely to preclude the evolution of diapause as an escape strategy (Southwood 1977) but, as long as enough favourable patches persist through the dry season to maintain populations at adequate levels, a migratory strategy which results in extensive dispersal in the rains will ensure that there is a high probability that they will contain populations irrespective of their location.

In some temperate-zone insects, migratory responses to reliable environmental cues of seasonal habitat deterioration provide the mechanism for escape to latitudes or to specific locations at which off-season survival is possible with or without diapause (reviews by Dingle 1982; Oku 1983). Many other species enter diapause without substantial movements and, for those which exploit ubiquitous host plants, many of the arguments regarding the low relative costs of emigration in relation to the large size and widespread distribution of the habitat patches may apply. Migratory strategies based on the genetic determination of flight capacity might be expected to evolve in such species and, in one of them, the univoltine weevil *Sitona hispidula* which feeds on clover, simple Mendelian inheritance of wing length has been demonstrated (Jackson 1928).

However, it is clear that the exploitation of generally available host plants in large patches does not always lead to the evolution of this type of migratory strategy. Temperate forests provide such habitats, but it appears that for species which exploit forest trees, the disadvantages of failing to disperse are outweighed by advantages of investing at least a proportion of their reproductive potential in the same location for as long as it remains favourable [e.g. *Choristoneura fumiferana* (Fisher and Greenbank 1979)], or of aggregation [e.g. in overcoming host-plant defences – bark beetles (Berryman 1976; Wood 1980) and possibly some Lepidoptera (Young 1983)].

The evolution of migratory strategies based on the maximisation of dispersal by flight in non-diapausing tropical species with ubiquitous host plants, has two important consequences. Where winds blow across the boundaries of their tropical and subtropical range, migrants will be carried into the temperate zone. In winter these individuals will be lost but in summer they are likely to encounter favourable habitats. As the adults of subsequent generations take to the wing, they will be dispersed and re-distributed by the local winds at that time. If these are predominantly towards higher latitudes, the result will be a continued extension of the range. In some regions of the northern hemisphere (e.g. southern China, south-eastern USA), there is a clear reversal in the direction of dominant winds from southerly and south-westerly in spring to north-easterly in autumn (D.E. Pedgley 1984, personal communication). A bias towards northward movements early in the season will therefore be reversed in the autumn when migrants will tend to be carried to lower latitudes. Many of these migrants which, because of the local winds they have encountered during flight, have remained in or returned to low latitudes during the summer, will regain regions where continuous generations can be maintained during the winter. Those which have been carried to higher latitudes may or may not do so, depending on the extent of their displacement in each generation and on the rapidity of the onset of habitat deterioration. Such a clear seasonal reversal of dominant winds is not apparent in other regions (e.g. Europe) or in the southern hemisphere but day-to-day changes in wind direction provide opportunities for northward and southward movements at all seasons.

Regular incursions into the temperate zone are characteristic of several tropical and subtropical species of Noctuidae, for example *Spodoptera frugiperda*, *Mythimna separata*, *Pseudaletia unipuncta*, *Alabama argillacea*, and *Autographa gamma* (Oku and Kobayashi 1978), all of which are polyphagous or feed on common and widely distributed host plants. *S. exempta* also extends its range seasonally into latitudes in South Africa where winter survival is not possible but similar northward movements are prevented by the Sahara and Arabian deserts. These incursions have been called pied piper migrations and have been referred to as non-adaptive (Dingle 1982) which is clearly misleading (Kennedy and Way 1979; Walker 1980), particularly as it seems highly probable that the extent of return movements in the autumn have been greatly underestimated for a number of reasons. Firstly, a reduction in the degree of synchrony of generations at the different latitudes through the summer, together with the wide dispersal of populations, will reduce the densities of flying moths at any one time to very low levels which are likely to be overlooked. Secondly, if meteorological disturbances capable of concentrating airborne moths are less frequent in autumn, populations will be at low densities, widely dispersed and therefore relatively inconspicuous. Finally, for pest species on which attention has been concentrated, populations at low densities in autumn have little or no agricultural importance and are therefore more likely to escape notice. One large-scale mark-recapture experiment has demonstrated autumn southward movement of *M. separata* over 800 km in China (Li et al. 1964).

A further indication that substantial return movements do occur in at least some of these species, is provided by recent work on *Pseudaletia unipuncta* which appears to have evolved the capacity to respond to environmental cues to achieve escape from high latitudes. The pre-reproductive period is prolonged in females and males from populations in southern Canada in response to increasing scotophase and/or low temperatures experienced in the pupal and early adult stages (Turgeon and McNeil 1983; Turgeon et al. 1983; Delisle and McNeil 1984). Prolonged pre-reproductive periods increase the number of nights over which moths can express their potential for migratory flight and thus displacement.

The second consequence of this type of migratory strategy is the inevitability of periodic concentration of flying moths by mesoscale meteorological disturbances (Sect. 1.3) resulting in locally intense oviposition and high larval densities. The adaptive significance of phase polyphenism in achieving rapid re-dispersal from these high densities has been discussed in relation to *S. exempta* (Sect. 2.5), and it is instructive that some of the migratory species of Noctuidae mentioned above as likely to have evolved similar strategies, show phase polyphenisms involving density-dependent responses very similar to those in this species (Table 2). The association between phase and flight capacity demonstrated in *S. exempta* has been investigated in only one of these species, *M. separata* (M.G. Hill 1984, personal communication), and his failure to obtain an effect can probably be attributed to the laboratory techniques and procedures he used. This

Table 2. Some effects associated with the density-dependent phase polyphenism in some non-dia-pausing migratory noctuid moths[a]

	Host plants	Larval pigmen- tation	Larval activity	Rate of larval develop- ment	Larval/ pupal fat reserves	Fecundity	Adult flight capacity
Spodoptera exempta	Oligophagous-grasses	+	+	+	+	0[1]	+ (♀)[2]
Mythimna separata	Polyphagous	+	+	+	+	0	– ? [3]
Autographa gamma	Polyphagous	+	+	+	+	+	?
Alabama argillacea	Malvaceae	+[4]	?	?	?	?	+ ? [4]
Pseudaletia unipuncta	Polyphagous						
Spodoptera frugiperda	Polyphagous	Density-dependent polyphenism 'assumed to occur' (Iwao 1962)					

[a] +, Higher/greater in crowded than in isolated individuals; 0, no difference; –, lower in crowded individuals. Data from review by Gruys (1970) and [1] U.R. Muhsin, unpublished results; [2] Woodrow and Gatehouse, in preparation; [3] M.G. Hill, personal communication; [4] Johnson et al. (1984)

aspect of the phase phenomenon in these species urgently requires investigation. Phase polyphenism was, of course, first described in locusts (Uvarov 1921) and its significance in their life-history strategies, to achieve emigration and displacement of populations into synoptic-scale wind convergences, where rainfall and thus favourable conditions for growth and development occur, is well-documented (Johnson 1969). However, populations of many locust species spend much of their time at low densities in the *solitaria* phase, occupying habitats in arid regions of the tropics. Although *solitaria* adults are known to migrate, nothing is known of the factors determining their flight capacity but the nature of the habitats of several species suggests that migratory strategies involving genetic determination of flight potential may be expected (as suggested by the results of Gunn and Hunter-Jones 1952).

The catholic feeding habits, voracity and mobility of many of the noctuid species discussed here make them formidable agricultural pests and the striking parallels between several of them and *S. exempta* suggest that they may have evolved similar migratory strategies. If this is confirmed, understanding these strategies and, in particular, appreciating the significance of low-density populations can be expected to improve the prospects of identifying the sources of migrants that initiate infestations and so of developing methods for long-range forecasting and more effective management of potentially damaging populations.

In the wider context, the main conclusion of this discussion is that stereotyped genetic determination of migratory capacity is likely to evolve not, as Ziegler (1976) suggested, when reliable environmental cues of habitat favourability are absent (adequate cues are probably always available), but when they are irrelevant because the balance between the costs of staying and leaving is in favour of the latter, even when local conditions remain favourable for growth and development. This condition is met when departing individuals have a high expectancy of reaching a new habitat, and/or when staying in the same location involves severe risks of mortality from natural enemies (including disease). In species subject to potentially heavy mortality from natural enemies, modulation of the tendency to disperse by a response to population density is to be expected, because this mortality is generally positively density-dependent.

Acknowledgements. The framework and impetus for much of the recent work on the African armyworm has been provided by Dr. Derek Rose and I have relied heavily on his, and Bill Page's generosity in allowing us free access to their ideas and unpublished observations and data. I have also benefitted greatly from several vigorous discussions with Dr. R.C. Rainey FRS Professor J.S. Kennedy FRS, and Dr. J.M. Cherrett both read this paper in draft and I am most grateful for their comments and suggestions.
   The work in this laboratory was carried out by several past and present research students and associates (whose contributions have been acknowledged in the text); my thanks to them all and particularly to Dr. Bill Parker, Dr. Alan Gunn, Pam Bowen, and Ann Vernon whose enthusiasm and insight we have lost by her tragic death in a car accident on 19th April 1986. We have been generously supported by studentships, grants, and facilities provided by SERC, TDRI (formerly COPR), DLCO-EA and the Kenya Agricultural Research Institute.

References

Aidley DJ (1974) Migratory capability of the African armyworm moth, *Spodoptera exempta* (Walker). E Afr Agric For J 40:202–203

Beck SD (1980) Insect photoperiodism, 2nd edn. Academic Press, New York

Berryman AA (1976) Theoretical explanation of mountain pine beetle dynamics in lodgepole pine forests. Environ Entomol 5:1225–1233

Blair BW (1972) An outbreak of the African armyworm *Spodoptera exempta* (Walker) (Lepidoptera: Noctuidae) in Rhodesia during December 1971 and January 1972. Rhod J Agric Res 10:159–168

Blair BW, Rose DJW, Law AB (1980) Synoptic weather associated with outbreaks of African armyworm *Spodoptera exempta* (Walker) (Lepidoptera:Noctuidae), in Zimbabwe during 1973 and 1976–77. Zimbabwe J Agric Res 18:95–110

Brown ES, Swaine G (1965) Virus disease of the African armyworm *Spodoptera exempta* (Walker). Bull Entomol Res 56:95–116

Brown ES, Swaine G (1966) New evidence on the migration of moths of the African armyworm *Spodoptera exempta* (Walker) (Lepidoptera:Noctuidae). Bull Entomol Res 56:671–684

Brown ES, Betts E, Rainey RC (1969) Seasonal changes in the distribution of the African armyworm, *Spodoptera exempta* (Walker) (Lepidoptera:Noctuidae), with special reference to Eastern Africa. Bull Entomol Res 58:661–728

Caldwell RL, Hegmann JP (1969) Heritability of flight duration in the milkweed bug *Lygaeus kalmii*. Nature 223:91–92

Delisle J, McNeil JN (1984) Calling behaviour of the armyworm *Pseudaletia unipuncta* (Haw.) (Lepidoptera:Noctuidae) under different photoperiodic regimes. 17th Int Congr Entomol. J Zool (Lond) 185:539–553
   20–26 August 1984, Hamburg. Abstract, p 453 (abstract only)

Den Boer MH (1978) Isoenzymes and migration in the African armyworm *Spodoptera exempta* (Lepidoptera:Noctuidae). J Zool 185:539–553

Dingle H (1968) The influence of environment and heredity on flight activity in the milkweed bug *Oncopeltus*. J Exp Biol 48:175–184

Dingle H (1980) Ecology and evolution of migration. In: Gauthreaux SA (ed) Animal migration, orientation and navigation. Academic Press, New York, pp 1–101

Dingle H (1982) Function of migration in the seasonal synchronisation of insects. Entomol Exp Appl 31:36–48

Dingle H, Arora G (1973) Experimental studies of migration in bugs of the genus *Dysdercus*. Oecologia (Berl) 12:119–140

Dingle H, Brown CK, Hegmann JP (1977) The nature of genetic variance influencing photoperiodic diapause in a migrant insect, *Oncopeltus fasciatus*. Am Nat 111:1047–1059

Faure JC (1943) Phase variation in the armyworm *Laphygma exempta* (Walker). Sci Bull Dep Agric For Un S Afr no 234

Fisher RA, Greenbank DO (1979) A case study of research into insect movement: spruce budworm in New Brunswick. In: Rabb RL, Kennedy GG (eds) Movement of highly mobile insects: concepts and methodology in research. North Carolina State University, Raleigh, pp 220–229

Fuseini BA, Kumar R (1975) Ecology of cotton stainers (Heteroptera:Pyrhocoridae) in southern Ghana. Biol J Linn Soc 7:113–146

Gatehouse AG, Hackett DS (1980) A technique for studying flight behaviour of tethered *Spodoptera exempta* moths. Physiol Entomol 5:215–222

Gruys P (1970) Growth in *Bupalus pinniarius* (Lepidoptera:Geometridae) in relation to larval density. Verh Rijksinst Naturbeheer 1:1–127

Gunn A, Gatehouse AG (1985) Effects of larval and adult food and water uptake on reproduction in the African armyworm, *Spodoptera exempta* (Walker) (Lepidoptera:Noctuidae). Physiol Entomol 10:53–63

Gunn DL, Hunter-Jones P (1952) Laboratory experiments on phase differences in locusts. Anti-Locust Bull 12:1–29

Haggis MJ (1984) Distribution, frequency of attack and seasonal incidence of the African armyworm *Spodoptera exempta* (Walker) (Lepidoptera:Noctuidae) with particular reference to Africa and southwestern Arabia. Report of the Tropical Development and Research Institute, London, L69

Hamilton WD, May RM (1977) Dispersal in stable habitats. Nature 269:578–581

Harrison RG (1980) Dispersal polymorphisms in insects. Annu Rev Ecol Syst 11:95–118

Iwao S (1962) Studies on the phase variation and related phenomena in some Lepidopterous insects. Mem Coll Agric Kyoto University no 84, pp 1–80

Jackson DJ (1928) The inheritance of long and short wings in the weevil *Sitona hispidula*, with a discussion of wing reduction among beetles. Trans R Soc Edinb 55:665–735

Johnson CG (1969) Migration and dispersal of insects by flight. Methuen, London

Johnson CG (1976) Lability of the flight system: a context for functional adaptation. In: Rainey RC (ed) Insect flight. RES symp 7. Blackwell, Oxford, pp 217–234

Johnson SJ, Hammond AM, Foil LD (1984) The correlation of phase variation in the cotton leafworm *Alabama argillacea* with its premigratory condition. 17th Int Congr Entomol, 20–26 August 1984, Hamburg. Abstracts, p 304 (abstract only)

Kennedy JS (1961) A turning point in the study of insect migration. Nature 198:785–791

Kennedy JS, Way MJ (1979) Summing up the conference. In: Rabb RL, Kennedy GG (eds) Movement of highly mobile insects: concepts and methodology in research. North Carolina State University, Raleigh, pp 446–456

Khasimuddin S (1977) On the occurrence of an aestivation/diapause phenomenon in the African armyworm *Spodoptera exempta* (Walker) (Lepidoptera:Noctuidae). E Afr Agric For J 42:350

Khasimuddin S (1981) Phase variation and off-season survival of the African armyworm, *Spodoptera exempta* (Walker) (Lepidoptera:Noctuidae). Insect Sci Appl 1:357–360

Li K-p, Wong H-h, Woo W-s (1964) Route of the seasonal migration of the Oriental armyworm moth in the eastern part of China as indicated by a three-year result of releasing and recapturing marked moths. Acta Phytophylacica Sin 3:101–110

Matthee JJ (1945) Biochemical differences between the solitary and gregarious phases of locusts and noctuids. Bull Entomol Res 36:343–371

Matthee JJ (1946) A study of the phases of the armyworm *Laphygma exempta* (Walker). J Entomol Soc South Afr 9:60–77

McCaffery AR, Page WW (1978) Factors influencing the production of long-winged *Zonocerus variegatus*. J Insect Physiol 24:465–472

Norris MJ (1965) Influence of constant and changing photoperiods on imaginal diapause in the red locust (*Nomadacris septemfasciata* Serv.). J Insect Physiol 11:1105–1119

Oku T (1983) Aestivation and migration in noctuid moths. In: Brown VK, Hodek I (eds) Diapause and life cycle strategies in insects. Junk, The Hague

Oku T, Kobayashi T (1978) Migratory behaviours and life cycle of noctuid moths (Insecta:Lepidoptera) with notes on recent status of migrant species in northern Japan. Bull Tohoku Natl Agric Exp Stn (Morioka) 58:97–209 (in Japanese: English summary)

Padgham DE (1983) The influence of host-plant on the development of the adult brown plant-hopper *Nilaparvata lugens* (Stal) (Hemiptera:Delphacidae) and its significance in migration. Bull Entomol Res 73:117–128

Parker WE (1983) An experimental study on the migration of the African armyworm moth, *Spodoptera exempta* (Walker) (Lepidoptera:Noctuidae). Thesis, University of Wales, Bangor

Parker WE, Gatehouse AG (1985a) The effect of larval rearing conditions on flight performance in females of the African armyworm *Spodoptera exempta* (Walker) (Lepidoptera:Noctuidae). Bull Entomol Res 75:35–47

Parker WE, Gatehouse AG (1985b) Genetic factors in the regulation of migration in the African armyworm moth, *Spodoptera exempta* (Walker) (Lepidoptera:Noctuidae). Bull Entomol Res 75:49–63

Pedgley DE, Reynolds DR, Riley JR, Tucker MR (1982) Flying insects reveal small-scale wind systems. Weather 37:295–306

Persson B (1981) Population fluctuations of the African armyworm, *Spodoptera exempta* (Walker) (Lepidoptera:Noctuidae), in outdoor cages in Kenya. Bull Entomol Res 71:289–297

Rainey RC, Betts E (1979) Continuity in major populations of migrant pests: the desert locust and the African armyworm. Philos Trans R Soc Lond B Biol Sci 287:359–374

Riley JR, Reynolds DR, Farmery MJ (1983) Observations of the flight behaviour of the armyworm moth, *Spodoptera exempta*, at an emergence site using radar and infra-red optical techniques. Ecol Entomol 8:395–418

Roff DA (1975) Population stability and the evolution of dispersal in a heterogenous environment. Oecologia (Berl) 19:217–237

Roff DA (1984) The evolution of wing polymorphism in insects. 17th Int Congr Entomol, 20–26 August 1984, Hamburg. Abstracts, p 357 (abstract only)

Rose DJW (1972) Dispersal and quality in populations of *Cicadulina* species (Cicadellidae). J Anim Ecol 41:589–609

Rose DJW (1979) The significance of low-density populations of the African armyworm, *Spodoptera exempta* (Walker). Philos Trans R Soc Lond B Biol Sci 287:393–402

Rose DJW, Dewhurst CF (1979) The African armyworm, *Spodoptera exempta* – congregations of moths in trees before flight. Entomol Exp Appl 26:346–348

Smith NG (1983) Host-plant toxicity and migration in the dayflying moth *Urania*. Fla Entomol 66:76–85

Southwood TRE (1962) Migration of terrestrial arthropods in relation to habitat. Biol Rev 37:171–214

Southwood TRE (1977) Habitat, the template for ecological strategies? J Anim Ecol 46:337–365

Southwood TRE (1978) Escape in space and time – concluding remarks. In: Dingle H (ed) Evolution of insect migration and diapause. Springer, Berlin Heidelberg New York, pp 277–279

Stinner RE, Barfield CS, Stimac JL, Dohse L (1983) Dispersal and movement of insect pests. Annu Rev Entomol 28:319–335

Tucker MR (1983) Light-trap catches of African armyworm moths *Spodoptera exempta* (Walker) (Lepidoptera:Noctuidae) in relation to rain and wind. Bull Entomol Res 73:315–319

Tucker MR, Pedgley DE (1983) Rainfall and outbreaks of the African armyworm *Spodoptera exempta* (Walker) (Lepidoptera:Noctuidae). Bull Entomol Res 73:195–199

Tucker M, Mwandoto S, Pedgley DE (1982) Further evidence for the windborne movement of armyworm moths, *Spodoptera exempta*, in East Africa. Ecol Entomol 7:463–473

Turgeon JJ, McNeil JN (1983) Modifications in the calling behaviour of *Pseudaletia unipuncta* (Lepidoptera:Noctuidae) induced by temperature conditions during pupal and adult development. Can Entomol 115:1015–1022

Turgeon JJ, McNeil JN, Roelofs WL (1983) Responsiveness of *Pseudaletia unipuncta* males to female sex pheromone. Physiol Entomol 8:339–344

Uvarov BP (1921) A revision of the genus *Locusta* L. (*Pachytylus* Fieb) with a new theory as to the periodicity and migrations of locusts. Bull Entomol Res 12:135–163

Walker TJ (1980) Migrating Lepidoptera: are butterflies better than moths? Fla Entomol 63:79–98

Whellan JA (1954) The African armyworm and its control. Rhodesia Agric J 51:414–427

Wood DL (1980) Approach to research and forest management for western pine beetle control. In: Huffaker CB (ed) New technology of pest control. Wiley, New York, pp 417–448

Young AM (1983) On the evolution of egg placement and gregariousness of caterpillars in the Lepidoptera. Acta Biotheor 32:43–60

Zera AJ, Innes DJ, Saks ME (1983) Genetic and environmental determinants of wing polymorphism in the waterstrider *Limnoporus canaliculatus*. Evolution 37:513–522

Ziegler JR (1976) Evolution of the migration response: emigration by *Tribolium* and the influence of age. Evolution 30:579–592

10 Dispersal in Aphids, A Problem in Resource Allocation

A. F. G. DIXON and M. T. HOWARD [1]

1 Introduction

Although weak and clumsy fliers, aphids have by riding the winds achieved amazing feats of dispersal (Elton 1925). The tendency to disperse is an adaptation (Johnson 1969) that has enabled aphids to spread the chance of survival in space, and to seek out and colonize plants that are of above average quality. Although the speed and direction of flight is usually governed by the wind, aphids nevertheless, by choosing when to fly (Dixon and Mercer 1983) and terminating their flight by actively flying downwards (Thomas et al. 1977), have some control over the distance they travel.

At the population level, the urge to disperse has been measured in terms of the proportion of a population that develops into alatae (Lamb and Mackay 1979); and at the individual level, in terms of whether an aphid flies before or after giving birth to offspring (Shaw 1970b), which has been attributed to environmentally-induced differences in size between individuals (Taylor 1975). Crowding during nymphal life affects the development of wings, flight musculature, fuel reserves and gonads. Consequently migrants have a well-developed flight apparatus, but small gonads: the oogenesis flight syndrome (Johnson 1969; Shaw 1970c). Flight behaviour is partly dependent on crowding experience during nymphal (Shaw 1970b,c) and early adult life (Dixon et al. 1968; Walters and Dixon 1982).

Because aphids reproduce parthenogenetically aphid populations are made up of clones, the individuals of which can be highly polymorphic (Dixon 1985). There can be marked differences in the flight behaviour of the various morphs especially in terms of host selection (Dixon 1971a) and temperature thresholds for flight (Walters and Dixon 1984). In addition to stating how dispersal affects an aphid's rate of increase, this paper will consider variation in flight behaviour, at the individual level within a particular morph and clone, in relation to dispersal in species of aphids of the subfamily, Aphidinae.

2 Cost of Dispersal

Alatae of aphids that show alary dimorphism take from 6% to 26% longer to reach maturity than the apterae (Fisher 1982; Gutierrez et al. 1971; Mittler 1958; Noda

---

[1] School of Biological Sciences, University of East Anglia, Norwich NR4 7TJ, United Kingdom.

Insect Flight: Dispersal and Migration
Edited by W. Danthanarayana
© Springer-Verlag Berlin Heidelberg 1986

1960; Thornback 1983) and although alatae may initially have a reproductive rate comparable to apterae (Dixon and Wratten 1971; Taylor 1975) alatae are more likely to have a lower overall reproductive rate (Wratten 1977). Alates also give birth to smaller offspring than apterae of the same species do (Dixon and Wratten 1971), which may be an adaptation of alates to the generally better quality of the plants they colonize than they leave. In the colonization of new habitats the initial rapid production of many offspring may be more important than their size at birth, and this could also have contributed to the evolution of small size of the offspring of alates.

Winged individuals of species that show alary dimorphism autolyse their wing muscles shortly after settling on a plant (Johnson 1953, 1957). This prevents further dispersal and possibly has the advantage of making a small amount of protein available for embryo development (Dixon 1971b). In addition, the brachypterous alatae of *Drepanosiphum dixoni*, which differ from the macropterae in lacking indirect flight muscles and not being able to fly are 32% more fecund (Dixon 1972). Therefore, the maintenance of wing muscles is costly, and by autolysing or not developing them, aphids free resources for growth and reproduction.

The shorter developmental time and greater reproductive rate of apterae, compared with alates, gives them a higher rate of increase. A first step in the evolution of apterousness was possibly a simple increase in fecundity with the appearance of brachypterous forms or as a result of wing-muscle autolysis. Later apterous development could then give the additional advantage of a shorter developmental time and lower temperature thresholds for development (Noda 1960).

Thus, in addition to the assumed high mortality suffered during dispersal, alate aphids incur other disadvantages that are more easily quantified. The development of a flight apparatus and/or its maintenance, is costly in terms of time and fecundity. If aphids fly then they may incur an additional cost in that their potential fecundity is further reduced (Burns 1971) and there is a further delay in the onset of reproduction. The combined effect is a marked reduction in reproductive potential and rate of increase.

3 Differential Dispersal

Variability in the tendency of individuals to disperse is associated with differences in the level of reproductive investment (number of ovarioles in their gonads) between individuals within a clone, and this is independent of their size. Those with few ovarioles take off more readily and at a steeper angle, delay wing-muscle autolysis longer and are more resistant to starvation, and have relatively more olfactory organs than aphids with many ovarioles (Dixon and Dharma 1980; Leather and Wellings 1981; Walters and Dixon 1983; Ward et al. 1983). Individuals of *Aphis fabae* fly for varying periods of time (Kennedy and Booth 1963) and although these authors did not relate flight duration to ovariole number the proportion flying for increasing lengths of time are similar to the proportions that have decreasing numbers of ovarioles (Walters and Dixon 1983).

Similarly, the proportions of alates of *A. fabae* that take off before giving birth, after giving birth or that do not fly at all (Shaw 1970b) may also be associated with reproductive investment in terms of ovariole number (Walters and Dixon 1983). The range and frequency distribution of reproductive investment is characteristic for a species, and even for a morph (Wellings et al. 1980) and is another aspect of the polymorphism of aphids.

Recent experiments have shown that these differences in flight behaviour between individuals of a clone result in differential dispersal. The experiments were done in a glasshouse vented to the exterior at one end via a fan, which caused air to flow through the glasshouse. An aphid population on a host plant (bird cherry sapling) was placed at the opposite end of the glasshouse from the fan, and between it and the source plant other aphid-free plants were positioned at a distance of 1 m from the source plant. Aphids were prevented from walking between plants by surrounding them with sticky traps. Emigrants of the bird cherry-oat aphid *(Rhopalosiphum padi)* were used, because they are obligate migrants and the individuals of a clone show a wide range of reproductive investment (Walters and Dixon 1983). In the first experiment the aphid-colonized oats, and in the second they were given a choice of oats or *Lolium perenne*, both of which are amongst the most preferred of the secondary host plants of this species of aphid (Leather and Dixon 1982). Each experiment lasted for 4 days, emigrants were collected from the secondary host plants daily and a sample taken from the bird cherry saplings at the beginning and end of each experiment.

Proportionately more aphids with high ovariole numbers colonized the secondary host plants in both experiments than were present on the primary host, bird cherry (Figs. 1 and 2; $\chi_1^2 = 20$, d.f.$_1 = 4, P < 0.001; \chi_2^2 = 21$, d.f. $= 2, P < 0.001$). This could not be attributed to a differential rate of departure from the primary host of emigrants of different ovariole classes because all show the same readiness to take off (Walters and Dixon 1983) and therefore as predicted there was no significant difference in the proportions of the various ovariole classes on the primary host at the beginning and end of each experiment ($\chi_1^2 = 4.9$, d.f. $= 4$, N.S.; $\chi_2^2 = 0$, N.S.). Two secondary hosts were offered in the second experiment in order to test the possibility that host species may influence the settling of individuals of different ovariole classes. However, there is

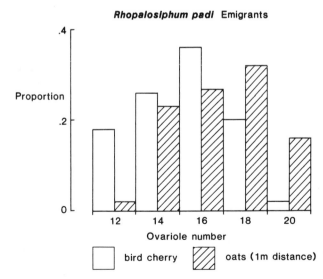

Fig. 1. The proportion of emigrants of the bird cherry-oat aphid of ovariole classes 12 to 20 leaving bird cherry and colonizing oats 1 m away (☐ n = 100; ▨ n = 56)

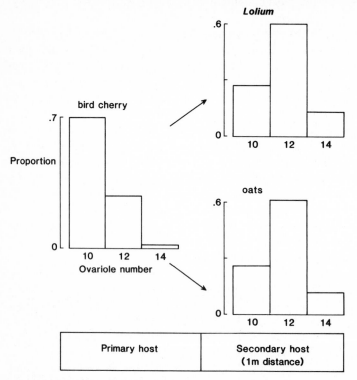

Fig. 2. The proportions of emigrants of the bird cherry-oat aphid of ovariole classes 10 to 14 leaving bird cherry and colonizing *Lolium* and oats 1 m away (Bird cherry n = 43; *Lolium* n = 30; oats n = 41)

no significant difference in the proportions of individuals of the different ovariole classes that colonized the two host plants ($\chi^2 = 0.02$, N.S.). Thus, in addition to the marked differences in flight behaviour of individuals with different reproductive commitments, there is evidence that this behaviour results in the aphids with the larger reproductive commitment mainly colonizing plants nearby, whereas those with a low reproductive commitment are assumed to colonize mainly plants at some distance.

The difference in the range of ovariole classes in the two experiments indicates the variability between clones. It is assumed that these differences are genetically determined and influence the fitness of the clones.

4 Discussion

In those species that show alary dimorphism it is either the tactile stimulation associated with crowding (Lees 1967; Shaw 1970a), deterioration in host quality (Dixon and Glen 1971), a combination of host quality and crowding (Watt and Dixon 1981), day length (Yagamuchi 1976), or a combination of these factors (Matsuka and Mittler 1978)

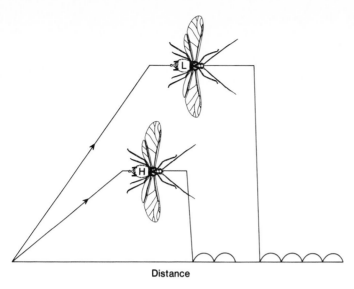

Fig. 3. The relationship between low (*L*) and high (*H*) reproductive investment and dispersal

that induces the development of winged forms. By responding to a number of stimuli rather than to one, aphids can possibly achieve a closer and more reliable tracking of environmental conditions. This enables them to produce the more fecund and faster developing apterae while the host plant is favourable and as it becomes unfavourable to switch resources to producing alatae that disperse to other plants.

Within a clone the alate aphids with small gonads have a greater urge to disperse, may fly longer and more frequently, and are better able to survive starvation than aphids with large gonads. The short-distance dispersers consist mainly of individuals with large gonads, and the long-distance migrants should consist mainly of individuals with small gonads (Fig. 3). Plants differ in quality and distribution both within and between seasons. In habitats that vary in quality between years clones of aphids with mixed reproductive strategies have a greater fitness than those that don't (Ward and Dixon 1984). The degree of differential dispersal shown by an aphid clone will probably reflect both the within and between year heterogeneity of the environment. In favourable environments individuals with a high reproductive investment, the 'risk takers' will be fitter than those with a low reproductive investment, the 'risk averse' and vice versa.

The variability in reproductive investment between individuals within a clone is programmed. That is, the intraclonal tactical diversity in flight and reproductive behaviour is an adaptation to living in, rather than a consequence of developing in, an heterogeneous environment. The proportion in each reproductive category can vary between clones, but is similar amongst the first and last offspring born to a mother (Dixon and Dharma 1980). However, if subjected to nutritional stress, a mother will give birth to proportionally more offspring with a low reproductive commitment, but nevertheless it will continue to produce offspring with a range of reproductive investments including the largest (Walters, Dixon, and Howard, unpublished).

Alary dimorphism and programmed intraclonal tactical diversity in reproductive investment is particularly characteristic of one of the ten subfamlies of the Aphididae, the Aphidinae, and may partly account for their success with 60% of aphid species belonging to this subfamily.

Acknowledgements. This work was supported by a Ministry of Agriculture Fisheries and Food grant to M.T. Howard. We are also indebted to Dr. Ted Evans for his helpful criticism of the manuscript.

References

Burns M (1971) Flight in the vetch aphid,*Megoura viciae* Buckton. PhD Thesis, University of Glasgow
Dixon AFG (1971a) The life-cycle and host preferences of the bird cherry-oat aphid, *Rhopalosiphum padi* L., and their bearing on the theories of host alternation in aphids. Ann Appl Biol 68:135–147
Dixon AFG (1971b) Migration in aphids. Sci Prog 59:41–53
Dixon AFG (1972) Fecundity of brachypterous and macropterous alatae in *Drepanosiphum dixoni* (Callaphididae, Aphididae). Entomol Exp Appl 15:335–340
Dixon AFG (1985) Structure of aphid populations. Annu Rev Entomol 30:155–174
Dixon AFG, Burns MD, Wangboonkong S (1968) Migration in aphids: response to current adversity. Nature 220:1337–1338
Dixon AFG, Dharma TD (1980) Number of ovarioles and fecundity in the black bean aphid, *Aphis fabae*. Entomol Exp Appl 28:1–14
Dixon AFG, Glen DM (1971) Morph determination in the bird cherry-oat aphid, *Rhopalosiphum padi* L. Ann Appl Biol 68:11–21
Dixon AFG, Mercer DR (1983) Flight behaviour in the sycamore aphid: factors affecting take-off. Entomol Exp Appl 33:43–49
Dixon AFG, Wratten SD (1971) Laboratory studies on aggregation, size and fecundity in the black bean aphid, *Aphis fabae* Scop. Bull Entomol Res 61:97–111
Elton CS (1925) The dispersal of insects to Spitsbergen. Trans R Entomol Soc Lond 1925:289–299
Fisher M (1982) Morph determination in *Elatobium abietinum* (Walk) the green spruce aphid. PhD Thesis, University of East Anglia
Gutierrez AP, Morgan DJ, Havenstein DE (1971) The ecology of *Aphis craccivora* Koch and subterranean clover stunt virus I. The phenology of aphid populations and the epidemiology of virus in pastures in south-east Australia. J Appl Ecol 8:699–721
Johnson B (1953) Flight muscle autolysis and reproduction in aphids. Nature 172:183
Johnson B (1957) Studies on the degeneration of the flight muscles of alate aphids I. A comparative study of the occurrence of muscle breakdown in relation to reproduction in several species. J Insect Physiol 1:248–266
Johnson CG (1969) Migration and dispersal of insects by flight. Methuen, London, 763 pp
Kennedy JS, Booth CO (1963) Co-ordination of successive activities in an aphid. The effect of flight on the settling responses. J Exp Biol 40:351–369
Lamb RJ, Mackay PA (1979) Variability in migratory tendency within and among natural populations of the pea aphid, *Acyrthosiphon pisum*. Oecologia (Berl) 39:289–299
Leather SR, Dixon AFG (1982) Secondary host preferences and reproductive activity of the bird cherry-oat aphid, *Rhopalosiphum padi*. Ann Appl Biol 101:219–228
Leather SR, Wellings PW (1981) Ovariole number and fecundity in aphids. Entomol Exp Appl 30:128–133
Lees AD (1967) The production of the apterous and alate forms in the aphid *Megoura viciae* Buckton, with special reference to the role of crowding. J Insect Physiol 13:289–318
Matsuka M, Mittler TE (1978) Enhancement of alata production by an aphid, *Myzus persicae*, in response to increases in daylength. Bull Fac Agric Tamagawa Univ 19, pp 1–7
Mittler TE (1958) Studies on the nutrition of *Tuberolachnus salignus* (Gmelin) (Homoptera, Aphididae) III. The nitrogen economy. J Exp Biol 35:626–638
Noda I (1960) The emergence of winged viviparous female in aphid VI. Difference in the rate of development between the winged and the unwinged forms. Jpn J Ecol 10:97–102

Shaw MJP (1970a) Effects of population density on alienicolae of *Aphis fabae* Scop. I. The effect of crowding on the production of alatae in the laboratory. Ann Appl Biol 65:191–196

Shaw MJP (1970b) Effect of population density on alienicolae of *Aphis fabae* Scop. II. The effects of crowding on the expression of migratory urge in the laboratory. Ann Appl Biol 69:197–203

Shaw MJP (1970c) Effect of population density on alienicolae of *Aphis fabae* Scop. III. The effect of isolation on the development of form and behaviour of alatae in a laboratory clone. Ann Appl Biol 65:205–212

Taylor LR (1975) Longevity, fecundity and size, control of reproductive potential in a polymorphic migrant, *Aphis fabae* Scop. J Anim Ecol 44:135–159

Thomas AAG, Ludlow AR, Kennedy JS (1977) Sinking speeds of falling and flying *Aphis fabae* Scopoli. Ecol Entomol 2:315–326

Thornback N (1983) The factors determining the abundance of *Metopolophium dirhodum* (Walk.), the rose grain aphid. PhD Thesis, University of East Anglia

Walters KFA, Dixon AFG (1982) Effect of host quality and crowding on the settling and take-off of cereal aphids. Ann Appl Biol 101:211–218

Walters KFA, Dixon AFG (1983) Migratory urge and reproductive investment in aphids: variation within clones. Oecologia (Berl) 58:70–75

Walters KFA, Dixon AFG (1984) The effect of temperature and wind on the flight activity of cereal aphids. Ann Appl Biol 104:17–26

Ward SA, Dixon AFG (1984) Spreading the risk, and the evolution of mixed strategies: seasonal variation in aphid reproductive biology. In: Engels W et al. (eds) Advances in Invertebrate Reproduction, vol 3. Elsevier, Amsterdam, pp 367–386

Ward SA, Wellings PW, Dixon AFG (1983) The effect of reproductive investment on pre-reproductive mortality in aphids. J Anim Ecol 52:305–313

Watt AD, Dixon AFG (1981) The role of cereal growth stages and crowding in the induction of alatae in *Sitobion avenae* and its consequences for population growth. Ecol Entomol 6:441–447

Wellings PW, Leather SR, Dixon AFG (1980) Seasonal variation in reproductive potential: a programmed feature of aphid life cycles. J Anim Ecol 49:975–985

Wratten SD (1977) Reproductive strategy of winged and wingless morphs of the aphids *Sitobion avenae* and *Metopolophium dirhodum*. Ann Appl Biol 85:319–331

Yagamuchi H (1976) Biological studies on the todo-fir aphid *Cinara todicola* Inouye with special reference to its population dynamics and morph determination. Bull Gov For Exp Stn (Tokyo) 283:1–102

11 Direction of Insect Migrations in Relation to the Wind

K. Mikkola[1]

1 Introduction

In recent decades, the views on the mechanics of long-range migrations of insects have diverged into two main lines: (1) French and White (1960), Shaw (1962), Hurst (1963, 1964), and Mikkola and Salmensuu (1965), working on Lepidoptera presumed downwind direction of migrations and analysed their data with the aid of wind trajectories which describe atmospheric tracks of air particles. Rainey (1963) showed that the displacement of locust swarms is predominantly downwind. A general theory was formulated on these lines by Kennedy (1961). (2) Baker (1978), however, has suggested that during migrations insects mainly use compass orientation. Thus, two controversial doctrines have been proposed. Nevertheless, it seems that even an individual species may use different migration and orientation mechanisms under different situations. In addition, regular and definite upwind migrations have been shown to occur in wasps and bumble bees (Mikkola 1978). The intention of this article is to document data on the role of wind as a factor affecting the direction of insect migrations. Examples come from the published and unpublished data gathered by the author in Finland and from the literature.

To discuss the problem at a more 'realistic' level, insect migrations may be compared with the movements of aeroplanes. Aeroplanes depart and land flying against the wind. From the classical studies of Kennedy (1940) we know that the takeoff of insects also regularly takes place against the wind (cf. e.g. Solbreck 1980). After the departure, an aeroplane may change its dependence on wind direction to compass or some other mechanical means of orientation. The track of migrating insects after takeoff is less well defined. This is the phase which is discussed in this article.

If an insect drifts in the wind, it is difficult to know if its direction is determined solely by the wind, or if the wind direction is the same as the compass or any other direction preferred by the insect. The former parallels the situation in which a balloon is carried by the wind, and the latter that in which a sailor uses midwind to reach his goal. Earlier authors such as Johnson (1969) have supported the downwind displacement thesis, but Baker (1978) claims that "to delegate displacement solely to prevailing wind (is) a strategy that has not yet been demonstrated for any animal". According to him, strong-flying migrants, even by night, choose such winds that are close to their preferred compass direction.

---

[1] University of Helsinki, Dept. of Zoology, P. Rautatiekatu 13, 00100 Helsinki 10, Finland.

Insect Flight: Dispersal and Migration
Edited by W. Danthanarayana
© Springer-Verlag Berlin Heidelberg 1986

2 Migrations with the Wind

In terms of the distance covered, downwind flights are the most spectacular migrations of insects. Single flights have been frequently cited in the literature (examples are given by Johnson 1969 and Pedgley 1982). There are extensive records on the occurrence of windborne insects in the United States (e.g., Pienkowski and Medler 1964; Rabb and Kennedy 1979; Raulston et al., Wolf et al., this volume), in Australia (Farrow 1975, this volume), China (Li et al. 1964), Japan (Oku 1983), Africa (Gunn and Rainey 1979), the United Kingdom (Johnson 1969; Johnson et al. 1962; Taylor 1974), and Northern Europe (Mikkola 1967). The windborne insect fauna has been analysed on fire ships or on weather ships (e.g. Asahina and Turuoka 1970; Haeseler 1974; Lempke 1962), on ships on transoceanic routes (e.g. Gressitt and Yoshimoto 1964), on oceanic islands (e.g. Fox 1978), and on mountainous ice fields (Burmann 1952; Edwards 1972).

2.1 Immigrations of Lepidoptera into Finland in 1946-1966 and 1972-1981

The Period 1946-1966. The 100 migrations observed during these 21 summers were compared to the daily weather patterns, the mean temperature of which exceeded the mean of 30 years by at least 2.5 centidegrees (Mikkola 1967). Warm days constituted 24.9% of all days, but 58 migrations of 100 occurred during warm days. Of these 58 migrations, 48 coincided also with movements of air currents. Statistical analysis of these data showed that the migrations and the occurrence of air currents were significantly correlated.

Air currents carrying migrants to Finland are regularly southern to southeastern (Fig. 1). Using wind trajectories it was shown that the South-Russian steppe area, south of 50 °N is the probable source of most migrations which are observed in Finland north of 60 °N. Baker (1978) referring to the above analysis wrote that "Large-scale back-tracking ... can rarely distinguish between conditions in the study area suitable for flight and conditions that would lead to immigration from reasonable sources". This aspect has, in fact, been considered by Mikkola (1967). The significant correlation of the migrations both with the air currents originating to the south of 50 °N and with warm days means that they are related to specific weather situations. Thus, neither warm stationary weather, air currents from the northern latitudes, nor colder ones from the south produce many migrants. It would seem that a warm air mass or a warm front arriving at the source area is a situation conducive to migrations and stimulates migrants to take off.

The above migrations, with the air currents moving towards the northern directions, probably end up all over in areas to the north of the steppe belt, the northern border of which is in eastern Europe roughly at 50 °N. Within the steppe belt, the migrations cannot be clearly linked to weather patterns. A difficulty experienced in analysing data from Central or Western Europe is that the potential source areas are situated at roughly similar distances from the observation points in many directions, i.e. in North Africa, the Near East and the Russian steppes.

The Period 1972-1981. Yearly migration records have been published in the reports of the Finnish Lepidopterological Society meetings since 1972 and in the German

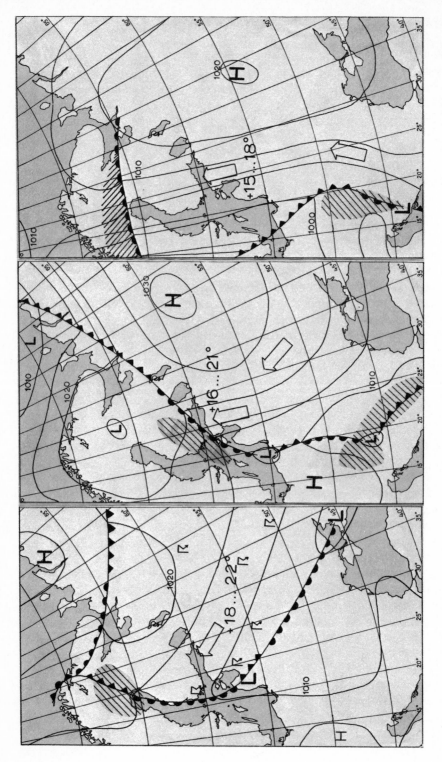

Fig. 1. Synoptic weather maps showing air currents which have carried many migrants to Finland: southwest or west of a region of high pressure lying over Russia, warm air is flowing to the northwest or north. (After Mikkola 1967)

journal *Atalanta* from 1979 onwards. Finnish data are mainly based on results from continuously operated light traps for moths, and on visual observations for butterflies.
The weather situations were analysed separately and the air flows were divided into
moderate or good with respect to migrations, taking into account the strength and
direction of the flow and the temperature. As shown in Fig. 2 good air currents were

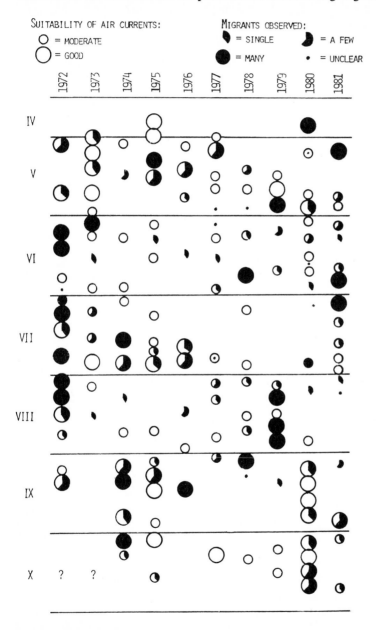

Fig. 2. Relation between immigrations observed and the suitability of air currents for migrations
in the years 1972–1981 in Finland. The years 1972 and 1973 lack data for October

frequent early and late in the season, i.e. towards the end of April/May and September/ October when they were recorded on 36 occasions. But the incidence of migrations during these periods was relatively low, with a mean number of migrations of 1.5 (i.e. between a single and several migratory species observed). The number of good air currents during mid-season (June–August) was low, 21 in all, but the incidence of migrations was clearly higher than during the other months, with a mean of 2.3, i.e. between several and many species recorded. A total of 71 moderate air currents was recorded for the entire season and the numbers of migrations recorded under the moderate wind conditions were low, giving a mean value of 0.6, i.e. either none or only a single species being observed. The most frequent immigrants into Finland during the period under study (1972–1981) are listed in Table 1.

In Fig. 2, it is seen that there are a few occasions where migrants have been observed without any obvious relationships to air currents. In these instances, only a single or a few species were involved, mostly during June–August when suitable air movement occurred just to the south or to the east of Finland. Also some of these individuals could have been local offspring of spring migrants. It would seem that for observations on lepidopteran migrations, Northern Europe is favourable, and Finland in particular, for there is no alternation of land and water masses nor high mountains on the migratory routes. Clearly most of the lepidopteran migrations to Finland occur when there are favourable air currents which are easily identifiable. In source areas, migrations may originate during a variety of warm-weather patterns, but only those insects taking off with suitable air currents are brought into Finland. The close correlation of immigrations with abnormally high temperatures shows that induction of flights in the source areas occurs under certain situations such as the approach of warm air fronts. The latter is often seen on the weather maps and gives reliability to the analyses; usually no data on the migrants in the source areas exist, but there are exceptions to this general rule (Mikkola and Salmensuu 1965; Mikkola 1971).

Table 1. The most frequent immigrants to Finland during the 10-year period 1972–1981[a]

Pieris brassicae (L.)	17	Herse convolvuli (L.)	6
Cynthia cardui (L.)	14	Ostrinia nubilalis (Hb.)	5
Agrotis ipsilon (Hfn.)	12	Orthonama obstipata (Fabr.)	5
Plutella xylostella (L.)	11	Loxostege sticticalis (L.)	5
Artogeia rapae (L.)	11	Pontia daplidice (L.)	4
Vanessa atalanta (L.)	10	Nycteola asiatica (Krul.)	4
Autographa gamma (L.)	9	Catocala adultera Men	4
Nomophila noctuella (D. & S.)	8	Ephesia fulminea (Scop.)	4
Colias hyale (L.)	7	Acherontia atropos (L.)	3
Macroglossum stellatarum (L.)	6	Spodoptera exigua (Hb.)	3

[a] The numbers denote immigrations presented in Fig. 1 (all species observed at least three times are included).

2.2 Aspects of Single Migrations

2.2.1 Change of Direction During Migration

The independence of directional changes in migrations from compass directions can be shown, if the direction of migration changes by at least 90° in tandem with the wind direction. Because only a few specific migrations have been observed at several geographical points, opportunities for such observations appear to be rare. Published work from Finland contains information on the flight direction changes in two species, *Spodoptera exigua* (Hb.), a medium- to strongly-flying night migrant (Mikkola 1967), and the strongly-flying *Pieris brassicae* (L.) which is a daytime migrant (Fig. 3a,b). During

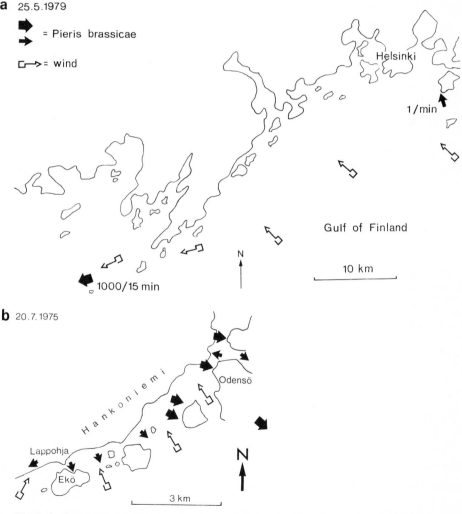

Fig. 3a,b. One spring (**a**) and one autumn (**b**) migration of *Pieris brassicae* (L.), the former a tailwind and the latter a headwind flight, both from the southern coast of Finland. In both cases the wind affected the direction of flight of the butterflies by roughly 90°

the migration of *S. exigua* in 1964, the flight direction changed three times from the main direction (NW) by about 130°, nearly reversing, corresponding to changes in the wind direction. This track did not correspond with compass orientation.

Especially during the spring, southeastern winds blowing over the Gulf of Finland give rise to a compensatory eastern or even northeastern wind at the Finnish coast. This is caused by temperature differences over the sea and land and friction. In two cases (Fig. 3a and Vepsäläinen 1968), with relatively weak winds, it was observed that the direction of immigrating *Pieris brassicae* changed with the wind direction, although maintenance of the original direction would have been quite possible under the prevailing wind conditions.

2.2.2 Changes to Sex Ratio During a Migration

In late 1972 numerous immigrant *Leucoma salicis* (L.) (Lymantriidae) arrived in SE Finland and proceeded up to the line of Åland Islands–Lahti-Joensuu in middle Finland. The prevailing winds were from the southeast. It is notable that during this migration the ♂:♀ sex ratio in the southeast of Finland ranged from 1.2:1 to 0:1 (i.e. 45%–100% ♀♀) (Fig. 4). In two cases, large numbers of *L. salicis* were observed to have drifted to the seashore: the proportions of females were 100% at Haapasaari in the extreme southeast, but 75% at Pellinki archipelago, some 60 km east of Helsinki. The sex ratio based on all the records to the east of Helsinki was 1:3 (73 ♂♂:214 ♀♀). In the Helsinki area and to the west of it, females were rare with a ratio of 1:0.02 (271 ♂♂: 5 ♀♀). An explanation of this phenomenon would be that most of the egg-laden females

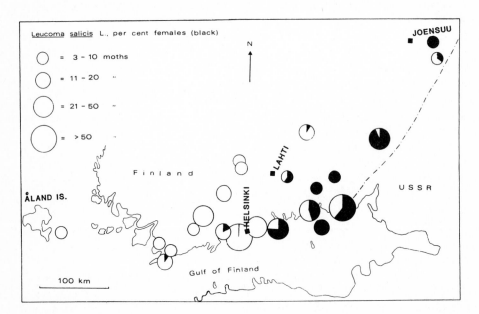

Fig. 4. Change of sex ratio in *Leucoma salicis* (L.) during a downwind migration to Finland in late June 1972

were disseminated en route, and only a few continued topass the Gulf of Finland. The subsequent very low proportion of females in the population would suggest that the males continued to migrate to a distance of 100–200 km greater than that achieved by females. In another example, during a migration of *Pontia daplidice* in 1970 (see below) the sex ratio changed during a flight from east to west from 50% to 70% males (Keynäs 1971), the drop in the proportion of females was presumed to be due to oviposition activity.

2.2.3 The Rate of Advance and Its Variations

The meteorological examination of data by Mikkola (1967) referred to above shows that lepidopteran migrations, including many flights of Pieridae, arrive in Finland quite regularly from a distance of about 1000 km, i.e. from south of 50°N latitude (this does not exclude migrations over short distances). Baker (1978), however, suggests "that even for ... *Pieris rapae* and *P. brassicae*, which have a relatively high incidence of migration throughout their lives, the vast majority of individuals will die within 300 land km (plus the distance of any overseas crossings) from their place of emergence". What could be the reasons for these conflicting conclusions?

Exact data on the rate of advance of lepidopteran migrations are scarce. Two accounts of migratory rates have been documented from Finland. In one instance, a migration of *Spodoptera exigua* occurred in August 1964 over a period of six nights from the southeastern border of Finland to Denmark (Mikkola and Salmensuu 1965), giving a rate of advance of 220 km per night. Prior to this, the swarm had travelled from Kazahstan, U.S.S.R., where there had been an explosive reproductive success, to Finland in 3 days or less – at a minimum rate of about 800 km day^{-1}. In the other instance, individuals of *Lymantria dispar* (L.) migrated from the southeastern suburbs of Moscow to Finland in July 1958 in three nights (Mikkola 1971) at a rate of travel of 350 km per night. In both these cases the rate of advance has been explained clearly in terms of air-current movement. Also several instances are known from the 1970s and 1980s in which the arrival of butterflies to southern and southeastern Finland has been followed by a steady spread of them to the north. From Table 2 it can be seen that the rate has been about 300 km day^{-1} for *Cynthia cardui* L., but 50 km for *P. brassicae* and still less for *Inachis io*. With *P. daplidice* in 1972, the rate of advance seems to have subsequently slowed down in Finland. A particularly notable example is the vast immigration of *C. cardui* and *Plutella xylostella* (L.) to the southeastern border of Finland in June 1978. The swarm gradually proceeded to the west, but subsequently advanced at an enormous speed to Lapland and onwards to Spitzbergen (Lokki et al. 1978). The rate of advance in *C. cardui* to the west, apparently without the aid of an air current, was about 70 km day^{-1}, but this increased to about 1000 km day^{-1} during the northbound migration.

It is also noteworthy that at the time of a mass immigration of *Pontia daplidice* (L.) to Finland in 1970, a distant steppe species, *Euchloe chloridice* (Hb.) also immigrated (Karvonen and Karvonen 1983). This was then the first time that this species had been seen in the Nordic countries, and a total of 18 individuals were observed. This is a clear example of the distant nature of this lepidopteran migration. There is meteorological

Table 2. Daily rate of advance during eight migrations of Papilionoidea observed in Finland in the years 1972–1984

Species[a]	Dates	Maximal distance (km)	Direction	Observation points/days	Rate of advance/ day (km)
1. *Cynthia cardui*	8–11 July 1972	960	N	3/3	320
2. *Cynthia cardui*	4– 5 September 1978	270	W	2/1	270
3. *Cynthia cardui*	24–25 June 1978	650	N	2/1	650
4. *Cynthia cardui*	24–25 June 1978	350	WNW	4/6	60
5. *Cynthia cardui*	26–27 July 1981	450	N	2/1	450
6. *Inachis io*	2–11 August 1983	330	NW	4/9	30–40
7. *Pontia daplidice*	8– 9 July 1972	250	N	2/1	250
8. *Pontia daplidice*	9–13 July 1982	360	N	2/4	90
9. *Pieris brassicae*	10–15 May 1975	250	N	2/5	50
10. *Pieris brassicae*	20–25 May 1984	250	N	2/5	50

[a] In one case (numbers 3–4), the rate of advance in two different directions during a single migration has been given, and in another (7–8), a migration has been divided into two stages. Observation points = successive observations along the migratory route.

evidence that the migration of the two species originated 2 days prior to their appearance in Finland, just north of the Black Sea and the Caspian Sea (Keynäs 1971). *E. chloridice* is hardly a permanent species in those areas, but it is not known if the immigrants were descendants of a spring migration or if they had migrated from further south. The rate of advance might have been of the order of 600 km day^{-1}. It should be noted that the first specimen of *P. daplidice* in SE Finland was caught at night at light (Hanski 1971). The advance of the swarm was terminated in southern Finland by a depression. During a migration 2 years later, the northernmost specimen was observed at Kuusamo (66 °N, see Table 2).

The question arises as to why there exist in a single species such big differences in the speed of advance during a migration. I propose that this is caused by two distinctly different phases of migration which may, however, pass through intermediate stages (see below). Now coming back to Baker's (1978) estimations of distances covered and the period of butterfly residence in gardens and countryside, his cross-country migration refers to something similar to the advance within Finland which occurred at a slow rate, as discussed above. This then would be a secondary migratory period of the butterflies. The primary one would be the long range migration which, as opposed to the cross-country migration, could be called above-country migration which might be partially synonymous with Taylor's above boundary layer migration (Taylor 1958, 1960, 1974). Because the distance from the source areas to Finland is simply too great to be covered by cross-country migrations, it seems that most immigrations to Finland represent above-country migrations. It is possible that such migrations occur high in the air and, therefore, without visual orientation with respect to the ground. Long-range butterfly migrations, on this basis, are largely dependent on the direction of the wind.

2.2.4 Flight in a Warm Air Mass Over a Cold One

One of the most exceptional records of a migrant in Finland in the last decade is that of a specimen of *Loxostege sticticalis* (L.) on August 10, 1977 in the most northwestern corner of Finland (69° 30'N) at Fjeld Harroaivi, some 400 km from the previous northernmost record of the species. The catch on August 10, 1977 is clearly connected to simultaneous captures of four specimens in the southernmost part of Finland. The weather maps provide an explanation for these finds. There was a region of high pressure over northeastern Russia, and to the southwest of it a strong and warm air current was directed from the northeastern side of the Caspian Sea towards the Kola Peninsula, to the east of Finland. Even at a height of 1500 m, the temperature was roughly 20 °C. It would seem that a migration of large numbers of *L. sticticalis* had taken place towards the Kola Peninsula. The warm air mass moved temporarily towards southern Finland, and on August 8th and 9th two specimens of *L. sticticalis* were recorded in SE Finland and on August 10th two more specimens in the south and southwest. According to the surface maps, warm air did not spread from the Kola Peninsula to Finnish Lapland. But at 1500 m a narrow tongue of it was pushed over Lapland. When the moth was caught on the Fjeld Harroaivi at a height of some 700 m, the temperature was 10 °C, but at 1500 m it was 10° warmer. This single catch is likely to be a representative of a larger number of moths which flew westwards from the Kola Peninsula higher up with the warm air. Such flights seem to be typical for *L. sticticalis*, since I have observed this species three times at heights of over 2000 m, sometimes in considerable numbers, at Vitoscha Mountain in Bulgaria and on the Altai and Khamar-Daban Mountains in Siberia.

The observation on *L. sticticalis* is paralleled by that made during the *S. exigua* migration in 1964 (Sect. 2.2.3). In the latter case, whilst the van of the main swarm was still in Finland, some 400 km to the northeast, a few specimens were encountered on the southeastern coast of Sweden. This supports the proposition that during its early phase of migration, the swarm would have continued to advance at a higher altitude over the southeastern border of Finland (Mikkola 1970). A recent publication on *S. exigua* (Aarvik 1981) records a capture from the same migration, on August 15, 1964, from southern Norway, a week after the swarm passed Finland. As with *C. cardui* and *P. daplidice* it would seem that above-country and cross-country migrations occur in *S. exigua* also.

2.2.5 Wind-Assisted Migrations of "Non-Migrant" Species

In my meteorological analysis of the migration data for the years 1946–1966 (Mikkola 1967), I listed migrations of species such as *Amphipyra perflua* (Fabr.) and *Catocala adultera* Men. which are not considered to be migrants, but have well-documented instances of migratory activity. Since then more information on several other so-called non-migrant species has been collected. One such example is a visual observation made on October 6, 1980 at 10.00 h at the bird observatory of Lågskär, to the south of Åland Islands. Starlings were seen capturing buff-coloured moths that were drifting in NNW-bound wind blowing about 40 km h^{-1}. Two specimens were collected and were identified as male *Colotois pennaria* (L.) (Geometridae), the females of which are known to fly only seldom. The source of these moths was back-tracked to the inner parts of southern Latvia, south of Stretpils, 250 km away. The moths had been over the Baltic Sea at dawn and continued their flight in the daytime, which is highly abnormal for this species. It is known that *C. pennaria* occurs regularly on a narrow strip on the southern coast of Finland, and it seems that several finds from more northern areas up to the towns of Vaasa and Iisalmi (Northern Savonia), 300–400 km to the north, were windborne specimens.

In late September and early October 1981, at the same observatory, 12 specimens of *Erannis defoliaria* (Cl.) (Geometridae) were collected in flight during the day. All were drifting with the wind to the NW and had crossed the Baltic Sea. The female of this nocturnal species is brachypterous. A similar species is *Phigalia pilosaria* (D. & S.), a spring-flying moth, the female of which is also brachypterous. This latter species regularly occurs in Finland only on the Åland Islands, but windborne males have been caught on ten occasions in southern Finland, always associated with warm southern air flows.

The examples cited above show that wind-associated migrations occur in a wide range of species which are not classified as migrants. These migrations are sporadic, and have been observed rarely on fire ships and at bird observatories, well outside their areas of regular occurrences. As it is unlikely that the males of *C. pennaria* or *Ph. pilosaria* would migrate in copula (*Erannis* males might do), the adaptive value of these migrations can hardly be anything other than the promotion of gene flow.

2.3 Downwind Migrations Cited in the Literature

The most well-known midwind migrant is the desert locust, *Schistocerca gregaria* (Forsk.) swarms of which reach the Intertropical Convergence Zone and the rains, by advancing with the wind (Rainey 1963), although Baker (1978) has been critical of this view. A similar continuously migrating windborne insect is the noctuid moth *Spodoptera exempta* (Wlk.) (Brown et al. 1969; Rose and Khasimuddin 1979; Tucker et al. 1982). According to the results of mark-recapture experiments carried out in China, *Mythimna separata* (Wlk.) and several other pests perform long-range migrations which proceed early in the season northwards with the prevailing wind (Li et al. 1964). *Heliothis zea* (Boddie) arrived in Texas and Arkansas with air currents from Mexico (Hartstack et al. 1982). Taylor et al. (1973) observed that the nighttime flight of *Autographa gamma* (L.) is predominantly downwind, but the daytime flight is not. In Kenya, the majority of noctuids, among them several migrant species, showed downwind displacement (Brown 1970). Mikkola (1968), analysing the data of the immigrant pyralid moth *Nomophila noctuella* (D. and S.) in Greenland, came to the conclusion that the moths had their origin in the southeastern parts of the United States. Later Munroe (1973) described this American taxon as a distinct species *Nomophila nearctica*, and Wolff (1975) determined the specimens from Greenland as *N. nearctica*, thus confirming the conclusions based on meteorological analysis of data. Kettlewell and Heard (1961) trapped a specimen of *N. noctuella* in England on March 10, 1960 which was found to be contaminated with a single radioactive particle, 9 μ in diameter, that originated in the French nuclear bomb test on February 13, in southern Algeria. The radioactive cloud from the test moved east and did not reach Europe until February 26–29, after the first global circuit, as suggested by meteorological data. As it appears that the radioactive particle was picked up by the moth in Africa and not in Europe, because it is unlikely that such a particle would have remained airborne for that long, it seems most likely that this moth was windborne over the 2000 km distance.

Of day-flying insects, heteropteran species of *Eurygaster* and *Aelia* migrate predominantly downwind (Brown 1965). The coleopteran *Hylobius abietis* (L.) takes off against the wind, but subsequent flight is with the wind (Solbreck 1980). The first flight of the neuropteran *Chrysoperla carnea* (Steph.) also occurs downwind (Duelli 1980). Roer (1968) considered the movements of marked *Aglais urticae* (L.) and other butterflies as downwind displacement, but Baker (1978) did not accept this result. Several small insects like aphids and microlepidopterans show regular downwind migrations (Johnson 1969; Delucchi and Baltensweiler 1979; Fisher and Greenbank 1979; Pedgley 1982).

3 Upwind Migrations

Kennedy (1940) has shown that flight against the wind is a compensatory reaction, a basic optomotor phenomenon in insects. The origin of upwind migrations may lie in the fact that such migrations accord well with the neurophysiology of insects. Nevertheless, upwind flight will be impossible in situations where wind speeds exceed flight speed. Upwind migrations are unlikely to occur in very small insects as even light winds often exceed their flight speeds. The well-understood situations in which insects fly upwind, when following the paths of pheromones or the odour of host plants or food, are not migratory and are not considered here.

3.1 Spring Migrations of Wasps and Bumble Bees

I have been observing the regular spring migrations of wasps (*Vespula* spp.) and bumble bees (*Bombus* spp.) on the Finnish southeast since 1975. These migrations invariably occurred in an upwind direction under definable wind conditions. The coast acts as a leading line (a continuous landmark) so that these hymenopters fly along the coast to ENE against winds from southeast and east and to the WSW against winds from the southwest (Mikkola 1978). On most occasions, the numbers flying in other directions did not exceed 2% of those migrating (cf. Baker 1978, p. 448). The greatest numbers flying were 23000 h^{-1} on a 800 m front for wasps and 900 h^{-1} on a 150 m front for bumble bees. The most numerous species were: *V. rufa* (L.) and *V. saxonica* (F.), and *B. lucorum* (L.) and *B. lapidarius* (L.), all the specimens examined being queens, mostly with undeveloped or slightly matured ovaries.

It was observed that when a willow *(Salix)* twig with blossoms was placed on a sandy beach, the immediate response of the migrating bumble bees was to stop and feed. Thus, there is a parallel between this migration and that of the cross-country migration of Lepidoptera, e.g. *Pieris rapae* (Sect. 2.2.3). It has also been observed that the wasps and the bumble bees may orientate upwind over the open sea (Mikkola 1984). The bumble bees were able to cross the Gulf of Finland, flying from Finland to Estonia. The first wasps were seen about 20 km from the Estonian coast after having flown about 50 km over the sea. The movement of the wasps as well as the bumble bees out to sea shows that the model developed to explain the presence of wasps in so-called insect drifts (Fig. 8 of Mikkola 1978) is applicable. Deterioration of the weather conditions may cause the wasps to drop on to the sea and subsequently to be returned by wave action to the original shore (Mikkola 1978).

There is one other record of a *mass* migration of wasps of the genus *Vespula (sensu lato)*. This is by Rudebeek (1965) in Sweden, but the dropping of the wind did not enable the determination of the upwind nature on this occasion. Beall (1942) observed, however, that the migration of *Polistes fuscatus* var *pallipes* LeP. in Canada is upwind. Also, of several observations on individual migrations of bumble bees in Great Britain, two occurred during the spring along the coastline upwind (see Mikkola 1978).

3.2 Coleoptera

Palmén (1944), studying Coleoptera in insect drifts, and particularly species floating from the Baltic countries to Finland, acknowledged that some of them may have been of local origin. My own collection of material as well as visual observations on individuals on upwind migrations indicate that many of them were actively migrating from local sources.

3.3 Odonata and Diptera

Together with the hymenopterans, individuals of Odonata and Syrphidae were observed to migrate upwind and along the Finnish southern coast (Mikkola 1978) and large-scale migrations of *Libellula quadrimaculata* L. have been recorded several times previously along the Finnish coastline. These migrations were identical to those of Hymenoptera, and occurred mostly along the southern coast against eastern or southeastern winds, but in one instance to the south against southern wind and on another occasion to the southwest against southwestern wind.

As noted by Fraenkel (1932), of the 23 migrations of Odonata known to him, 15 proceeded upwind. A more recent account of such a migration is that by Kumerloeve (1970) when nine species of Odonata were migrating upwind in September 1969 along the coast of Montenegro, Yugoslavia. Unfortunately, Dumont and Hinnekint (1973) in their thorough review on mass migration of Odonata did not pay attention to the question of wind direction, though the optical effects were covered. It would seem that wind direction has a primary effect and landmarks that provide a leading line only a secondary one, though the latter may obscure the former. But pure landmark-oriented migrations do certainly occur (Kiauta 1964; Dumont and Hinnekint 1973).

The best documented reports on the migration of Syrphidae again show the same pattern: upwind orientation, often modified by the leading-line effects of coast (e.g. Owen 1956; Jeekel and Overbeek 1968; Sutton 1969; Burmann 1978). On the isolated island Jussarö, off the southern coast of Finland, I observed a massive migration of

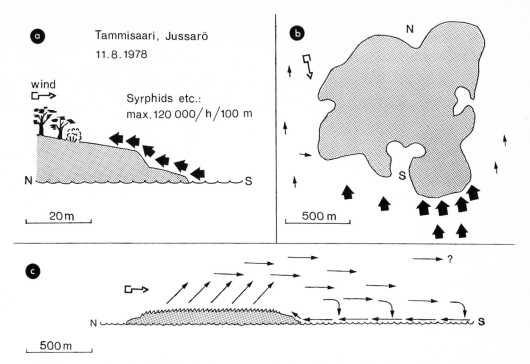

Fig. 5a–c. A migration of Diptera in the outer archipelago of southern Finland. **a** From the side; **b** from above; **c** an interpretation of the whole

Diptera on August 11, 1978, the mechanisms of which are relevant (cf. Fig. 5). Low over the southern cliffs of the island, a steady stream of flies were arriving from the sea to the south. A sample showed that the most numerous flies were syrphids of the species *Syrphus ribesii* (L.), *Episyrphus balteatus* (DeG.), and *Platyceirus* sp., but also flies belonging to other families were participating. Counts over a strip 100 m wide showed that about 120000 flies were arriving per hour. The wind was from the NNW at a force of 4 (Beauf scale; about 7 m s^{-1}) and clearly too strong for any sustained upwind flight over the cool sea. Observations made on a boat showed that a few of these insects were flying to the east and to the west of the island also (Fig. 5b). It is postulated that the flies departed from the island, were carried by the wind towards the open sea, and on descending because of the cooler conditions, made visual contact with the wave pattern, and returned to the island flying upwind (Fig. 5c). As the movement of the flies was seen several hundred metres to the south of the islands and, as most of the flies moving to the east and the west of the island were also flying up wind, it is unlikely that there was any visual orientation with respect to the island itself.

3.4 Lepidoptera

During the migration of Syrphids referred to above (Fig. 5), individual Large Whites *(Pieris brassicae)* were seen drifting with the NNW wind, towards the sea. The interesting point about this movement is that they were trying to head northwards upwind, towards the land, but for long periods were being forced towards the sea by the wind, giving the impression that they were flying backwards. That Large Whites do clearly head upwind was established during observations, made with the aid of a fast motor boat, on July 20, 1975 (Fig. 3b). The butterflies were crossing a 7 km strip at the rate of 7000 individuals per hour. The sea-breeze effect changed the wind direction around the peninsula and the direction of the butterfly movement also changed accordingly. Nevertheless, there was a distinct difference between the main direction of migration and the wind direction, possibly caused by the visual cues provided by the islands. After the downwind flights during the spring migration, the orientation of the Large Whites changes. During the first few days in May 1975 they flew roughly downwind, i.e. in the opposite direction to that of the wasps and the bumble bees (Sect. 3.1) which were heading east, but subsequently the flight direction was reversed.

Observations on large-scale migrations of *Cynthia cardui* (L.) and *Autographa gamma* (L.) by Palmén (1946) in the same locality showed that these moths also followed exactly the same route as that of the wasps and the bumble bees, i.e. along the coast to the east and against the weak southeastern wind. Williams (1958, 1970) notes that for many species of Lepidoptera, including *Autographa gamma*, a good proportion of the visually observed migrations is often downwind or upwind at an angle to the wind; his compilation of 400 migrations of *V. cardui* includes 123 examples downwind, 73 upwind, and 81 across-wind, with the rest of the examples in intermediate directions. Williams (1970, p. 172) also notes that there is much bias in many of the above observations and there is "probably a biassed observational error, as observers are more likely to take notice of cases where the winds are with the flight ("obviously the cause"), or against it ("how curious"), than merely across". As discussed above, upwind orientation may be difficult to observe because of the leading-line effect. One other notable example of an upwind migration is that of the hesperiid *Parnara guttata* Brem. and Grey in Japan in 1970 (Miyashita 1970).

3.5 Formicidae

Flights of ants are usually classified as swarming, rather than as migrations. But undoubtedly these flights lead to dispersal of populations and re-location of their genetic material. Swarming probably does not lead to migrations of the same scale as those of the foregoing examples; the distances travelled during *Vespula* and *Bombus* migrations seem to range from several tens to a few hundred kilometres, whereas the range of the swarming flights of Formicidae appear to be at most a few kilometres. According to Dr. R. Rosengren (oral comm.) males and females of the *Formica rufa* group tend to fly downwind. The following observations show that in *Camponotus* the orientation is mainly upwind. I made the first observations on the swarming of *Camponotus herculeanus* (L.) at Tammisaari, SW Finland, on June 23, 1979 at 1400–1700 h. The wind direction was SW at a force of 2 (Beauf. Scale; 2.3 m s^{-1}) and the temperature was 24.3 °C. Ants, including a pair in copula, flew predominantly towards the SW, with frequent turns. On June 4, 1984, the wind was from the SSE, at a force of 2 (Beauf. Scale; 2.3 m s^{-1}) and the temperature was 23.3 °C. Data on the swarming of *C. her-*

culeanus were collected by two persons, the observer facing downwind, using vertically-directed 10 × 40 binoculars, the other keeping a notebook. The wind direction was determined at a height of 20–30 m, the height at which the ants flew, with the aid of windblown seeds used as markers. The flight directions were determined from the visual field of the binoculars using watch-dial distribution, 12 being upwind and 6 being downwind. The time of observation was 1830 h (GMT + 2 h + 1 h summertime). Numbers of males flying during five periods, of 1 min each were recorded with respect to the wind direction. Of 123 individuals recorded 83 flew upwind and 7 downwind. The numbers of females were too low to make reliable conclusions.

4 Migrations Independent of Wind Direction

As indicated previously (Sect. 3) at least some migrations of Odonata may proceed along rivers and other leading lines, probably independent of the wind direction (see Dumont and Hinnekint 1973). There are several examples, however, where it is claimed that the direction of migration is compass-orientated. The monarch butterfly, *Danaus plexippus* (L.) and several other Lepidoptera are said to have compass orientation in their seasonal migrations (see Baker 1978; Gibo 1984; Johnson 1969; Williams 1970). It has also been suggested that wasps may resort to compass orientation when wind calms down (Mikkola 1978). Compass orientation may have a considerable significance during certain migrations (Baker 1978), but this may not necessarily occur in the majority of such migrations. There would be variations caused by the geographical location; for instance, in Finland, migrations seem to be predominantly downwind, and compass migration, perhaps more frequent in some species in the south, may not occur regularly at the northern latitudes.

There are also records of seasonal migrations, the directions of which are independent of physical environmental factors (e.g. Beebe 1949; Gatter 1981; Lack and Lack 1951; Pruess and Pruess 1971; Sutton 1966).

5 Discussion

It is clear that wind plays a significant role in insect migrations. Downwind as well as upwind migration ensures that the track of the migrants are kept straight, since wind fields are relatively steady. This ensures movement away from the current breeding areas and minimizes the risk of the migrants returning thereto. A return would occur if the migrating insects were randomly oriented with respect to direction. With regard to downwind migrations, the carrying power of the aircurrents makes long-range migrations and the occupation of wider geographical areas feasible. For very minute insects downwind flights may be essential even for changes of biotopes in restricted areas as well.

The importance of the direction of the migratory flights is seen from the fact that upwind migrants, which do not have the advantage of the carrying power of the wind and must compensate for any displacement caused by the wind, nevertheless fly upwind. This is seen in the extreme form in the wasps *(Vespula)* and bumble bees *(Bom-*

bus). They show regular, large-scale migrations against the wind, and no data exist on downwind flights. In my view, upwind migrations are much more common than is known, for they are probably diffuse and difficult to be spotted on homogeneous terrains and occur in more conspicuous form along leading lines, especially on coasts. Upwind flights occur only when there is visual contact with the ground or water below and, therefore, occur invariably during the daytime and at low levels. As described earlier both downwind and upwind flights may take place during a single migration, the former at higher and the latter at lower altitudes, as in *Danaus plexippus* (Kennedy 1951).

It would seem that even seasonal migrants may use wind for orientation during their migrations. The Large White butterflies, *Pieris brassicae*, arrive in the spring flying with the warm southern to southeastern winds, heading downwind, probably with no visual contact with the ground (Sect. 2.2.1); there is considerable evidence that subsequently, in the late summer, the orientation changes so that the butterflies fly upwind. This change must be an integral part of the seasonal migrations of such species, since the warm winds suitable for migratory activity in the northern latitudes are largely uni-directional. The exact cause of the upwind flights, and the significance of sun orientation emphasised by Baker (1978) need further clarification. Clear changes of course occurred in *Spodoptera exigua* during its immigration to Fennoscandia in 1964 (Mikkola and Salmensuu 1965); the main directions of flight were NW and SW, corresponding to air currents, but there was no likely compass-orientation mechanism in this nocturnal flier.

The presently available data suggest that long-range migrations, in many instances, can be divided into two phases: (1) a primary, mostly above-country and resource-independent phase and (2) a secondary, mostly cross-country and resource-dependent phase. The first phase would correspond with the teneral migration as defined by Johnson (1969) which occurs prior to feeding, mating, and egg-laying activities and which usually terminates after a long-range displacement. The second phase would be a more local component of a migration, with feeding, mating, and oviposition activities fitted-in when necessary (cf. Baker 1978). The costs of long-range dispersal and resource finding would thus be temporarily divided. This sounds reasonable, since both could hardly be met effectively at the same time. That these two phases have not been separated previously may explain the many conflicting views about insect migrations, especially those of butterflies. It is not appropriate, however, to try to apply these concepts rigidly to every migration. Thus, it seems that upwind migrations, in general, lack the first phase, and the change from the first phase to the second is not necessarily abrupt, but gradual. Of the known examples, the clearest transition from the first to the second phase occurs during the seasonal spring migrations of butterflies.

References

Aarvik L (1981) The migrant moth *Spodoptera exigua* (Hübner) (Lepidoptera, Noctuidae) recorded in Norway. Fauna Norv Ser B 28:90–92
Asahina S, Turuoka Y (1970) Records of the insect visits to a weather ship located at the Ocean Weather Station "Tango" on the Pacific, V. Insects captured during 1968. Kontyu 38:318–330
Baker RR (1978) The evolutionary ecology of animal migration. Hodder and Stoughton, London, 1012 pp

Beall G (1942) Mass movement of the wasp, *Polistes fuscatus* var. *pallipes* LeP. Can Field Nat 56: 64–67

Beebe W (1949) Insect migration at Rancho Grande, north-central Venezuela. General account. Zoologica (NY) 34:107–110

Brown ES (1965) Notes on the migration and direction of flight of *Eurygaster* and *Aelia* species (Hemiptera, Pentatomoidea) and their possible bearing on invasions of cereal crops. J Anim Ecol 34:93–107

Brown ES (1970) Nocturnal insect flight direction in relation to the wind. Proc R Entomol Soc Lond Ser A Gen Entomol 45:39–43

Brown ES, Betts E, Rainey RC (1969) Seasonal changes in distribution of the African armyworm, *Spodoptera exempta* (Wlk.) (Lep., Noctuidae), with special reference to Eastern Africa. Bull Entomol Res 58:661–728

Burmann K (1952) Wanderfalter auf Gletschern. Z Wien Entomol Ges 37:101–103

Burmann K (1978) Syrphiden-Wanderungen in Gebirge. Beobachtungen aus Nordtirol (Österreich) (Insecta: Diptera, Syrphidae). Ber Naturwiss Med Ver Innsb 65:129–137

Delucchi V, Baltensweiler W (ed) (1979) Dispersal of forest insects: evaluation, theory and management implications. Mitt Schweiz Entomol Ges 52:125–342

Duelli P (1980) Adaptive dispersal and appetitive flight in the green lacewing *Chrysopa carnea*. Ecol Entomol 5:213–220

Dumont HJ, Hinnekint BON (1973) Mass migration in dragonflies, especially in *Libellula quadrimaculata* L.: a review, a new ecological approach and a new hypothesis. Odonatologica (Utr) 2: 1–20

Edwards JS (1972) Arthropod fallout on Alaskan snow. Arct Alp Res 4:167–176

Farrow RA (1975) Offshore migration and the collapse of outbreaks of the Australian plague locust (*Chortoicetes terminifera* Walk.) in South-East Australia. Aust J Zool 23:569–595

Fisher RA, Greenbank DO (1979) A case study of research into insect movement: spruce bud worm in New Brunswick. In: Rabb, Kennedy (1979), see below, pp 220–229

Fox KJ (1978) The transoceanic migration of Lepidoptera to New Zealand – a history and a hypothesis on colonisation. New Zea Entomol 6:368–380

Fraenkel G (1932) Die Wanderungen der Insekten. Ergeb Biol 9:1–238

French RA, White JH (1960) The diamond-back moth outbreak of 1958. Plant Pathol (Lond) 9: 77–84

Gatter W (1981) Insektenwanderungen. Kilda, Greven, 94 pp

Gibo DL (1984) Flight behaviour in migrating butterflies. Abstr 6.3.8, 17th Int Congr Entomol Hamburg

Gressitt JL, Yoshimoto CM (1964) Dispersal of animals in the Pacific. In: Gressitt JL (ed) Pacific basin biogeography. 10th Pacif Sc Congr Hawaii 1961, pp 283–292

Gunn DL, Rainey RC (1979) Strategy and tactics of control of migrant pests. Philos Trans R Soc Lond B Biol Sci 287:245–488

Haeseler V (1974) Aculeate Hymenopteren über Nord- und Ostsee nach Untersuchungen auf Feuerschiffen. Entomol Scand 5:123–136

Hanski I (1971) (The mass migration of *Pontia daplidice* in the summer 1970) (in Finnish) Kymenlaakson Luonto 12:5–9

Hartstack AW, Lopez JD, Muller RA, Sterling WL, King EG, Witz JA, Eversull AC (1982) Evidence of long range migration of *Heliothis zea* (Boddie) into Texas and Arkansas. Southwest Entomol 7:188–201

Hurst GW (1963) Small mottled willow moth in southern England, 1962. Meteorol Mag Lond 92: 308–312

Hurst GW (1964) Meteorological aspects of the migration to Britain of *Laphygma exigua* and certain other moths on specific occasions. Agric Meteorol 1:271–281

Jeekel CAW, Overbeek H (1968) A migratory flight of hover-flies (Diptera, Syrphidae) observed in Austria. Beaufortia 15:123–126

Johnson CG (1969) Migration and dispersal of insects by flight. Methuen, London, 763 pp

Johnson CG (1971) Comments at a meeting. Proc R Entomol Soc Lond 36:33–36

Johnson CG, Taylor LR, Southwood TRE (1962) High altitude migration of *Oscinella frit* L. (Diptera:Chloropidae). J Anim Ecol 31:373–383

Karvonen J, Karvonen E (1983) *Pontia chloridice* in Finland (Lepidoptera, Pieridae). Notulae Entomol 63:67–68

Kennedy JS (1940) The visual responses of flying mosquitoes. Proc Zool Soc Lond 109:221–242

Kennedy JS (1951) The migration of the desert locust (*Schistocerca gregaria* Forsk.) I. The behaviour of swarms. II. A theory of longrange migration. Philos Trans R Soc Lond B Biol Sci 235:163–290

Kennedy JS (1961) A turning point in the study of insect migration. Nature 189:785–791

Kettlewell HBD, Heard MJ (1961) Accidental radioactive labelling of a migrating moth. Nature 189:676–677

Keynäs K (1971) The migration of *Pontia daplidice* into Finland in the summer 1970. Circ Finn Lep Soc 12:5

Kiauta B (1964) Over een trekvlucht van *Libellula quadrimaculata* L. (Odonata-Libellulidae). Levende Nat 67:59–63

Kumerloeve H (1970) Massenzug von Libellen im montenegrischen Küstenland im Spätsommer 1969

Lack D, Lack E (1951) Migration of insects and birds through a Pyrenean pass. J Anim Ecol 20: 63–67

Lempke BJ (1962) Insects captured on the light-ship "Noord-Hinder". Entomol Ber (Amst) 22: 101–111

Li KP, Wong HH, Woo WS (1964) Route of the seasonal migration of the oriental armyworm moth in the eastern part of China as indicated by a three-year result of releasing and recapturing of marked moths. Acta Phytophyl Sin 3:101–110

Lokki J, Malmström KK, Suomalainen E (1978) Migration of *Vanessa cardui* and *Plutella xylostella* (Lepidoptera) to Spitsbergen in the summer 1978. Notulae Entomol 58:121–123

Mikkola K (1967) Immigrations of Lepidoptera, recorded in Finland in the years 1946–1966, in relation to aircurrents. Ann Zool Fenn 2:124–139

Mikkola K (1968) Transportation by air of *Nomophila noctuella* Schiff. (Lep., Pyralididae) to Greenland from North America in 1954. Entomol Medd 36:509–510

Mikkola K (1970) The interpretation of long-range migrations of *Spodoptera exigua* Hb. (Lepidoptera: Noctuidae). J Anim Ecol 39:593–598

Mikkola K (1971) The migratory habit of *Lymantria dispar* (Lep.: Lymantriidae) adults of continental Eurasia in the light of a flight to Finland. Acta Entomol Fenn 28:107–120

Mikkola K (1978) Spring migration of wasps and bumble bees on the southern coast of Finland. Ann Entomol Fenn 44:10–26

Mikkola K (1984) Migration of wasp and bumble bee queens across the Gulf of Finland (Hymenoptera: Vespidae and Apidae). Notulae Entomol 64:125–128

Mikkola K, Salmensuu P (1965) Migration of *Laphygma exigua* Hb. (Lep., Noctuidae) in northwestern Europe in 1964. Ann Zool Fenn 2:124–139

Miyashita K (1970) An observation on the migration of *Parnara guttata* Bremer et Grey (Lepidoptera: Hesperiidae) in 1970. Jpn J Ecol 23:251–254

Munroe E (1973) A supposedly Cosmopolitan insect: the celery webworm and allies, genus *Nomophila* Hübner (Lepidoptera: Pyralidae: Pyraustinae). Can Entomol 105:177–216

Oku T (1983) (Annual and geographical distribution of crop infestation in northern Japan by the Oriental armyworm in special relation to the migration phenomena). Misc Publ Tohoku Natl Agric Exp Stn 3:1–49

Oku T, Koyama J (1976) Long-range migration as a possible factor caused the late-summer outbreak of the Oriental armyworm, *Mythimna separata* Walker, in Tohoku district, 1969. Jpn J Appl Entomol Zool 20:184–190

Owen DF (1956) A migration of insects at Spurn Point, Yorkshire. Entomol Mon Mag 92:43–44

Palmén E (1944) Die anemohydrochore Ausbreitung der Insekten als zoogeographischer Faktor. Ann Zool Soc Zool Bot Fenn Vanamo 10:1–262

Palmén E (1946) Ein auffallender Massenflug von *Phytometra gamma* L. und *Pyrameis cardui* L. (Lep.) in Südfinnland. Ann Entomol Fenn 12:122–131

Pedgley DE (1982) Windborne pests and diseases: meteorology of airborne organisms. Horwood, Chichester, 250 pp

Pienkowski RL, Medler JT (1964) Synoptic weather conditions associated with long-range movement of the potato leafhopper, *Empoasca fabae*, into Wisconsin. Ann Entomol Soc Am 57: 588–591

Pruess KP, Pruess NC (1971) Telescopic observation of the moon as a means for observing migration of the army cutworm, *Chorizagrotis auxiliaris* (Lepidoptera:Noctuidae). Ecology 52:999–1007

Rabb RL, Kennedy GG (ed) (1979) Movement of highly mobile insects: concepts and methodology in research. North Carolina State University Press, Raleigh, NC

Rainey RC (1963) Meteorology and the migration of desert locusts. Tech Notes Wild Met Org 54:1–115

Roer H (1968) Weitere Untersuchungen über die Auswirkungen der Witterung auf Richtung und Distanz der Flüge des kleinen Fuchses (*Aglais urticae* L.) (Lep. Nymphalidae) im Rheinland. Decheniana 120:313–334

Rose DJW, Khasimuddin S (1979) Wide-area monitoring of the African armyworm, *Spodoptera exempta* (Walker) (Lepidoptera: Noctuidae). In: Rabb, Kennedy, see above, pp 212–219

Rudebeck G (1965) On a migratory movement of wasps, mainly *Vespula rufa* (L.), at Falsterbo, Sweden. Proc R Entomol Soc Lond Ser A Gen Entomol 40:1–8

Shaw MW (1962) The diamond-back moth migration of 1958. Weather 17:221–234

Snow DW, Ross KFA (1952) Insect migration in the Pyrenees. Entomol Mon Mag 88:1–6

Solbreck C (1980) Dispersal distances of migrating pin weevils *Hylobius abietis*, Coleoptera: Curculionidae. Entomol Exp Appl 28:123–131

Sutton SL (1966) South Caspian Insect Fauna, 1961 II. Migration, status and distribution of certain insect species in Northern Persia. Trans R Entomol Soc Lond 118:51–72

Sutton SL (1969) A migration of Syrphid flies at Spurn. Naturalist (Leeds) 909:51–53

Taylor LR (1958) Aphid dispersal and diurnal periodicity. Proc Linn Soc Lond 169:67–73

Taylor LR (1960) The distribution of insects at low levels in the air. J Anim Ecol 29:45–63

Taylor LR (1974) Insect migration, flight periodicity and the boundary layer. J Anim Ecol 3:225–238

Taylor LR, French RA, Macaulay EDM (1973) Low-altitude migration and diurnal flight periodicity; the importance of *Plusia gamma* L. (Lepidoptera: Plusiidae). J Anim Ecol 42:751–760

Tucker MR, Mwandoto S, Pedgley DE (1982) Further evidence for windborne movement of armyworm moths, *Spodoptera exempta*, in East Africa. Ecol Entomol 7:463–473

Vepsäläinen K (1968) The immigration of *Pieris brassicae* L. (Lep., Pieridae) into Finland in 1966, with a general discussion of insect migration. Ann Entomol Fenn 34:223–243

Williams CB (1958) Insect migration. Collins, London

Williams CB (1970) The migrations of the painted lady butterfly, *Vanessa cardui* (Nymphalidae), with special reference to North America. J Lepid Soc 24:157–175

Wolff NL (1975) *Nomophila*-arterne (Lepidoptera, Pyralidae) fra det nordatlantiske område. Entomologiske Meddelelser 43:129–135

12 Flight Strategies of Migrating Monarch Butterflies (*Danaus plexippus* L.) in Southern Ontario

D. L. GIBO[1]

1 Introduction

Most migrating butterflies fly relatively close to the ground, presumably to avoid the full force of the wind and to better control their courses (Baker 1978; Walker 1980; Walker and Riordan 1981; Walker TJ 1985). In North America, the monarch butterfly, *Danaus plexippus* L., is a well-known exception. During the late summer and fall migration, individuals of this species regularly fly high above the ground (Beall 1941a,b; Gibo and Pallett 1979; Gibo 1981; Schmidt-Koenig 1984; Lugger 1980; Urquhart 1960) and often proceed by soaring flight (Gibo 1981; Gibo and Pallett 1979; Schmidt-Koenig 1979; Schmidt-Koenig 1985). The migrating *D. plexippus* can reach surprising altitudes, presumably by circling in thermals. Glider pilots have observed the butterflies flying more than 1 km above the ground (Gibo 1981). In addition to being the only North American butterfly known to migrate at high altitudes, *D. plexippus* also accomplish a remarkable feat of navigation. Each fall most of the eastern North American population make their way to a small group of overwintering sites in the mountains of central Mexico (Brower 1977; Calvert et al. 1979; Brower 1984; Urquhart and Urquhart 1976, 1977, 1978).

It is not clear how migrating *D. plexippus* accomplish their annual, long-distance migration. Because soaring flight is the most energy-efficient method of flight possible (Pennycuick 1975), particularly for insects, where the difference between resting metabolism and flight metabolism is so great (Zebe 1954), the butterflies have the potential to travel great distances with little effort (Gibo and Pallett 1979). However, soaring cross-country presents unique problems in navigation. Not only must the soaring butterflies spend much of their flight time at relatively high altitudes fully exposed to the wind drift, but they must also spend a great deal of time soaring in circles in thermals in order to gain, or maintain altitude. Since thermals drift downwind, any *D. plexippus* circling in them will also drift downwind. Even when the butterflies are flapping or gliding between thermals, their relatively slow airspeeds, approximately 5 m s^{-1} for powered, cruising flight (Urquart 1960), and 2.6 to 3.6 m s^{-1} for gliding and soaring flight (Gibo and Pallett 1979), restrict their ability to compensate for crosswinds or make significant progress in most headwind conditions. The fact that *D. plexippus* are occasionally found hundreds of kilometres offshore over the Atlantic during the late

[1] Department of Zoology, Erindale College, University of Toronto, Mississauga, Ontario, L5L 1C6, Canada.

Insect Flight: Dispersal and Migration
Edited by W. Danthanarayana
© Springer-Verlag Berlin Heidelberg 1986

summer (Urquhart and Urquhart 1979b) is strong evidence that high-altitude, soaring flight entails serious risks.

Previous studies of the late summer migration of *D. plexippus* have given a contradictory picture of the effectiveness of the butterfly's directional control. Most studies indicated that in eastern North America *D. plexippus* migrated in a wide range of wind conditions and their directional control was relatively poor (Beall 1941a,b; Kanz 1977; Urquhart 1960; Urquhart and Urquhart 1977, 1978, 1979a). However, it was generally agreed that the butterflies tended to fly SW more often than in other directions. In contrast, Schmidt-Koenig (1979, 1985) found that migrating *D. plexippus* had fairly precise control of their direction. For example, in the first study, he investigated five locations in eastern and southeastern North America and he found that the mean bearings of the migrants at each location had relatively little scatter and could not be distinguished from the local Great Circle routes to the overwintering sites. He also reported that *D. plexippus* could be seen attempting to compensate for crosswinds by advancing (rotating) their headings. Appropriate use of this manoeuvre, termed crabbing because the animal appears to be flying sideways, could allow the butterflies to maintain a particular course in a wide range of crosswind conditions. Finally, an extensive mark, release and recapture program combined with field studies, indicated that the general movement of many of the *D. plexippus* leaving the southern Ontario region of Canada was often SE towards the east coast or S towards the Gulf states. The butterflies then flew W, finally entering Mexico along a broad front extending from Brownsville, Texas, NW to the Big Bend area (Urquhart 1960; Urquhart and Urquhart 1977, 1978).

In summary, the literature describes *D. plexippus* as a butterfly that migrates in a wide range of wind conditions, frequently flies well above the ground, fully exposed to wind drift, often proceeding by soaring in circles while drifting downwind, and, nevertheless, performs an amazing feat of long-distance migration. In addition, the butterfly has either poor directional control and a general preference to fly in a SW direction, or has good directional control and a strong preference for the shortest route to the overwintering sites. In an attempt to reconcile these views and to determine the flight strategies of migrating *D. plexippus* in southern Ontario, Canada, a 3 year field study was carried out in Mississauga, Ontario.

2 Methods

The flight behaviours of migrating *D. plexippus* were observed in open habitats in Mississauga, Peel County, Ontario, Canada in 1978, 1979, and 1981. Migrating *D. plexippus* were observed each year from late August until mid- to late September. Individual observation periods ranged from 15 min to over 2 h. Most observations were made between 10:00 AM (EST) and 3:00 (EST), corresponding to the daily peak in migratory activity (Urquhart 1960; Gibo and Pallett 1979). Field equipment consisted of 7 × 35 field glasses, compasses, wind-speed indicators, thermometers, notebooks, and tape recorders.

Observations were restricted to *D. plexippus* that passed through an imaginary 'window' extending approximately 30 m to either side of the observer and upward to a maximum of approximately 300 m, the limit at which the butterflies were visible through field glasses. During each observation period the sky was scanned with the field glasses every few minutes to detect high-flying individuals. For this study, *D. plexippus* that flew past and continued on until lost to sight were classified as migrants. In contrast, individuals that flew past and then paused to investigate, or alight on flowers, bushes, trees or any structure, were classified as foragers. Analysis of the direction of flight of *D. plexippus* classified as foragers showed that they tended to fly upwind. For example, in 1978, 34 butterflies were classified as foragers. A Rayleigh test showed that the mean direction of the foragers did not differ from random, with n = 34, mean angle = 94° ± 71°, r = 0.23, and $P > 0.05$. A Rayleigh test (Batchelet 1981) also showed that the wind directions for the sample did not differ significantly from random, with n = 34, mean angle = 46° ± 65°, r = 0.28 and $P > 0.05$. Examination of the data showed that 24 of the 34 were flying within 20° of the upwind direction. Consequently, individuals classified as foragers were not included in further analysis.

Vanishing bearings were recorded for each migrating *D. plexippus*. Any approaching butterflies that were flying within 10 m of the ground were usually detected 50 m or more away. As a result, it was often possible for the observer to move left or right as appropriate, to reduce parallax error. If the butterflies were flying in a straight line, the observer also attempted to determine the bearings of their headings, that is, the direction of alignment of their bodies. Two methods were used to determine headings. If the migrating *D. plexippus* flew past within 1.5 m of the ground, the observer would attempt to sight along its body, in the direction it was 'facing', to any conspicuous object in the distance, such as a tree. The observer would then continue to track the migrant until it vanished. After the vanishing bearing had been recorded, the observer would then take the bearing for the object. The second method was used for migrants that flew overhead and within approximately 100 m of the ground. As the *D. plexippus* flew overhead, the observer would glance in the direction the butterfly was 'facing' and pick out any conspicuous object. The observer would then continue to track the migrant until it vanished. As above, the bearing for heading was obtained after the vanishing bearing. When the observer was using field glasses, the second method required some practice to be able to relocate the butterflies after glancing away. Finally, headings for *D. plexippus* that were obviously rotated to the left or right, but flew by either too far to one side or too fast to obtain a good measurement, were simply recorded as rotated left or right of their course. Standard Polar statistical methods (Batchelet 1981) were used to calculate mean angle and mean vectors for the flight directions and headings.

The flight altitude above the ground was estimated for each migrant. Under ideal viewing conditions the field glasses made it possible to distinguish *D. plexippus* flying as high as approximately 300 m. However, on most days haze, bright back lighting from the sky or reduced light due to overcast conditions, brought the viewing ceiling down to approximately 150 to 200 m.

The flight of migrants was classified as either flapping, soaring or gliding. Individuals in flapping flight often proceeded by a flight mode that has been termed cruising flight (Urquhart 1960), that is, bouts of flapping alternating with brief glides. Butterflies in simple gliding flight descended while holding their wings outspread. Since the glide ratio of *D. plexippus* is in the range of 3:1 to 4:1 (Gibo and Pallett 1979), the descent was moderately steep. An individual was classified as soaring if it was gliding, that is, not flapping its wings, while either maintaining or gaining altitude.

The flight was further classified as either straight, circling or adjusting. Migrants in straight flight continued directly away from the observer until lost to sight. Circling flight was associated with soaring in thermals and the circles were generally 1–2 m in diameter. It should be emphasized that soaring flight did not require that the butterflies fly in circles; migrating *D. plexippus* often soared directly across a thermal. Flight was classified as adjusting when the *D. plexippus* continuously changed their headings, perhaps in response to variable crosswinds, or after entering weak or disorganized thermals. The ground tracks of adjusting butterflies ranged from nearly straight, to scalloped, to S turns.

The preferred direction was defined in this study as the mean direction adopted by migrating *D. plexippus* that were not experiencing significant wind drift. This direction was calculated by two different methods. Preferred direction A was the mean direction of all individuals that were flying straight and had their courses and headings aligned. As these butterflies were neither drifting nor compensating, it was assumed that they were in complete control of their direction of flight. Preferred direction B was the mean direction of all individuals flying in a straight line and flying within 3 m of the ground, irrespective of their headings. In this case it was assumed that low-flying individuals also had complete control of their courses.

The weather was recorded at the beginning of each observation period and whenever conditions changed during the observation period. The wind direction and wind velocity at 1.5 m, air temperature at 1 m, indications of thermals, type of cloud cover and amount of cloud cover were all noted. The wind direction at the altitudes of high-flying butterflies was determined, when possible, by noting the direction of drift of circling individuals. Otherwise, the wind directions aloft were assumed to be approximately the same as at 1.5 m. Thermal activity was indicated by any of the following: soaring butterflies, soaring birds, thistledown drifting upward, cumulus clouds, and frequent wind shifts. Variable winds were a good indicator of thermal activity because, as thermals drifted through or near the observation site, the complex air circulation of these convection cells usually caused temporary and erratic shifts in wind direction and velocity.

The Great Circle Route to the overwintering site was calculated using a commercially available hand calculator programmed for aviation calculations. The data entered for the calculations were the latitude and longitude of both Mississauga, Ontario, Canada, and the latitude and longitude of the overwintering sites in Central Mexico. The values for Mississauga were lat 43° 32' and long 79° 40'. The values used for the overwintering sites, taken from Brower (1985), for site alpha, were lat 19° 40', long 100° 17'. This gave a local Great Circle Route of 222° 40'.

3 Results

During a total of approximately 2500 min of observations, the flight behaviours for 575 migrating *D. plexippus* were recorded. Most migratory activity was observed between 11:00 A.M. (EST) and 3.00 P.M. (EST). On one occasion, however, a few mi-

Table 1. Mean direction, preferred direction A and preferred direction B of migrating *D. plexippus* in southern Ontario [a]

Group	Mean				
	N	Angle	Direction	A.D.	r [b]
Mean direction	575	222°	SW	47°	0.66 ***
Preferred direction A	24	234°	SW	29°	0.87 ***
Preferred direction B	49	232°	SW	46°	0.68 ***

[a] Abbreviations; A.D. = angular deviation; r = length of mean vector.
[b] Rayleigh test significance levels: *** $P < 0.001$.

grants were seen before 8:00 A.M., and on another occasion, some were seen after 5:30 P.M. Migration occurred in winds from each of the eight major sectors of the compass and with wind velocities, at 1.5 m, ranging from near calm to 6.7 m s^{-1} (24 km h^{-1}). Although most migration occurred when there was at least some sky visible, some migration occurred under complete overcast conditions. Six individuals were even observed migrating during periods of light rain. Most migration occurred when air temperatures were above 15 °C, however, some flights were observed when the air temperature was 11 °C. Migrating *D. plexippus* in southern Ontario clearly tolerate a wide range of weather conditions.

Table 1 shows that in southern Ontario the mean vanishing bearings for all migrants, including those circling in thermals, was 222°. This is almost identical to the bearing for the local Great Circle Route (222° ·40°) to the Site Alpha overwintering area in central Mexico. Table 1 also shows that preferred direction A and preferred direction B differed by only 2°. However, because the 95% confidence limits (Batchelet 1981) for preferred directions A and B were 14° and 16°, respectively, neither could be distinguished from the 222° ·40° Great Circle Route. If the migrants flew the route indicated by either of these two bearings, they would cross from the USA into Mexico in the region of Eagle Pass, Texas, an area well within the broad front of entry delineated by Urquhart and Urquhart (1978) and where large flights of migrants have been reported (Urquhart 1960). In any case, the preferred direction is clearly SW. In this paper, preferred direction B, 232°, the one with the larger sample size, will be used for comparison to other mean directions.

Once the preferred direction was known, the eight major wind bearings could be classified according to their effect on this direction. Accordingly, a NE wind was a pure tailwind, an E wind was a combination of a tailwind and left crosswind, a SE wind was a pure left crosswind, a S wind was a combination of a headwind and left crosswind, a SW wind was a pure headwind, a W wind was a combination of a headwind and a right crosswind, a NW wind was a pure right crosswind, and a N wind was a combination of a tailwind and a right crosswind combination. Table 2 shows that the eight wind conditions were associated with an array of mean directions. Four of the eight wind bearings, NE, E, SE, and S winds, were associated with SW to WSW mean directions. Since three of the four, E, SE, and S winds, were partial or pure left crosswinds, the SW to WSW mean directions of the butterflies flying in these winds indicated that they

Table 2. Flight behaviour of migrating *D. plexippus* for each wind bearing

Wind bearings	Mean direction				Frequency of *D. plexippus*				
					At each altitude range				
	N	Mean angle	A.D.[a]	r[a]	< 3 m	3–29 m	= > 30 m	Soaring	Soaring in circles
NE	64	229° (SW)	26°	0.90 ***[b]	0.109	0.328	0.563	0.734	0.641
E	150	247° (WSW)	32°	0.84 ***	0.087	0.260	0.653	0.693	0.307
SE	131	257° (WSW)	30°	0.86 ***	0.061	0.550	0.389	0.489	0.229
S	15	237° (WSW)	57°	0.51 *	0.533	0.333	0.133	0.200	0.200
SW	35	143° (SE)	68°	0.30 *	0.171	0.343	0.486	0.371	0.086
W	20	128° (SE)	52°	0.59 ***	0.200	0.500	0.300	0.600	0.400
NW	107	153° (SSE)	33°	0.83 ***	0.028	0.327	0.645	0.636	0.411
N	53	185° (S)	26°	0.90 ***	0.038	0.491	0.472	0.717	0.528
Total and weighed mean frequencies	575				0.089	0.383	0.527	0.605	0.352

[a] A.D. = angular deviation; r = length of mean vector.
[b] Rayleigh test significance levels: * $P < 0.05$; *** $P < 0.001$.

were somehow compensating for wind drift. In contrast, winds from the W, NW, and N, were associated with large deviations from the preferred SW direction. As winds from W, NW, and N were either partial or pure right crosswinds to a SW course, the butterflies were apparently experiencing a large amount of wind drift.

The SE mean direction observed for *D. plexippus* migrating in a SW wind strongly suggested that at least some of the butterflies had rotated their headings towards the S, counterclockwise from their preferred SW direction, otherwise they could not have been drifting SE. Examination of the data showed that 23 of the 35 observations took place from 9:59 to 11:29 A.M. on August 27, 1979, a 90 min period with unusual weather conditions. The SW wind was very light, ranging from near calm to $1.4 \, \mathrm{m \, s^{-1}}$ ($5 \, \mathrm{km \, h^{-1}}$) and relatively weak thermals were present. The 23 migrants had a mean course of $128° \pm 60°$ or SE. A Raleigh test showed that the mean direction was significant, with n = 23, r = 0.44, and $P = 0.01$. At the end of the 1.5 h observation period the wind went through a few fluctuations and then abruptly backed to the SE. The migrants observed immediately after the wind shift were flying SW. The remaining 12 individuals flying with a SW wind were observed on different days and had a mean direction of $233° \pm 76°$. However, a Rayleigh test showed that the observed mean direction was not significant, with n = 12, r = 0.11, and $P > 0.85$. Apparently, the *D. plexippus* observed during the morning of August 27 were migrating, while those seen during most periods of SW winds were doing something else.

The altitude observations were grouped into three flight levels, < 3 m, $3-29$ m, and $= > 30$ m. Avoiding unfavourable winds by flying near the ground was not a common strategy. Table 2 shows that for all wind bearings, except S, over 50% of the migrants were flying higher than 3 m above the ground. In N, NE, E, SW, and N winds, approximately 50% to 65% of the *D. plexippus* were flying at least 30 m above the ground. The tendency of migrating *D. plexippus* to fly well above the ground in E, SE, and S winds, was consistent with the view that the butterflies compensate for left-crosswind conditions. Although these high-flying individuals were fully exposed to the wind, the mean direction of each group showed only slight drift to the right (westward). Even when the wind was a pure left crosswind, that is, from the SE, over 90% of the migrants flew well above the ground. Despite this exposure to wind drift, the mean direction for this group, $257°$, only deviated $25°$ from the $232°$ preferred direction. Right crosswinds presented a different picture. The mean directions of the three groups of butterflies flying in right crosswinds, that is, W, NW, and N winds, were shifted S to SE, or $49°$ to $104°$ from the $232°$ preferred direction. The fact that most of the migrating *D. plexippus* observed flying in right-crosswind conditions were also flying at high altitudes was consistent with the view that the butterflies were not compensating. Finally, the altitude grouping provided evidence that SW winds can elicit two different behaviours. None of the *D. plexippus* migrating during the morning of August 27, 1979, were flying less than 3 m above the ground. Six were between 3 and 29 m and 17 were flying at least 30 m above the ground. In contrast, 6 of the 12 non-migrating individuals flying during other periods of SW winds were flying less than 3 m above the ground and six were flying within $3-29$ m of the ground. None were flying at the highest flight level.

Soaring flight was observed for all eight wind bearings. Table 2 shows that soaring was the dominant mode of flight, except when the wind was from the SE, S or SW.

The frequency of total soaring activity and the frequency of soaring by circling in thermals both reached maximum values when the winds were NE, pure tailwinds. As the winds swung clockwise to the E, SE, and S, the frequencies of these behaviours declined. In contrast, the high frequencies of soaring and circling in thermals observed in winds from the N, NW or W, were further evidence that the *D. plexippus* made little or no effort to avoid drift in right crosswinds. Examination of the data for the SW headwind condition showed that the August 27, 1979, group was distinct. Twelve of the 23 observed migrating during the 90 min period of SW winds were soaring. Nine of the 12 were soaring directly across thermals and three were drifting NE while circling in thermals. In contrast, only 1 of the 12 non-migrating *D. plexippus* observed during the other periods of SW winds was soaring. Finally, because a total 202 or 35.2% of all *D. plexippus* observed were drifting downwind, while soaring in circles in thermals, these circling individuals were a major component of the migrating population. Surprisingly, the mean direction for the circling individuals, $222° ± 62°$, was identical to the mean course for the population as a whole. The Rayleigh test showed that the mean direction was significant, with n = 202, r = 0.41, and $P < 0.001$. Apparently, natural selection has adjusted the frequency at which the butterflies circle in thermals for each wind direction to produce a bias towards the SW.

The role of crabbing in crosswind compensation was examined for each wind direction by comparing the mean direction and mean headings of migrating *D. plexippus* observed to be flying in a straight line (i.e. not circling or adjusting). Table 3 shows that when the wind was a NE tailwind, most of the migrants were flying downwind. Although five (24%) of this group were seen to have their headings rotated to the left of their course, their mean rotation was less than $20°$ and the mean heading was only shifted to SSW. Table 3 also shows that migrating *D. plexippus* clearly compensated for left crosswinds, that is, E, SE, and S winds, by crabbing left. When winds swung to the E, the tailwind/left-crosswind condition, the mean direction of the butterflies was SW and 61 (75%) of them were seen to have headings rotated to the left. The amount of rotation to the left of the $232°$ preferred direction that was necessary to compensate for the average, left-crosswind component of an E wind was $53°$, resulting in a mean heading of S. In SE winds, the pure, left-crosswind condition, left rotation was seen in 75 (87%) of the migrants. However, the rotation of the mean heading from the preferred direction was slightly less than in E winds, $43°$, and the mean course was slightly drifted right to WSW. Winds from the S, the headwind/left-crosswind condition, were associated with a SW course. Unfortunately, only one heading measurement was recorded for this wind direction, a left rotation from the preferred direction of $24°$.

Table 3 also shows that in right-crosswind conditions, that is, W, NW, and N winds, migrating *D. plexippus* did not appear to compensate for drift by crabbing. The mean headings of the butterflies, ranging from $230°$ to $244°$, were very close to the preferred $232°$ direction. Since most of the butterflies were flying well above the ground and soaring, the resulting wind drift was considerable. Although most of the butterflies for which headings were observed were seen to be rotated to the right of their flight direction, this was unlikely to be a compensating manoeuver. They were simply holding a SW heading in a right crosswind and being drifted left (southward). However, for some individuals, the situation may have been more complicated than it appeared. Inspection of the data showed that approximately 15% of the migrants flying in right-crosswind

Table 3. Mean direction and mean heading of straight flying, migrating D. plexippus

Wind bearings	Mean direction				Mean heading				Number with heading rotated[a]	
	N	Mean angle	A.D.[b]	r[b]	N	Angle	A.D.	r	Left	Right
NE	21	224° (SW)	37°	0.79 ****c	12	207° (SSW)	46°	0.68 **c	5	–
E	81	235° (SW)	27°	0.89 ***	48	180° (S)	35°	0.81 ***	61	–
SE	86	247° (WSW)	26°	0.90 ***	50	190° (S)	27°	0.89 ***	75	–
S	10	219° (SW)	42°	0.73 ***	1	208° (SSW)	–	–	1	–
SW	30	156° (SSE)	65°	0.35 *	13	167° (SSE)	33°	0.83 ***	1	12
W	11	175° (S)	54°	0.55 *	8	238° (WSW)	44°	0.70 *	–	7
NW	58	164° (SSE)	39°	0.77 ***	46	230° (SW)	33°	0.83 ***	3	50
N	21	196° (SSW)	21°	0.93 ***	9	244° (WSW)	47°	0.67 *	1	11
Totals	318				187				142	80

[a] Rotation of heading does not necessarily mean crosswind compensation – see text.
[b] A.D. = angular deviation; r = length of mean vector.
[c] Rayleigh test significance levels: * $P < 0.05$; ** $P < 0.01$; *** $P < 0.001$.

conditions had W headings and SW directions. Consequently, a minority of the individuals actually may have been compensating for right crosswinds by crabbing right.

Finally, Table 3 shows that in SW winds, the *D. plexippus* migrating on August 27, 1979, did adopt a preferred direction that was south to southeast. Examination of the data showed that 11 of the 13 observations for which headings were recorded were seen during the 90 min period on August 27. The 11 individuals had a mean heading of approximately $162° \pm 31°$ or SSE, quite different from the preferred SW course. A Raleigh test showed that the observed mean direction was significant, with n = 11, r = 0.85, and $P < 0.001$. The left-rotated headings of the August 29 group resulted in a mean direction of $128° \pm 61°$, or SE. A Raleigh test showed that this was significant, with n = 23, r = 0.44, and $P < 0.05$. The directions of the remaining two butterflies, flying during other periods of SW winds, were 170° and 240°. Unfortunately, the exact preferred course for the August 27 group cannot be determined by either of my methods because it is impossible for the butterflies flying in a crosswind to have their course and headings aligned, and none of the migrants were observed flying within 3 m of the ground.

4 Discussion

Flight strategies of *D. plexippus* migrating in southern Ontario appear to be a response to two main selection pressures. These are (1) selection for minimizing the effort of the migration and (2) selection for rapid escape from the higher latitudes. It is obvious that minimizing the effort of migration is adaptive, as the butterflies must arrive at the overwintering sites in sufficiently good condition and with sufficient reserves of fat to survive the winter, mate in the spring, and, for the females, to make a long return flight to at least as for north as the Gulf states (Brower 1977; Brower 1985; Urquhart and Urquhart 1979a). Furthermore, because nectar sources at the overwintering sites are inadequate, the butterflies cannot replenish their fat reserves before departing in the spring (Brower 1977; Brower 1985). Escaping from the high latitudes as quickly as possible is also adaptive because periods of weather favourable for migration in southern Ontario occur less and less frequently after the third week in September as cool, wet conditions become the norm. In addition, days with favourable temperatures are usually accompanied with S to SW winds making migration difficult. Butterflies that delay departure until late in the season could easily become trapped in Ontario by the deteriorating weather conditions. These two selection pressures are potentially in conflict. Selection for minimizing energy expenditures would favour migrating *D. plexippus* adopting a SW course, flying in tailwinds and proceeding largely by soaring. Selection for rapid evacuation would favour migration in as wide a range of weather conditions as possible and taking advantage of every opportunity to move south.

Both selection pressures appear to have shaped the way migrating *D. plexippus* respond to each wind condition. A NE tailwind was the most favourable situation and probably the one wind condition in which selection for minimizing energy expenditures and selection for rapid escape favour similar behaviours, flying at high altitudes and soaring. Flying well above the ground maximizes ground speed by exposing the butterflies to the stronger winds aloft, while soaring in thermals, which also requires high-altitude flight, reduces energy expenditures. Accordingly, in this wind condition, almost 90% of the butterflies were flying higher than 3 m above the ground and maximum frequencies were recorded for *D. plexippus* soaring and circling in thermals. In contrast, left-crosswind conditions, E, SE, and S winds, if not compensated for, could have been very unfavourable for the butterflies. If the butterflies flying in E winds spent much of their flight circling in thermals, they would have drifted W, delaying their retreat from the higher latitudes. Similarly, frequent periods of circling in thermals on SE or S winds would have resulted in the butterflies drifting NW or N, bringing them to higher latitudes, where the season is more advanced than their departure point. As a result, selection for rapid evacuation from the higher latitudes has apparently favoured migrating *D. plexippus* reducing their exposure to drift in left crosswinds. This was accomplished by selection for individuals that spent less time flying high, less time soaring, less time circling in thermals and that resisted drift during straight flight by crabbing left. Right crosswinds (W, NW, N) present the butterflies with different problems. For two of the conditions, NW and N winds, any drift will be towards lower latitudes. However, in NW winds, the pure right crosswind, the drift resulted in a mean direction towards the SSE (Table 2), almost perpendicular to the preferred SW direction. Consequently, the migrating butterflies would be making little progress towards the Eagles Pass area of Texas. Although the migrants did minimize energy expenditures by flying high and soaring for most of their flight, this behaviour did not reduce the cost of migration to the Eagles Pass area, just the cost of reaching a lower latitude. Apparently, for NW winds, selection for rapid evacuation of the higher latitudes has dominated selection for minimizing energy expenditures. The N wind as a tailwind/right-crosswind condition, appeared to be similar to the NE-wind condition in that both selection pressures favoured the same behaviours. Selection for rapid evacuation favoured high-altitude flight to exploit the higher-velocity winds aloft. Selection for minimizing energy expenditures also favoured high-altitude flight and soaring. Although the resulting southward drift meant that their mean direction, 185° (Table 2), was rotated approximately 47° off course, the butterflies were still getting closer to the Eagles Pass area of Texas and preceeding largely by low-cost soaring.

Two separate strategies appear to have been favoured when SW headwinds prevail. During most periods with SW winds, the *D. plexippus* did not migrate. This strategy would reduce energy expenditures because making significant progress in most headwinds would require energy-intensive, low-level, flapping flight. In contrast, when SW winds were very light and thermals were present, the butterflies switched to a second strategy. They rotated their headings southward, apparently adopting a more southward preferred direction. By altering their preferred direction and rotating their headings to the left, they had effectively changed a pure SW headwind into a headwind/right-crosswind combination. As a result, they could fly high and exploit thermals during a brief period of unusual weather. As with W and NW winds, selection for rapid evacuation of the higher latitudes seemed to be the main selection pressure favouring this behaviour. The predominately SE drift resulting from the rotation of their headings meant that the butterflies were moving to a lower latitude, but not getting any closer to the Eagles Pass area.

Given the complexity of the flight strategies, exhibited by migrating *D. plexippus*, it is not surprising that different observers have reported different flight behaviours.

Depending on wind direction, the mean directions of butterflies in southern Ontario can range from WSW to SSE. The frequencies of butterflies observed at different flight levels, and the frequencies of soaring and circling individuals will depend on both wind direction and whether or not thermals are present. Any soaring individuals that are circling in thermals will be observed to drift downwind. If foragers are also considered as migrants, these individuals will usually be seen flying upwind. As thermals pass by, the behaviour of the butterflies will change and individuals flying in different sections of the convection cell will often have different mean directions and headings. Migrating *D. plexippus* in southern Ontario will also show an asymmetric response to left and right crosswinds to their preferred SW direction. The butterflies will be very effective at compensating for left crosswinds, but will largely ignore right crosswinds. Although these flight behaviours all appear to be a response to only two main selection pressures, the result can appear quite chaotic, giving the impression of poor directional control. Finally, the asymmetric response to left and right crosswinds has an interesting implication. If this behaviour persists for much of their migration across eastern North America, then it could account for the large number of individuals flying S to SE migration routes, as determined by mark and recapture studies (Urquhart 1960; Urquhart and Urquhart 1977, 1978). Although the mean direction of the migrants observed in this study was SW, not S to SE, this may be only part of the story. Because *D. plexippus* can fly much higher than 300 m (Gibo 1981), the upper limit for observations in this study, the mean direction of an unknown proportion of the migrants was not recorded. These high-flying individuals would have been exposed to the higher velocity winds aloft and should have experienced much more drift than the butterflies flying at lower altitudes. If the butterflies flying above 300 m persist in not compensating for right crosswinds, the tendancy to drift S to SE would be greatly enhanced. Another possibility is that the migrating population encounters regions where right crosswinds are more common as they move through the northeastern USA.

Acknowledgements. I thank Cathy Neal for her help in making the observations and Lincoln Brower, Jody McCurdy, and Lena Chamely for reading and commenting on the manuscript. This research has been supported, in part, by grants from the Bickle Foundation and from the Natural Science and Engineering Research Council of Canada.

References

Baker RR (1978) The evolutionary ecology of animal migration. Holmes and Meier, New York

Batschelet E (1981) Circular statistics in biology. Academic Press, London

Beall G (1941a) The monarch butterfly, *Danaus archippus* Fab. I. General observations in southern Ontario. Can Field Nat 55:123–129

Beall G (1941b) The monarch butterfly, *Danaus archippus* Fab. II. The movement in southern Ontario. Can Field Nat 55:133–137

Brower LP (1977) Monarch migration. Nat Hist 86:40–53

Brower LP (1985) New perspectives on the migration biology of the monarch butterfly, *Danaus plexippus* L. In: Migration: mechanisms and adaptive significance. Contrib Mar Sci, Suppl, vol. 27

Brower LP, Calvert WH, Christian J, Hedrick LE (1977) Biological observations on an overwintering colony of monarch butterflies (*Danaus plexippus*, Danaidae) in Mexico. J Lepid Soc 31: 232–242

Calvert WH, Hedrick LE, Brower LP (1974) Mortality of the monarch butterfly (*Danaus plexippus* L.): avian predators at five overwintering sites in Mexico. Science 204:847–851

Gibo DL (1981) Altitudes attained by migrating *Danaus p. plexippus* (Lepidoptera:Danaidae), as reported by glider pilots. Can J Zool 59:571–572

Gibo DL, Pallett MJ (1979) Soaring flight of monarch butterflies *Danaus plexippus* (Lepidoptera: Danaidae) during the late summer migration in southern Ontario. Can J Zool 57:1393–1401

Kanz JE (1977) The orientation of migrant and non-migrant monarch butterflies, *Danaus plexippus* (L.). Psyche (Camb Mass) 84:120–141

Lugger O (1980) On the migration of the milkweed butterfly. Proc Entomol Soc Wash 1:156–158

Pennycuick CJ (1975) The mechanics of flight. In: Farmer DS, King JR, Parks KC (eds) Avian biology, vol 5. Academic Press, New York, pp 1–75

Schmidt-Koenig K (1979) Directions of migrating monarch butterflies (*Danaus plexippus*; Danaidae; Lepidoptera) in some parts of the eastern United States. Behav Processes 4:73–78

Schmidt-Koenig K (1985) Migration strategies of monarch butterflies (*Danaus plexippus* (L.); Danaidae; Lepidoptera). Contrib Mar Sci, Suppl, vol 27

Urquhart FA (1960) The monarch butterfly. University of Toronto Press, Toronto

Urquhart FA, Urquhart NR (1976) The overwintering site of the eastern population of the monarch butterfly (*Danaus p. plexippus*: Danaidae). J Lepid Soc 30:73–87

Urquhart FA, Urquhart NA (1977) Overwintering areas and migratory routes of the monarch butterfly (*Danaus p. plexippus*, Lepidoptera:Danaidae) in North America with special reference to the western population. Can Entomol 109:1583–1589

Urquhart FA, Urquhart NA (1978) Autumnal migration routes of the eastern population of the monarch butterfly (*Danaus p. plexippus* L.; Danaidae; Lepidoptera) in North America to the overwintering site in the Neovolcani Plateau of Mexico. Can Entomol 111:15–18

Urquhart FA, Urquhart NA (1979a) Vernal migration of the monarch butterfly (*Danaus p. plexippus*, Lepidoptera:Danaidae) in North America from the overwintering site in the neo-volcanic plateau of Mexico. Can Entomol 111:15–18

Urquhart FA, Urquhart NR (1979b) Aberrant autumnal migration of the eastern population of the monarch butterfly, *Danaus p. plexippus* (Lepidoptera:Danaidae) as it relates to the occurrence of strong westerly winds. Can Entomol 111:1281–1286

Walker TJ (1980) Migrating Lepidoptera: are butterflies better than moths? Fla Entomol 63:79–98

Walker TJ (1985) Butterfly migration in the boundary layer. Contrib Mar Sci, Suppl, vol 27

Walker TJ, Riordan AJ (1981) Butterfly migration: are synoptic-scale wind systems important? Ecol Entomol 6:433–440

Zebe E (1954) Über den Stoffwechsel der Lepidopteran. Z Vgl Physiol 36:290–317

13 Interactions Between Synoptic Scale and Boundary-Layer Meteorology on Micro-Insect Migration

R. A. FARROW[1]

1 Introduction

Insect migration over long range is associated with particular weather patterns. In temperate latitudes, the eastward movement of cold fronts and the associated poleward surges of warm air are the most important (Johnson 1969). Less migratory activity is reported in the cold polar air moving towards the equator in the wake of fronts. Such movements are, however, important to the maintenance of source populations of spring migrations at lower latitudes and may contribute towards substantial 'return' movements in autumn.

Direct measurement of the distribution and abundance of insects in the atmosphere was not possible until radar was applied to the study of migration. This has led to a quantitative understanding of the effects of the atmospheric environment on migrating insects in terms of parameters such as track, speed and orientation of individuals and variations of their density with height, distance and time (Drake 1981). These parameters have been related to wind vectors and to wind, temperature, and humidity profiles and fields. Although the size range of insect migrants occupies a continuum, it has become convenient to divide migrants into 'macro' (> 1.5 cm) and 'micro' (< 1.5 cm). Macro-insects are characterised by their high-flying speed of $2-5$ m s^{-1}, relative to the wind. Micro-insects generally fly at speeds of less than 1 m s^{-1}, and their distribution and displacement are necessarily controlled by atmospheric processes to a much greater extent than in macro-insects. Most radar investigations on long distance migration have, until recently, been limited to observations on the relatively large macro-insects because they are readily detectable as individuals by 3-cm wavelength entomological radars (Drake 1982). Examples include noctuid moths and grasshoppers, both of which have been studied in Australia (Drake and Farrow 1983; Drake and Farrow 1986).

More recently, Farrow and Dowse (1984) developed aerial sampling techniques which could identify insects under observation by radar. These samples revealed an unsuspected abundance of micro-insects which were too small to be seen as individuals by conventional radar, although they outnumbered macro-insects by an average of 1000:1 (Farrow 1982). From measurements of the variations in composition of micro-insect populations by day and night in the surface and upper air, Farrow (1982) demonstrated that there is continuity of upper-air populations from day to night. The present paper discusses the ways in which diel changes in the structure of the planetary

---

[1] CSIRO, Division of Entomology, G.P.O. Box 1700, Canberra, A.C.T. 2601, Australia.

Insect Flight: Dispersal and Migration
Edited by W. Danthanarayana
© Springer-Verlag Berlin Heidelberg 1986

boundary layer interact with broader-scale weather patterns to influence micro-insect migration. The interactions will be illustrated using the results from a 5 day sampling period in spring 1983. This was part of a longer series of samples taken at intervals between March 1982 and February 1984 to identify micro-insects and to quantify their migrations over a site in central western New South Wales.

2 Boundary-Layer Processes

2.1 Structure

For the purposes of this discussion, the atmosphere can, in biological terms, be divided into three layers. Closest to the ground is the biological boundary layer where an insect can control its flight movements because its flying speed is greater than the wind speed (Taylor 1974). For micro-insects, this layer is rarely more than a few metres deep by day, although it may deepen at night when the surface air becomes calm. Migration over long range by the slow-flying micro-insects is obviously very limited at this level.

The next layer is the planetary boundary layer in which the wind flow is slowed by friction with the earth's surface. By day, this layer is relatively deep due to the effects of convective mixing and may extend to 1000 m or more. Although wind speeds tend to fluctuate due to the effects of updrafts and downdrafts, they are, on average, less than the geostrophic speed and tend to increase progressively with altitude. By night, when convection ceases, the planetary boundary layer becomes very shallow with virtually calm air at the surface and a strong wind shear developing between the surface and the geostrophic flow at 100–300 m. Under cloudless conditions this system develops rapidly with the cessation of thermal convection about an hour before sunset. Although micro-insects have little control over their migration in this layer, opportunities for long-range migration increase with altitude due to the effects of wind shear. By day, micro-insects may only need to ascend a few metres to reach a significant wind flow, but insects taking off at dusk, or later, must ascend actively to reach a significant wind flow which may be as high as 50 m.

The third layer, the geostrophic, extends upwards to the tropopause and its winds exhibit a relatively constant velocity that is proportional to the pressure gradient, and it offers the greatest opportunities for insect displacement. By day, access to this layer is limited because of its altitude, but at night it is more easily reached at 200–400 m.

2.2 Wind Effects

Vertical motion of air may exceed $1–2$ m s^{-1} by day and provides considerable assistance to the ascent of micro-insects, possibly to the top of the planetary boundary layer at more than 1000 m altitude. When micro-insect populations are sufficiently dense to be observed by radar by day, they are often concentrated in the walls of convective cells and appear to be entrained in columns of ascending air (Fig. 1a). These cells, when well-defined, are typically about 1 km in diameter and may be aligned downwind, particularly at higher wind speeds; they move in the direction of the

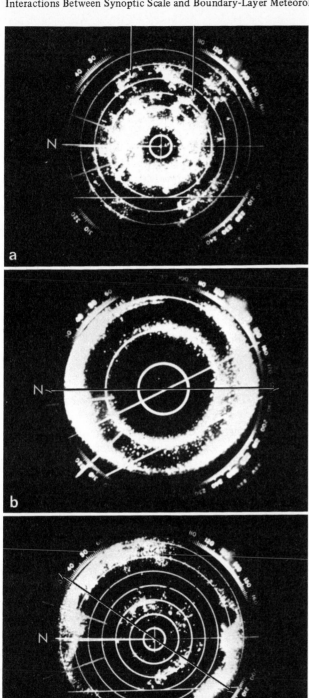

Fig. 1a–c. Photograph of PPI (plan-position indicator) of CSIRO Entomological Radar showing migrant micro-insects: **a** concentrated in the walls of convective cells, indicated by *arrows*, at 1203 h on October 12, 1978. Near the centre of the display, at low altitude, densities are relatively uniform. **b** Concentrated in two layers at 150–250 m altitude at 0556 h on October 3, 1978. The non-uniformity of the ring of targets indicates mutual orientation at 90/270°, perpendicular to arrow. **c** Moving upwards in a layer by convective lift at 0939 h on October 3, 1978. Scattered targets at the centre of the display indicate morning takeoff. Micro-insects in the layer are now orientated at 130/310°, perpendiculars to arrow

mean wind, typically at speeds of $5-20$ km h^{-1}. The passage of cells is marked by periodic wind gusts and small fluctuations in air temperature at the surface.

By night, when the air is horizontally stratified, there is no vertical mixing, except in the narrow zones of convergence of passing cold fronts. No assistance is provided to micro-insects ascending out of the surface boundary layer. Dense concentrations of micro-insects have been observed by radar in layers as little as 20 m thick which appear to correspond to the top of the planetary boundary layer at $100-300$ m altitude. On one occasion, a double layer was present after dawn in the geostrophic flow at $150-200$ m (Fig. 1b). This was later displaced to a higher altitude by thermal convection (Fig. 1c) and eventually broken up by the increasing intensity of convective turbulence. The non-uniform representation of the layers indicates a significant level of mutual alignment of the migrants that is strongly associated with wind direction.

Wind vectors in the planetary boundary layer can be estimated from pilot balloon ascents. During the study period September $20-24$, diel variation in surface and upper-wind velocity at 200 m (Fig. 2a), shows a fivefold difference in velocity between the surface and upper air each night, even when upper-wind velocities attained 80 km h^{-1}. Examination of the wind profile for 2222 h on September 23 in the northerly flow, just prior to the passage of a cold front, shows a peak velocity of $18\,\mathrm{m\,s^{-1}}$ ($64.8\,\mathrm{km\,h^{-1}}$) at 142 m (Fig. 2b) about 60 m below the sampling height. The profile illustrates the development of a supergeostrophic flow or low-level jet. Insects flying in this layer have the potential to achieve much greater net displacements than those flying at a few tens of metres above or below them. Jets are most frequently encountered in the warm air-flows ahead of cold fronts and their causes have been well-documented in the great plains of the USA (Blackadar 1957). Their latitudinal extent in Australia is not well-known, but it is thought that they occur over a wide area of the inland of the continent. The jet was not present at 1916 h and the time of its appearance over the study site is unknown. A jet was not detected in the previous frontal system on the night of September 20, although the profile (Fig. 2b) indicated an uneven wind shear between 70 and 140 m altitude.

2.3 Temperature Effects

By day, when the planetary boundary layer is convectively mixed, air temperature declines with altitude at the adiabatic lapse rate of 1 $^\circ$C per 100 m. With daily maxima averaging $23^\circ-27\,^\circ$C at 1.5 m during the study period, the temperature at 500 m would reach $18^\circ-22\,^\circ$C by mid-afternoon and the air above 500 m could be below the flight threshold of many migrants for much of the day. In mid-summer, when surface air temperatures may attain maxima of 35 $^\circ$C or higher at the study site, micro-insect migrants may ascend to the top of the planetary boundary layer at more than 1000 m, but it is unlikely that they ever reach the geostrophic layer by day. On cloudless nights, the air near the ground cools by radiation and a temperature inversion rapidly forms once thermal convection ceases. Temperature declines at about the adiabatic lapse rate in the geostrophic layer above the inversion and a substantial layer of warm air favourable for migration exists in the geostrophic flow at night.

A radiosonde was attached to the kite to measure temperature at the sampling height on two occasions during the study period and marked inversions were detected

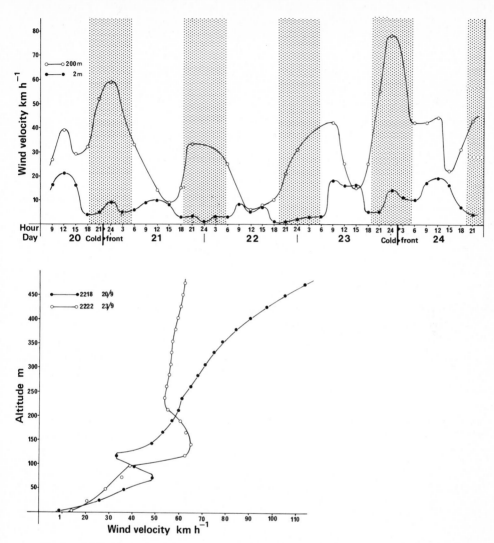

Fig. 2. a Diel variation in surface and upper (200 m) wind velocity September 20–24, 1983. **b** Wind profiles over the Trangie study site at 2218 h on September 20 and 2222 h on September 23, 1983

(Fig. 3). The temperatures recorded were 2°–3 °C less than the daily maximum surface temperature, corresponding to the local lapse rate, and, on the first occasion, the inversion was detected an hour before sunset. Temperature differences of 6°–7 °C were recorded between the sampling height and the surface during the night, although the peak temperature at the top of the inversion was unknown. This is likely to be higher than indicated by the local lapse rate because of the movement of air aloft from warmer areas to the north. For technical reasons we were unable to operate tethered and free-release radiosondes together, to measure temperature changes with both height and time simultaneously. Drake (1984) in a series of measurements of temperature

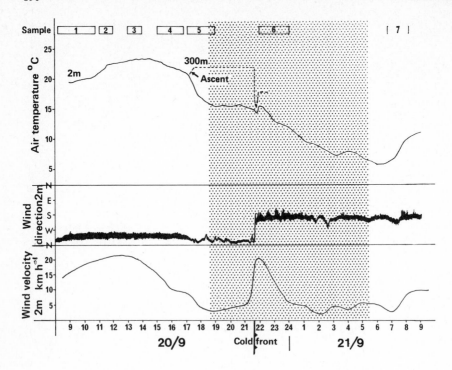

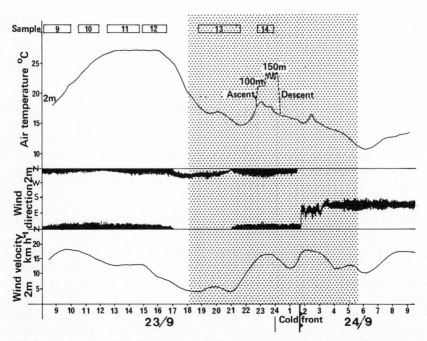

Fig. 3. Comparisons of surface temperature from a screen thermograph (*solid line*) and upper-air temperature from a radiosonde attached to kite (*broken line*) and surface-wind speed and direction at two insect sampling periods

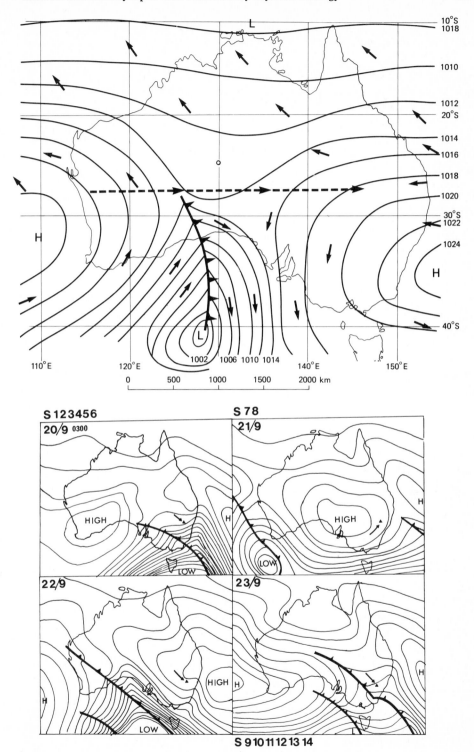

Fig. 4. a Chart of a generalised synoptic situation. **b** Daily synoptic charts (1500 h) for the period September 20–24, 1983. *Arrow* indicates wind direction over research site

profiles at night with free-release radiosondes, found that the top of the inversion was located approximately at the top of the zone of wind shear at 150–250 m, i.e. at about the sampling height used in the present study. Temperatures above the inversion decline by 1 °C per 100 m and would fall below 20 °C at an altitude of approximately 500 m. On September 20 there appeared to be little mixing between surface and upper air, until the immediate vicinity of the front where the temperature dropped by 6.5 °C in 8 min (Fig. 3a). It can also be seen that the inversion persisted in the westerly flow in the wake of the front (Fig. 3a), because the temperature at 300 m rose to 18 °C which was 3 °C higher than at the surface. On September 23, vertical mixing was more pronounced in the vicinity of the front than on September 20. This is apparent from the violent fluctuations of temperature in the upper air and the rise in temperature at the surface during the night (Fig. 3b). The radiosonde was not operating during the passage of the front because of the excessive turbulence of the upper winds after 2400 h.

3 Synoptic Meteorology

A synoptic chart of a generalised meteorological situation is presented in Fig. 4a. In the southern hemisphere, wind circulates anticlockwise around the anticyclones and accelerates in the zone of convergence associated with the cold front. Regular eastward displacement of the systems causes the winds to rotate at any given site in either a clockwise or anticlockwise direction depending on whether the site is to the north or to the south of the anticylonic ridge, while the wind speed increases in the vicinity of the front in the zone of low pressure (Fig. 2). The temperature of the airflows also varies in a systematic manner, the northerly airflows being 10°–15 °C warmer than the southerly flows with a sharp discontinuity at the frontal interface (Fig. 3).

The real synoptic situation for the study period is shown in Fig. 4b. Cold fronts crossed the study area on the nights of September 20/21 and 23/24, and were preceded by strong northerly airflows which were largely confined to the upper air at night because of the lack of mixing (Fig. 2).

4 Micro-Insect Migrants

Seasonal conditions were extremely favourable for insect development following mild wet conditions during the winter and an abundant growth of pasture. Samples of insects flying at 150–300 m were taken systematically from a tow net suspended from a kite (Farrow and Dowse 1984) over a 5-day period from September 20–24. The kite operates in winds of 4 m s^{-1} or more, so during periods of light winds or calms, when opportunities for long-range movement are limited, no samples can be obtained. Samples from the surface boundary layer were collected with a net mounted on a vehicle and analysed in the same way. Taxa were sorted into orders and those of economic importance were identified to species.

The volume of air sampled at each insect catch was estimated from the wind velocity at the sampling height, and the catch number was converted to absolute density and flux by the method of Farrow and Dowse (1984). While the radar provides an almost instantaneous measure of the number of insects in a volume of 30000000 m^3, aerial sampling takes an hour to provide an estimate of insect density in a volume averaging 10000 m^3 (depending on wind speed), over a limited height range so it cannot be used to provide density profiles.

Five aerial samples were obtained prior to the passage of the front on September 20 and three in its wake the next morning. No samples were obtained when the high pressure cell became established on the 21 and 22 as winds were very light. Six samples were obtained ahead of the second front on the 23 and five in its wake the next day. Aerial densities were substantially higher in the northerly airflow ahead of the front than in the southerly flow in its wake (Fig. 5). Densities at 200–300 m altitude in the northerly flow averaged 0.0844 ± 0.0500 m^{-3} (n = 11) as against 0.0055 ± 0.0024 m^{-3} (n = 8) in the southerly flow, suggesting that there was a net southward displacement of micro-insects at this period.

In spring and autumn, lack of micro-insects in southerly airflows is associated with two factors: (1) relative coldness of the southerly airflow, which at $12°$-15 °C, is probably below the flight threshold of many migrants, (2) relatively small source populations to the south of the study site compared to the north, due to the lower day-degree accumulation in the former areas, causing insects to emerge progressively later with increasing latitude in spring. During the preceding summer and autumn, sampling revealed that northerly airflows still carried many more insects than the southerly flows, although emergence differentials did not occur and temperatures were well above any limiting thresholds, suggesting that there are other features of the northerly airflows that make them more favourable for flight than airflows from other directions.

The densities of micro-insects flying in the biological boundary layer decline progressively after sunset with the cooling of the surface air (Farrow 1982), whereas the densities of those flying in the upper air at 150–300 m are maintained or increased (Fig. 5). Recruitment from the surface boundary layer cannot account for such changes because ascent to the upper air is confined to within an hour of sunset in macro-insects (Drake 1984) and is not known to be different in micro-insects. Cooling of air below the inversion itself provides a mechanism for explaining the increase in aerial density at 200 m observed on September 23 between 1830 h and 2400 h. If the insects migrating at this time had originally taken off by day, they would have been distributed over a wide altitudinal range by late afternoon, due to the effects of convective turbulence. Descent may be initiated by a variety of factors including the onset of darkness, but progressive cooling of air above the temperature inversion during the night would progressively force micro-insects to descend towards the top of the inversion.

Whatever the cause of the initial descent, it is proposed that further descent below the inversion is inhibited by the decline in temperature towards the surface, which steepens as the night progresses, and that micro-insects remain flying actively, in the absence of convective lift, in the warm air of the inversion. At 1600 h, aerial density averaged 0.019 m^{-3}: if it is assumed this was extended over a height range of 500 m, this represents a total of 9.67 insects in a square metre column of air. If these insects were progressively concentrated into a layer 20 m thick (as previously observed on radar, Fig. 1) during the course of the night, without further recruitment or loss, density would increase to 0.483 m^{-3}, as against the observed value of 0.581. These estimates do not take account of quantitative changes due to the effects of spatial variations in

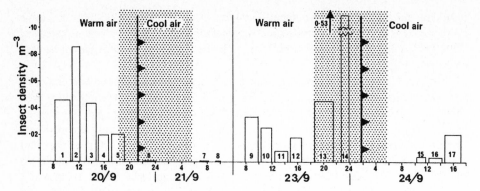

Fig. 5. Density of micro-insect migrants at 150–300 m during a sequence of samples September 20–24, 1983

the size of source populations. On September 23, for example, a difference of 2 h between sampling periods would, if wind speed was constant at 15 m s^{-1}, be equivalent to a separation of 100 km between source areas for insects taking off at the same time. With take-off and ascent to the upper air limited to within an hour after sunset, the high densities observed at 2330 h on September 23 could have also been affected by take-off of a large population of micro-insects at least 4 h earlier, localised 150–250 km north of the study site. It should be noted that the samples collected at night and by day on September 23 comprised the same range of species, and were not limited to nocturnal species (Farrow, unpublished data).

The night populations are also affected by the rate of fall-out in aerial populations, particularly once recruitment into the upper air has ceased about an hour after sunset. Major changes would be expected to occur in the late afternoon with the cessation of thermal convection and at the onset of darkness. It is not yet possible to determine the proportion of the population that is held up in the nocturnal inversion and to assess the rate of fall-out of micro-insect populations during night flight in favourably warm airflows, although once these turn cold with the passage of a front, fall-out, as indicated by the present study (Fig. 5), was dramatic and immediate and involved 99% of the population. The effect of the low-level jet on the night of September 23/24 was such that micro-insects which were overtaken by the cold front over the study site at about 0100 h could, if they had taken off prior to the cessation of thermal convection, have originated from areas 300 km or more north of the study site.

5 Conclusions

The characteristic pattern of synoptic weather of temperate latitudes of Australia results in a net-southward displacement of migrating micro-insects. This net-poleward displacement can only be sustained by high rates of population growth in the northern source areas because the northward movements involve very low densities of micro-insects. Nevertheless, such movements are essential to the founding of populations in the northern source areas and to their maintenance in the long term. In addition, the rotation of wind trajectories with the passage of weather systems, ensures an extensive redistribution of aerial populations in all directions over inland Australia, although the extent of such movements is modified by the lightness of winds and limiting temperatures in some directions. This interpretation of migration and redistribution of migrant insect populations avoids the necessity of requiring specific 'return' migrations in autumn to compensate for the ubiquitous poleward migrations of spring and at other periods of the year.

Despite their relatively feeble flight capabilities, micro-insects have been shown capable of controlling their flight behaviour by migrating at night in the absence of convective or convergent atmospheric processes. The extend of this control is supported by the observation that micro-insects are capable of some degree of orientation with respect to wind direction. Concentration occurs in zones of maximum wind velocity at the top of the nocturnal inversion where the air is warmest and may remain at a temperature favourable for flight throughout the night. Under clear air conditions in continental climates, rapid establishment of a nocturnal inversion in the planetary boundary layer appears to suppress evening descent of the kind observed in temperate maritime climates (Taylor 1974) and the nocturnal distribution of micro-insects comes to resemble the type reported in macro-insects by Drake (1984).

References

Blackadar AK (1957) Boundary layer wind maxima and their significance for the growth of nocturnal inversions. Bull Am Meteorol Soc 35:283–290

Drake VA (1981) Quantitative observation and analysis procedures for a manually operated entomological radar. CSIRO Aust Div Entomol Tech Pap no 19:41 pp

Drake VA (1982) The CSIRO entomological radar: a remote sensing instrument for insect migration research. In: Wisbey LA (ed) Scientific instruments in primary production. Australian Scientific Industry Association, Melbourne, pp 63–73

Drake VA (1984) The vertical distribution of macro-insects migrating to the nocturnal boundary layer: a radar study. Boundary-Layer Meteorol 28:353–374

Drake VA, Farrow RA (1983) The nocturnal migration of the Australian plague locust *Chortoicetes terminifera* (Walker) (Orthoptera: Acrididae): quantitative radar observations of a series of northward flights. Bull Entomol Res 73:567–585

Drake VA, Farrow RA (1986) A radar and aerial-trapping study of an early spring migration of moths (Lepidoptera) in inland New South Wales. Aust J Ecol 10:223–235

Farrow RA (1982) Aerial dispersal of microinsects. In: Lee KE (ed) Proc 3rd Australas Conf Grassl Invert Ecol, 30 Nov–4 Dec 1981, Adelaide. S.A. Govt Printer, Adelaide, pp 51–55

Farrow RA, Dowse JE (1984) Method of using kites to carry tow nets in the upper air for sampling insects and its application to radar entomology. Bull Entomol Res 74:87–95

Johnson CG (1969) Migration and dispersal of insects by flight. Methuen, London

Taylor LR (1974) Insect migration, flight periodicity and the boundary layer. J Anim Ecol 43:225–238

14 Derelicts of Dispersal: Arthropod Fallout on Pacific Northwest Volcanoes

J. S. EDWARDS[1]

1 Introduction

The exceptional capacity for aerial dispersal of insects and certain other terrestrial arthropods by active or by passive means has profound implications for their ecology, development, and evolution. It may well have been the exploitation of temporally and spatially patchy palaeozoic environments by means of aerial dispersal that underlay the evolutionary emergence of pterygotes (Wigglesworth 1976) and in due course their immense variety and their colonization of diverse, sometimes extreme, environments. But for their dispersive success, there is a price to pay. One consequence of aerial dispersal by very small animals is the element of indeterminacy in trajectory; they are largely at the mercy of weather and topography. Given the limited ability to control pathway in comparison with larger animals, such as birds, it is a necessary consequence that the process of dispersal is wasteful in the sense that a fraction, sometimes a minute one, of dispersers may terminate travel in a favorable habitat.

The emphasis of most studies of dispersal is on the factors leading to initiation of flight, entry into the air column, and pathway. This work addresses a particular aspect of the termination of dispersal in focusing on the arthropod-fallout fauna of alpine regions, the great biomass of arthropods for which travel terminates in the wrong places. A marine counterpart to this phenomenon is exemplified by Heydemann's (1967) estimate of 4.5 billion insects per summer day drifting over the North Sea from a 30-km-wide coastal strip, and comparable data reviewed by Bowden and Johnson (1976).

The phenomenon of arthropod fallout on snow fields seems first to have been noted, with characteristic acuity, by the founder of modern biogeography, Alexander von Humboldt, during his attempted ascent in 1802 of Chimborazo, then the world's highest known peak, and is now documented by a diverse, but mainly anecdotal, literature concerning the presence of diverse insects and spiders on alpine snowfields. Mani (1968) has assembled many such accounts in his pioneer work on alpine entomology and Swan (1963), recognizing the prevalence of the phenomenon, applied the term *aeolian zone* to regions where animal communities depend for their subsistence on wind-transported organic material. We have attempted quantitative estimates of arthropod fallout in montane central Alaska (Edwards and Banko 1976) and on volcanic peaks in the Pacific Northwest (e.g., Mann et al. 1980), while Papp (1978) and Spalding (1979) have made similar observations in California. These communities exemplify what Hutchinson

---

[1] Department of Zoology, NJ-15, University of Washington, Seattle, WA 98195, USA.

Insect Flight: Dispersal and Migration
Edited by W. Danthanarayana
© Springer-Verlag Berlin Heidelberg 1986

(1965) categorized as the *allobiosphere*, which is dependent on productivity arising elsewhere in the biosphere. The organisms of the ocean depths lie at one extreme of this spectrum, and high alpine communities at the other.

Our current studies of arthropod fallout on the volcanoes of the Pacific Northwest have had the objective of characterizing the arthropod fallout and the resident-arthropod communities of these alpine islands for which the fallout provides a substrate. The two principal foci of our studies have been Mount Rainier (4392 m), a dormant strato-volcano with permanent snowfields down to 2000 m, surrounded by forest and agricultural lands, and more recently the barren mineral surfaces in the blast zone within 7 km of the crater of Mount Saint Helens (2550 m) following the eruption of May 1980.

2 Arthropod Fallout on Mount Rainier and Associated Alpine-Arthropod Communities

Van Dyke (1919) and Melander (1921) first recorded the presence of insects on the snowfields of Mt. Rainier, and Scudder (1963) has drawn attention to the occurrence of several Heteroptera stranded on snow in the Pacific Northwest. Our studies, beginning in 1970, have shown that windborne insects may be found on snow at all times of the year, with minima in December through March. With the resumption of spring growth in the lowlands from mid-April on, increasing numbers of arthropods are carried by predominantly SW winds to the snowfields with a maximum in mid- to late June, and a slowly decreasing fallout rate through late summer and fall, punctuated occasionally by massive numbers of ants and of aphids. The highest densities occur at altitudes between 2000 and 3000 m; but material has been collected at all altitudes up to the summit at 4392 m, where syrphids, aphids, lygaeids, and Lepidoptera have been taken. Over 200 insect species, representing 105 families from 14 orders, and 23 species of spider from 8 families, have so far been recognized in the fallout fauna.

As a specific example of arthropod fallout, the composition of 40 random square meter samples taken from a snowfield on Mt. Rainier at 2500 m, June 1975 is shown in Table 1.

With the exception of the Plecoptera, the predominant contributors to the snow-field-fallout fauna are from terrestrial habitats. The Plecoptera, which figure prominent-

Table 1. Numbers and biomass of fallout insects on Muir snowfield, Mt. Rainier, June 1975 based on 40 random m² quadrats at 2500 m altitude

Order	Number of individuals	Biomass dry wt (mg)
Homoptera	397	6.74
Heteroptera	358	1.95
Diptera	147	2.03
Hymenoptera	45	2.49
Plecoptera	9	3.10
Coleoptera	7	0.53
Total	963	26.84
m^{-2}	24.07	6.71

ly in the data shown in Table 1 are associated with local meltwater streams, but the
majority of the fallout appears to be derived from lowland habitats. In general, our
sampling over a decade indicates that the orders most abundantly and diversely represented in the fallout are those known to be conspicuous dispersers, i.e., Homoptera
such as Aphididae, Cicadellidae, Heteroptera such as Lygaeidae, parasitic Hymenoptera,
Diptera, particularly Nematocera and Syrphidae, and Coleoptera, notably Coccinellidae
and Chrysomelidae. The overall representation of groups in the fallout, based on a total
of 1772 specimens taken from random 1 m^2 quadrats throughout the summer of 1975
are listed in Table 2.

Table 2. Taxonomic composition of fallout
arthropods, Mount Rainier (for details, see
text)[a]

	%
Homoptera	29.9
Heteroptera	22.3
Hymenoptera	21.9
Diptera	11.6
Coleoptera	5.0
Araneae	5.5
Lepidoptera	4.9

[a] Minor constituents included Plecoptera,
Trichoptera, Neuroptera, and Ephemeroptera.

The fallout arthropods strewn over many acres of alpine snowfield provide a resource
for birds as well as predatory or scavenging arthropods. Several species of bird, on
Mt. Rainier notably water pipits *(Anthus spinoletta)*, rosy finches *(Leucosticte tephro-
cotis)*, and horned larks *(Eremophila alpestris)*, but also ptarmigan *(Lagopus leucurus)*
and occasionally corvids, exploit the inactive and clearly visible arthropods. Two small
beetles, a carabid *(Bembidion* sp.) and a staphylinid also forage on the snow during
daylight hours, as do lycosid spiders. After nightfall, an arthropod guild composed of
five species of carabid, a grylloblattid *(Grylloblatta* n. sp.) and a phalangid *(Leiopilio
glaber)* scavenge on the snowfields, leaving shelter in shattered bedrock at the margins
of snowfields and glaciers with the onset of dark, reaching peak numbers at about midnight and declining until 3–4 a.m. (Mann et al. 1980; Edwards 1982). Comparable
nocturnal foragers have been observed on snowfields throughout the Washington Cas-
cades and on subalpine lava fields in the Mackenzie Pass area of Oregon (Edwards, unpublished). Further south in California, Papp (1978) described a similar nocturnal guild
dependent on arthropod fallout from the central Sierra Nevada. The general conclusion
to be drawn is that the resource provided by the stranding of dispersing arthropods on
snowfields is sufficiently reliable to have generated a foraging/scavenging guild in alpine
areas around the world. The broadest taxonomic diversity within the guild appears to
be reached in the alpine and subalpine zone of the Pacific Northwest.

3 The Role of Arthropod Fallout in the Recolonization of the Blast Zone of Mount Saint Helens

On May 18, 1980, a massive eruption destroyed the cone of Mount Saint Helens. Within hours, as the eruption and consequent debris flows subsided, the area to the north of the breached crater became a vast landscape of pyroclastic deposits, devoid of life, and initially, but only temporarily, hostile to living organisms. We were thus provided with an opportunity to monitor the process of recolonization of the bare mineral surfaces. Since 1980 we have monitored the composition and quantity of arthropod fallout in the blast zone at altitudes around 1000 m, using standard pitfall-trap arrays and fallout collectors. At least 70 insect families from 17 orders have been borne on the wind to our collection sites. Diptera and Homoptera predominate numerically in the fallout, while Diptera and Coleoptera contribute the greatest diversity, with at least 40 and 20 families, respectively.

The major dipteran component particularly in the first 2 years after the eruption comprised Chironomidae and Culicidae, doubtless reflecting the high productivity of freshwater bodies rich in organic materials as a result of the impact of the eruption outside the blast zone. Otherwise, the faunal composition resembled that described for Mt. Rainier. Within the first two summers 43 species of spider arrived as ballooners on the bare mineral surface of the blast zone. The efficacy of ballooning as a dispersal mechanism is emphasized by the presence in this remote site of three introduced European species, *Theridion bimaculatum* (Linne), *Euoplognatha ovata* (Clerk), and *Lepthyphantes tenuis* (Blackwell) none of which occurred in Washington state before about 1950 (Crawford 1985). Thirty-six of the immigrant spiders in the blast zone are characteristically western lowland species and must have traveled at least 30 km on the prevailing westerly winds to reach the sampling site. The occurrence of ballooning at characteristic stages of life history are reflected in the spider samples; for example, many juvenile *Pardosa* spp. were taken at pitfalls, but no adults, while *Erigone* spp. were all adults or penultimate instars. During the first 4 years it is doubtful that any spiders survived to reproduce, although they probably persisted for variable times by feeding on windborne prey.

4 Habitat Characteristics and the Survival of Insects After the Eruption

In addition to the destruction of life within the blast zone, a major impact of the eruption for insects was the lethal effect of the fine-textured tephra which acted as a desiccant (Edwards and Schwartz 1981). Within the blast zone the combined effect of wind and precipitation has served to stabilize surfaces, on which scavengers, such as carpenter ants (*Camponotus* sp.), and predators, such as salticid spiders, were found taking prey from the arthropod fallout within 8 weeks of the eruption.

Wind action has a dual significance in the recolonization of Mt. St. Helens. It is not only the principle vehicle for transport of fauna and flora, but it also serves to generate viable microhabitats by removing finer fractions of tephra from the surfaces of pyroclastic debris, to create the so-called desert pavement. Stones and pebbles remain at the surface, eventually forming a protective stabilized layer. The desert pavement so formed creates a potential microhabitat, where windspeed is damped, where shade from insolation is available, and where as a result water balance can be sustained in viable ranges for arthropods (Fig. 1).

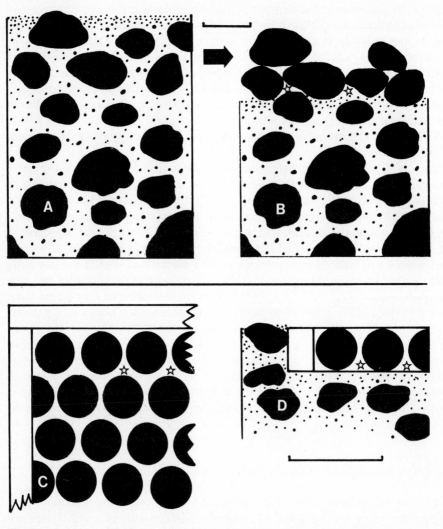

Fig. 1. A–B Formation of desert pavement on Mount Saint Helens. Wind action removes the finer fractions from the surface shown in profile in **A**, leaving a layer of small stones at the surface, depicted in **B**. Spaces (*stars*) under the stones provide a sheltered environment where organic fallout is trapped and where arthropods can survive. *Scale:* 2 cm. **C–D** Plan and elevation, respectively, of fallout collector used to estimate quantity and composition of organic fallout on Mount Saint Helens. Golf balls are enclosed in a wooden frame and supported on a fine, nylon-mesh base. The close-packed golf balls (shown slightly separated in diagrams) simulate the texture of desert pavement and provide sheltered spaces (*stars*) where organic fallout, and predatory or scavenging arthropods accumulate. The collectors are set in the mineral surface as shown in **D**. *Scale:* 5 cm

5 Quantitative Estimate of Arthropod Flux

A prime objective of our studies on arthropod fallout in the blast zone has been to develop a technique that enables us to monitor realistically the arthropod flux in these habitats. Standard techniques, such as pitfall and sticky traps, are variously selective (Southwood 1966), and do not reflect the natural situation in that the trapped arthropods cannot depart as they might from a natural surface. An effective simulation of the desert pavement was devised to function as fallout collectors. They were composed of a square 0.25 m^2 frame backed by nylon mesh which contain close-packed golf balls (Fig. 1C, D). Arrays of these collectors were set in the desert pavement and sampled at approximately 10-day intervals. The seasonal pattern of organic fallout for 1984 is shown in Fig. 2.

We recognize two principal components in the arthropod fallout: the potential survivors and colonists among predators and scavengers, e.g., carabids and staphylinid beetles and spiders, and the diverse fauna composed of herbivores, parasitoids, and nonfeeding adults that either constitute potential prey for predators and scavengers, or are decomposed to yield nutrient to the mineral surface. On the basis of data from the fallout collectors, we estimate that ca. 10 mg m^{-2} day^{-1} dry wt. biomass of arthropod accrues on the surface. The seasonal accumulation would, thus, on the basis of our estimates, be on the order of 1 kg ha^{-1}. These are preliminary figures derived from sets of data that are not yet fully analyzed, and so must be regarded as provisional and tentative, but they serve to convey the order of magnitude of the arthropod fallout as a nutrient source.

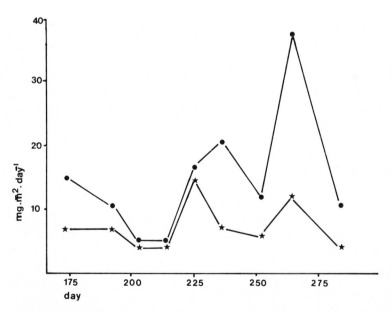

Fig. 2. Arthropod (*stars*) and total organic (*circles*) fallout at representative site in blast zone of Mount Saint Helens estimated by golf ball-fallout collectors as described in text. Days numbered consecutively from Jan. 1, 1984

Carabids were certainly the first survivors and colonizers among the insects in the blast zone of Mount Saint Helens. *Bembidion* (*incertum* grp.) were among the first immigrants to arrive at pitfall traps and probably established overwintering populations by the third year after the eruption. Of at least 15 carabid species arriving at the trapping sites, seven are now capable of sustaining permanent populations. *Bembidion planatum*, for example, which first appeared in fallout traps during 1981 has established large local populations yielding larvae throughout the summer, and large numbers of adults appear in pitfall traps in spring at the time of thaw. These resident populations depend entirely or in major part on windborne-arthropod fallout, since no other resident small arthropods, e.g., Collembola or mites, have established populations. Scavengers, such as Dermaptera, arrive in the fallout every summer, but we have no evidence that they have yet established permanent populations.

6 Arthropod Dispersal and Primary Succession

The classical concepts of primary succession (Clements 1916; Grime 1979) emphasize plants as pioneers of mineral surfaces. Herbivores follow the pioneers, and these in due course support predation and parasites. But on the bare mineral surfaces created by the eruption of Mt. Saint Helens, we find a reversal of the pattern outlined above. Instead of a sequence initiated by colonizing microorganisms, followed by plants on which herbivores could establish, and following them, predators and scavengers, we have found an inverted sequence, with the initial colonizer role occupied by predators and scavengers subsisting on windborne arthropods. An undetermined fraction of the arthropod fallout yields nutrients, such as phosphorus and fixed nitrogen, which constitute a potential source for primary-plant colonists. The misdirected products of arthropod dispersal may thus play a role in the establishment of vegetation, and thus to the diversification of the ecosystem. Of course, the establishment of pioneer plants in the blast zone, in this case the subalpine lupine *Lupinus lepidus* provides foci for herbivores and with them the diversification of the insect fauna; aphids, Lepidoptera, hymenopteran parasitoids, syrphids, and coccinelids quickly occupy the newly vegetated habitats.

My objective in this essay has been to emphasize the importance of the products of arthropod dispersal as a source of substrate for alpine and subalpine arthropods, and as a source of nutrient in developing soils. The fallout fauna of the alpine snowfields and pyroclastic surfaces provide annual testimony to the capacity of terrestrial arthropods for dispersal and thus for colonization of the patchy habitat spread beneath them.

Acknowledgments. Many people have aided in fieldwork and identification in the studies reviewed above. In particular I acknowledge Drs. D.H. Mann and S. Leffler, Carrie Becker, Rod Crawford, Merrill Peterson, and Rick Sugg.

Supported in part by a contract from the National Park Service for work on Mount Rainier, and a grant from N.S.F. for work on Mount Saint Helens.

References

Bowden J, Johnson CG (1976) Migrating and other terrestrial insects at sea. In: Cheng L (ed) Marine insects. Elsevier/North-Holland, Amsterdam, pp 97--117

Clements FE (1916) Plant succession. An analysis of the development of vegetation. Carnegie Inst
    Publ 520
Crawford RL (1985) Mt. St. Helens and spider biogeography. Proc Wash State Entomol Soc 46:
    700–702
Crawford RL, Edwards JS (1985) Ballooning spiders as a component of arthropod fallout on snow-
    fields of Mt. Rainier, Washington (ms submitted for publication)
Edwards JS (1982) Habitat, behavior and neurobiology of an American grylloblattid. In: Ando H
    (ed) Biology of the Notoptera. Kashiyo-Insatsu Co, pp 19–28
Edwards JS, Banko PC (1976) Arthropod fallout and nutrient transport: a quantitative study of
    Alaskan snowpatches. Arct Alp Res 8:237–245
Edwards JS, Schwartz LM (1981) Mt. St. Helens ash: a natural insecticide. Can J Zool 59:714–715
Glick PA (1939) The distribution of insects, spiders and mites in the air. US Dept Agric Tech Bull
    673, 150 pp
Grime JP (1979) Plant strategies and vegetation processes. Wiley, London, 222 pp
Heydemann B (1967) Der Überflug von Insekten über Nord- und Ostsee nach Untersuchungen auf
    Feuerschiffen. Dtsch Entomol Z 14:185–215
Hutchinson GE (1965) The ecological theater and the evolutionary play. Yale University Press,
    New Haven
Mani MS (1968) Ecology and biogeography of high altitude insects. Series Entomologica Junk, The
    Hague, 527 pp
Mann DH, Edwards JS, Gara RI (1980) Diel activity patterns in snowfield foraging invertebrates on
    Mount Rainier, Washington. Arct Alp Res 12:359–368
Melander AL (1921) Collecting insects on Mount Rainier. Smithsonian Rep 1921:415–422
Papp RP (1978) A noval aeolian ecosystem in California. Arct Alp Res 10:117–131
Scudder GCE (1963) Heteroptera stranded at high altitudes in the Pacific Northwest. Proc Entomol
    Soc B C 60:41–44
Southwood TRE (1966) Ecological methods, with particular reference to the study of insect popu-
    lations. Methuen, London
Spalding JB (1979) The aeolian ecology of White Mountain Peak, California: windblown insect
    fauna. Arct Alp Res 11:83–94
Swan LW (1963) Aeolian zone. Science 140:77–78
Van Dyke EC (1919) A few observations on the tendency of insects to collect on ridges and moun-
    tain snowfields. Entomol News 30:241–244
Wigglesworth VB (1976) The evolution of insect flight. In: Rainey EC (ed) Insect flight. R Entomol
    Soc Symp 7, pp 225–269

15 Ecological Studies Indicating the Migration of *Heliothis zea, Spodoptera frugiperda,* and *Heliothis virescens* from Northeastern Mexico and Texas

J. R. Raulston[1], S. D. Pair[2], F. A. Pedraza Martinez[3], J. Westbrook[2], A. N. Sparks[2], and V. M. Sanchez Valdez[4]

1 Introduction

The negative implications of crop pest-control technologies which evolved as a result of the development of effective pesticides are generally recognized today (Knipling 1982). The concept of integrated pest management has resulted as an effort to optimize our ability to protect our agriculture and concurrently alleviate our reliance on heavy pesticide use for such protection. However, Joyce (1982) succinctly points out that in order to effectively manage a pest population by either chemical or other means, the distribution of that population must be defined on a temporal and spatial basis. In this context, the role of insect movement must be considered in relation to the establishment of effective pest-management systems.

In the United States, the three species which are the subject of this paper, the corn earworm (CEW), *Heliothis zea* (Boddie), the fall armyworm (FAW), *Spodoptera frugiperda* (J.E. Smith), and the tobacco budworm (TBW), *H. virescens* (F.), annually invade a variety of crops and inflict heavy economic losses over large geographical areas. Among the major crops attacked by one or more of these insects are: corn, cotton, soybeans, sorghum, tomatoes, peanuts, and forage crops. In recent years, concentrated research efforts have been and are being made to establish pest-management systems to suppress these key pests. As part of this research thrust, efforts are underway to determine the ecological and environmental factors that influence the development of their populations within specific areas and the chronological and spatial interrelationships of populations between geographical areas.

2 Distribution

The three subject species are widely distributed in the North and South American continents and the adjacent Caribbean Islands. Luginbill (1928) indicated the FAW to be of tropical origin since no diapause mechanism exists which would allow the insect to normally survive the winters beyond the subtropical regions of southern Texas and

---

[1] Subtropical Crop Insects Research Unit, USDA, ARS, P.O. Box 1033, Brownsville, TX 78520, USA.
[2] Insect Biology and Population Management Research Laboratory, USDA, ARS, Tifton, GA 31793, USA.
[3] SARH, INIA, C.A.E. "Las Adjuntas", Apdo. Postal No. 3, S. Jimenez, Tamaulipas, Mexico.
[4] SARH, INIA, C.A.E. "Las Huastecas", Apdo. Postal C-1, Tampico, Tamaulipas, Mexico.

Insect Flight: Dispersal and Migration
Edited by W. Danthanarayana
© Springer-Verlag Berlin Heidelberg 1986

Florida. However, this species redistributes itself throughout the U.S. each year from its overwintering sites (Snow and Copeland 1969). The insect progressively extends its range northward reaching the northern states and Canada in August.

Hardwick (1965) indicates the CEW is also distributed throughout much of North and South America and the Caribbean Islands. Both the CEW and the TBW exhibit diapause mechanisms which allow them to overwinter further north than is possible for the FAW. Indeed, Snow and Copeland (1971) concluded that CEW overwintering occurred to a latitude of approximately 45°N in the U.S. Although winter survival is possible at the above-mentioned latitude, most of the overwintering studies that have been conducted in the U.S., even at latitudes as low as 30°N, indicate that actual survival of diapausing CEW is normally less than 5% (Blanchard 1942; Phillips and Barber 1929; Barber 1941; Stadelbacher and Pfrimmer 1972; Slosser et al. 1975; Stadelbacher and Martin 1980; Eger et al. 1983). These overwintering studies suggest that population buildup of the CEW at its more northerly ranges is probably a result of adult movement rather than overwintering.

The TBW distribution patterns within the U.S. have not been as clearly defined as that of the FAW or CEW. However, it is one of the most serious pests of cotton throughout the southern and western regions of the U.S. and occurs through much of Mexico, Central and South America, and the Caribbean Islands. One of the primary causes of the demise of the cotton industry in the Tampico area of the State of Tamaulipas, Mexico, was the inability to control this insect (Lukefahr 1970).

3 Study Region

Since most of the references cited previously indicate that continuous development of reproductive populations normally does not occur above the approximate latitude of 26°N, we have outlined a study region comprising northeastern Mexico and south Texas as shown in Fig. 1.

Raulston and Houghtaling (1984) give an ecological description of this region. Briefly, the study area is bounded on the north by the Lower Rio Grande Valley (LRGV) of Texas at a latitude of 25° 54'N; on the west by the Sierra Madre Oriental Mountains; on the south by Ciudad Mante and Tampico (22° 16'N); and to the east by the Gulf of Mexico (Fig. 1). Climatologically, the area is classified as semi-arid with annual precipitation rates averaging from 680 to 760 mm yr^{-1}. Precipitation maxima normally occur in September with smaller peaks occurring in May and June. Minimum mean monthly temperatures ranging from ca. 15° to 19 °C occur in January and February with maximum means ranging from 27° to 32 °C occurring between May and August.

Three major crops (corn, sorghum, and cotton) are grown in this region that sustain damage and potentially develop large populations of one or more of the species we are studying. The most important crop from the standpoint of production of populations of the CEW and FAW is corn. Between 200000–300000 ha of irrigated corn is produced in the combined northeastern section of the State of Tamaulipas and the LRGV. The crop is planted in February and matures in late May and June. An area in the central region of the State of Tamaulipas along the Soto la Marina River bounded to the west by Abasolo and to the east by Soto la Marina also produces ca. 15000 ha of

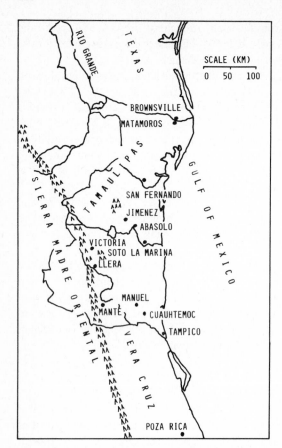

Fig. 1. Region of northeastern Mexico and Lower Rio Grande Valley in which ecological studies with *Heliothis zea, Spodoptera frugiperda*, and *H. virescens* were performed

irrigated corn. This crop is normally planted in late January and early February and matures in May ca. 2–3 weeks earlier than the northernmost corn area. The southern area of the State of Tamaulipas produces less than 5000 ha of corn, and the planting dates are variable depending primarily on available rainfall. Corn in this area can be found at various stages of maturity the entire year.

Northern Tamaulipas is the primary sorghum-producing area of Mexico with over 500000 ha of dryland planted to the crop annually in early February. The LRGV also produces ca. 200000 ha of sorghum which is planted in late February. The other crop of major importance in producing populations of these insects is cotton. This crop in recent years has been primarily planted in the LRGV with ca. 120000 ha planted annually in late February and early March.

4 Population Dynamics

4.1 Trap Observations

Pheromone traps (Hartstack et al. 1979) for all three species were placed at six locations within the study region and at Poza Rica, Vera Cruz, ca. 315 km south of Tampico. These locations listed from north to south are: (1) LRGV; (2) San Fernando; (3) Abasolo; (4) Padilla; (5) Llera; and (6) Cuauhtemoc (Tampico area). Tables 1–3 list 1983 capture data for Julian dates 1–120.

Table 1. Incidence of 1983 corn earworm male capture in pheromone traps located in the Lower Rio Grande Valley (LRGV) of Texas and northeastern Mexico[a]

10-Day period	Capture/indicated 10-day period (% total capture/trap)						
	LRGV	San Fernando	Abasolo	Padilla	Llera	Cuauhtemoc	Poza Rica
1	0	6.5	0.1	1.8	3.3	9.1	21.5
2	2.2	0	0.2	3.3	2.4	5.6	15.1
3	0.1	0	0.6	1.4	2.8	6.5	10.5
4	1.6	7.9	1.2	1.9	2.8	7.6	16.2
5	2.6	4.2	0.7	1.2	1.2	12.8	9.0
6	4.8	1.5	6.0	3.7	1.2	7.3	8.8
7	14.0	2.3	5.2	16.9	0.7	2.6	6.9
8	7.6	12.5	4.6	17.0	2.9	9.8	3.2
9	11.6	6.9	7.4	6.9	1.7	17.4	2.6
10	18.0	17.9	7.5	6.1	4.7	10.5	1.7
11	14.4	19.5	37.6	20.0	21.2	8.4	2.6
12	24.0	20.8	29.2	19.9	55.1	2.5	2.0
Total capture/trap	1464.0	636.5	888.0	2051.0	575.0	99.0	1219.2

[a] Data are from Julian day 1–120.

Table 2. Incidence of 1983 tobacco budworm male capture in pheromone traps located in the Lower Rio Grande Valley (LRGV) of Texas and northeastern Mexico[a]

10-Day period	Capture/indicated 10-day period (% total capture/trap)						
	LRGV	San Fernando	Abasolo	Padilla	Llera	Cuauhtemoc	Poza Rica
1	0	0.7	0.1	0.3	4.4	0.6	10.9
2	0.5	0.2	0.5	0.5	14.7	0	12.6
3	0	0	0.1	0	11.8	0.4	4.7
4	0	0	0.5	0.5	16.2	0.7	6.6
5	0.9	0	1.4	1.0	1.5	0.6	10.4
6	1.0	0.2	3.2	2.8	5.9	1.7	13.7
7	7.6	5.0	6.2	14.3	0	1.0	10.7
8	6.0	17.2	7.5	22.8	1.5	14.4	3.6
9	9.5	14.3	4.9	7.1	7.4	13.6	0
10	13.4	23.1	10.1	13.3	32.4	27.7	4.7
11	20.5	31.5	21.4	22.3	4.4	23.9	12.8
12	40.3	7.7	44.2	15.4	0	14.4	9.2
Total capture/trap	338.4	640.0	1109.0	1458.0	68.0	343.8	42.2

[a] Data are from Julian day 1–120.

Table 3. Incidence of 1983 fall armyworm male capture in pheromone traps located in the Lower Rio Grande Valley (LRGV) of Texas and northeastern Mexcio[a]

10-Day period	Capture/indicated 10-day period (% total capture/trap)						
	LRGV	San Fernando	Abasolo	Padilla	Llera	Cuauhtemoc	Poza Rica
1	0	No capture	7.1	1.1	2.4	12.1	7.6
2	8.3	No capture	9.0	1.1	4.8	10.4	33.3
3	0.6	No capture	20.8	0.1	9.7	5.0	17.1
4	1.4	No capture	17.6	0.1	6.5	10.7	19.7
5	4.3	No capture	1.6	14.3	1.6	13.5	8.6
6	4.9	No capture	7.5	6.4	4.0	10.6	4.5
7	21.9	No capture	3.9	16.9	2.4	6.4	3.4
8	17.5	No capture	2.0	33.4	0	17.9	2.7
9	7.0	No capture	0	2.9	8.1	3.9	1.1
10	2.3	No capture	6.3	15.0	60.5	7.5	0.5
11	2.1	No capture	0	8.7	0	1.1	1.3
12	4.7	No capture	24.3	0	0	0.9	0.3
Total capture/trap	46.9	No capture	255.0	909.0	124.0	337.0	1555.3

[a] Data are from Julian day 1–120.

4.1.1 Corn Earworm

The percent of CEW occurring within the first 40 days was generally progressively higher moving from north to south (Table 1). The highest capture peak at the southern-most location (Poza Rica) occurred within this 40-day period. However, the initial peaks at the more northern locations (LRGV to Padilla) did not occur until days 70–80 with the exception of San Fernando which had an initial peak at day 40. Although CEW capture at Cuauhtemoc was relatively low during this 120-day period, the trends in capture peaks at this location tended to be intermediate between the more souther-ly and northerly locations. Llera, which is bounded to the west by the Sierra Madre Oriental Mountains and to the north by a major escarpment (the Cuesta de Llera), does not appear to follow either a spatial or temporal pattern in relation to the other trap locations. Low levels of capture were observed at this location throughout the ob-servation period, and peak captures were then noted between days 110–120. Follow-ing the initial peaks, a second, larger-peak capture period occurred at the northern locations (LRGV-Padilla) between days 110–120. This adult peak was probably the result of emergence from the corn crop.

4.1.2 Tobacco Budworm

Of the trap locations observed, Llera and Poza Rica appeared to have the lowest TBW populations averaging only 68 and 42.2 per trap, respectively, for the 120-day ob-servation period (Table 2). Discounting these two locations, no major initial peaks were noted at the other trap sites; however, increases in trap captures were noted at all locations between days 70–80. Numbers of captured TBW males continued to increase at LRGV and Abasolo throughout the remainder of the observation period; however, at San Fernando, Padilla, and Cuauhtemoc, capture began to decline after days 100–110. Capture at Abasolo and Padilla was the highest of the locations for the 120-day period, even though no major cultivated hosts for this insect are grown within this area.

4.1.3 Fall Armyworm

For the 120-day trapping period, FAW captures were lowest at LRGV, the northernmost location, and highest at Poza Rica, the southernmost location (Table 3). Peak capture occurred at Poza Rica by day 20; however, at LRGV, the capture peak did not occur until day 70. Occurrence of FAW-capture peaks within the central region tended to be temporally more variable than those of CEW and TBW. However, increased capture of FAW within the central area was noted by day 50. Interestingly, no FAW capture was observed at San Fernando.

These trap data indicate a general temporal and spatial relationship of populations within the region. The populations of CEW and FAW in the southernmost location (Poza Rica) peaked earliest. These trap-capture peaks were associated with the developing corn crop in the Poza Rica area. The low-level captures occurring in the northern areas, prior to population development on wild or cultivated hosts, conceivably originated and dispersed northward from the southern areas. The later peaks then resulted from populations developing in local corn fields.

4.2 Heliothis spp. Diapause Induction and Termination

A chronological relationship between *Heliothis* spp. emergence from diapause and initial pheromone-trap captures was observed. Larvae of both CEW and TBW were collected at LRGV, Abasolo, and Cuauhtemoc from corn, pidgeon pea, and various wild hosts in the fall and winter months of 1982 and held for determination of diapause induction. Larvae were placed on artificial diet (Shaver and Raulston 1971) and held outside until pupation. After 10 days, numbers of diapaused pupae were determined by the presence of larval eyespots (Phillips and Newson 1966). Diapausing pupae were subsequently placed in 1.8-cm plastic tubes fitted on one end with a cotton plug and on the other end with a 113-g plastic cup covered with a fine-mesh screen. The plastic tube was then inserted into the soil to a depth of 2.5–5.0 cm. Subsequent emerging adults ascended the tube and were contained above ground in the plastic cup. Data in Table 4 indicate the incidence of diapause at the above locations. Low levels of CEW diapause were observed at LRGV between days 250–280. The percent of pupae entering diapause began to increase by day 290 and peaked at 31.3% by day 310.

Subsequently, lower levels of diapause were indicated (< 10%) through day 340 when the last pupae were observed from that area. CEW larval collections at Abasolo and Cuauhtemoc were sporadic; however, a similar pattern was observed at these locations with peak diapause occurring at Abasolo on day 310. Interestingly, higher levels

Table 4. Diapause induction of corn earworm and tobacco budworm at Lower Rio Grande Valley (LRGV), Abasolo, and Cuauhtemoc, 1982

Pupation date	No. pupae observed			Percent entering diapause		
	LRGV	Abasolo	Cuauhtemoc	LRGV	Abasolo	Cuauhtemoc
			Corn earworm			
240	50			0		
250	62			1.6		
260	44			0		
270	–			–		
280	132	79	56	2.3	17.9	23.2
290	130			10.0		
300	76			18.4		
310	67	85	55	31.3	46.6	20.0
320	67	38	17	7.5	5.3	11.8
330	54	84		7.4	7.7	
340	41			9.8		
350	–	92	15	–	8.3	6.6
360	–		55	–		5.5
			Tobacco budworm			
240	8			0		
250	39			5.1		
260	58			3.4		
270	–			–		
280	–			–		
290	–			–		
300	21			76.2		
310	24			91.7		
320	6			33.3		
330	3		11	33.3		54.5
340	–			–		
350	–			–		
360	–		35	–		54.3

of diapause were observed both at Abasolo and Cuauhtemoc earlier than at LRGV, with 17.9% and 23.2% diapause, respectively, being observed on day 280.

The incidence of diapause in the TBW followed a similar temporal pattern to the CEW at LRGV; however, the percent of the population entering diapause was higher compared to the CEW. The peak incidence of diapause occurred on day 310 with 91.7% of the pupae entering diapause. No TBW larvae were observed in the Abasolo area and only two collections were made in the Cuauhtemoc area between days 330 and 360. Approximately 55% of the larvae pupating in this period entered diapause. Raulston and Houghtaling (1984) show that the period when peak diapause is occurring in the southern area of the region, as observed in this test, is also the time when peak CEW and TBW populations were occurring on wild-host plants in the area. These authors also suggest that diapausing populations in the LRGV are relatively low due to a lack of host plant availability during the period of diapause induction.

Low levels of CEW and TBW emergence from diapause occurred at LRGV between days 40-80 with the peak occurring on day 90 (Table 5). Similar patterns of CEW

Table 5. Emergence from diapausing corn earworm and tobacco budworm pupae at Lower Rio Grande Valley (LRGV), Abasolo, and Cuauhtemoc, 1983

Julian date	Percent of total emergence/location				
	Corn earworm			Tobacco budworm	
	LRGV	Abasolo	Cuauhtemoc	LRGV	Cuauhtemoc
10	0	0	0	0	0
20	0	0	3.2	0	0
30	0	0	3.2	0	6.7
40	4.3	3.0	0	2.9	0
50	2.2	3.0	0	0	0
60	2.2	15.1	0	2.9	6.7
70	2.2	0	6.5	5.8	40.0
80	4.3	6.0	6.4	2.9	0
90	56.5	57.2	45.1	45.7	6.7
100	13.1	15.1	25.8	28.6	0
110	15.2	0	3.2	11.5	6.7
120	0	0	3.2	0	6.7
130–160	0	0	0	0	26.7

emergence occurred at Abasolo and Cuauhtemoc where emergence peaks also occurred on day 90. From 45.1% to 57.2% of the total emergence from diapause occurred at all three locations within this period.

Comparing the CEW diapause emergence peak at LRGV to trap captures at that location (Table 1), it is evident that the initial-capture peak which occurred on day 70 was not in synchrony with the peak emergence of moths from diapause. Further, since searches for larval populations on wild and cultivated hosts in the area prior to day 70 proved unfruitful, it would appear that the increase in the moth population indicated at day 70 did not originate locally. A similar pattern was observed at Abasolo, where high numbers of CEW were being captured by day 60 or 30 days prior to the diapause emergence peak. At Cuauhtemoc, a CEW-capture peak occurred during the peak emergence from diapause period (day 90); however, significant capture had been occurring in this area for the entire period previous to day 90, all of which could not have been associated with adults emerging from diapause. Further, corn fields were observed in this area as early as day 25 which were supporting CEW larval populations, and these developing populations were probably responsible for the trap captures observed in the area.

TBW-trap captures at LRGV began to increase by day 70 (Table 2), or 20 days prior to the peak emergence of moths from diapause in the area. As with the CEW, no larval infestations were observed in the area prior to that period which would produce the high moth population indicated by the trap captures. The TBW-diapause emergence peak at Cuauhtemoc occurred on day 70 or 10 days prior to the initial increased trap capture in that area (Tables 2 and 5).

Similar asynchronies between trap capture and emergence of overwintering CEW have been observed at other locations in the southern United States. Stadelbacher and

Pfrimmer (1972) observed a 33-day hiatus between initial trap capture and emergence from diapause at Stoneville, Mississippi, and Hartstack et al. (1982) observed a 19-day difference between their initial trap peak and the mean emergence from diapause at College Station, Texas. In view of the favorable wind flow from south to north at the time of their peak-trap capture, Hartstack et al. (1982) concluded that this adult population possibly originated in Mexico.

Raulston et al. (1982) also indicated that wind-flow patterns conducive to aerially transporting insects from south to north along the Gulf Coast of Mexico frequently occur within our study region. This phenomenon is further illustrated in Fig. 2 which shows successive 5-h nocturnal trajectories that would be followed by insects taking flight from Jimenez, Tamaulipas, Mexico, on days 67–70, 1983, at 1800 h CST and subsequently being transported by winds at 500 m above ground level. Data in Tables 1 to 3 show that initial increases in adult trap capture of CEW, TBW, and FAW occurred during this period. A maximum northward displacement of ca. 200 km over the initial 5-h period occurred on day 69; however, in all instances, a minimum-northward displacement of ca. 150 km and a maximum-northward displacement of ca. 320 km occurred within the three trajectory periods.

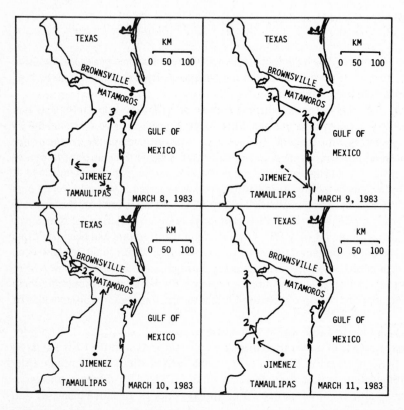

Fig. 2. Successive 5-h nocturnal trajectories at 500 m above ground level for insects taking flight at 1800 CST from Jimenez, Tamaulipas, Mexico

4.3 Development of Larval Populations

As indicated earlier, wild host-plant observations were made in 1983 at LRGV for evidence of infestation by *Heliothis* spp. These observations were initiated on day 28 and terminated on day 111. Four *Heliothis* spp. wild host plants, viz., wild tobacco *(Nicotiana rapanda), Abutilon trisulcatum, Bastardia viscosa*, and *Verbena* spp., were observed.

Graham and Robertson (1970) and Graham et al. (1972) indicate that wild tobacco was the most important wild host plant of *Heliothis* spp. in the LRGV; however, during the dry conditions in the spring of 1983 populations of this plant were extremely low. Also very low numbers of two other host species attractive to *Heliothis* spp., *A. trisulcatum* and *B. viscosa* were noted, and the only host plants found to be relatively abundant were the *Verbena* spp. No *Heliothis* eggs or larvae were found on any of these plants during our observations.

Heliothis larval infestations were also monitored on three wild hosts (*B. viscosa, A. umbellatum,* and *A. lignosum*) in the San Fernando area between days 81–118. Data listed in Table 6 show both CEW and TBW were developing on wild hosts within this area by day 81. Indeed, the highest rates of infestation were observed on days 81–82; however, by day 118 these host plants had matured and were no longer supporting *Heliothis* larval populations. Although the magnitude of these wild host-plant populations or their actual influence on *Heliothis* population dynamics is not known, Raulston and Houghtaling (1984) observed that *B. viscosa* and *A. trisulcatum* are prevalent along roadsides, in fallow fields, and in fields recently cleared of native vegetation. In 1981, they also observed initial *Heliothis* spp. larval populations on *B. viscosa* near San Fernando on day 77 with the peak occurring on day 93. Interestingly, these *Heliothis* larval infestations are well synchronized with the initial trap-capture peaks at San Fernando and occur prior to the peak emergence from diapause. Thus, considering that no major cultivated or wild-host reservoirs are available within the area earlier for de-

Table 6. Occurrence of *Heliothis* spp. larvae on wild-host plants in San Fernando area, 1983

Julian date	Plant species	No. plants observed	Av. No. larvae/plant		% Larvae parasitized	
			Corn earworm	Tobacco budworm	Corn earworm	Tobacco budworm
81–82	*Bastardia viscosa*	50	0.34	0.36	47.1	27.8
82	*Abutilon umbelatum*	10	0	0.50	–	0
	A. lignosum	20	0	0.35	–	14.3
97	*B. viscosa*	50	0.20	0.20	90.0	60.0
103	*B. viscosa*	32	0.03	0.19	0	16.7
	A. umbelatum	20	0	0.40	–	0
	A. lignosum[a]	12	0	0	–	–
118	*B. viscosa*[a]	ca. 50	0	0	–	–
	A. umbelatum[a]	ca. 20	0	0	–	–
	A. lignosum[a]	ca. 20	0	0	–	–

[a] Plants were senescent and were no longer attractive.

velopment of *Heliothis* populations, those adults responsible for these initial infestations on wild hosts are likely to be migrants.

Observations of CEW and FAW larval infestations on corn were initiated on day 71 in the LRGV and Abasolo areas. Combined observations were made on the irrigated corn produced in the Rio Bravo area of northeastern Mexico and LRGV. Data in Table 7 indicate that CEW and FAW infestations on pre-tassle corn in the LRGV were low, with the peak infestations for both species occurring between days 111 and 120. The major occurrence of mature larvae for both species on pre-tassle corn also occurred during this period at LRGV. The initial incidence of CEW larvae on pre-tassle corn in the Abasolo area, between days 81–90, was low and no larvae were observed on two pre-tassle corn fields between days 121–130. The FAW infestation on pre-tassle corn at Abasolo peaked between days 121–130. By this period (days 121–130) much of the corn at Abasolo was in the silk stage, and CEW infestations on this age of corn averaged 32.3% (Table 8). In comparison, FAW infestations on silking corn at this time was low, which may indicate differences in preference between the two species for the various corn stages. A similar preference pattern was observed at LRGV.

An initial increase of CEW larvae on fruiting corn occurred at LRGV between days 121–130 (Table 8). An increase subsequently occurred after day 151 on later planted corn which probably was a result of reproduction by adults represented by the initial infestation. CEW infestation during this period was 66% (days 161–170). The initial increase of FAW larval infestation on fruiting corn at LRGV occurred between days 131–140 and was larger than the initial CEW infestation; however, the subsequent FAW infestation did not build up to the same magnitude of that of the CEW. A similar pattern was observed at Abasolo; however, in this region, initial infestations on silking corn were higher than those observed at LRGV and did not develop to the magnitude observed later at LRGV. This may be partly due to the fact that crop maturity occurred earlier and over a shorter period of time at Abasolo which did not allow for the development of a second generation on fruiting corn within that area.

Table 7. Corn earworm (CEW) and fall armyworm (FAW) infestation of pre-tassle corn in northeastern Mexico and Lower Rio Grande Valley (LRGV) of Texas (spring-early summer observations)

Julian date	No. fields	Total larvae	Infestation level (%)		Frequency of larval stages (%)					
					CEW			FAW		
			CEW	FAW	Small	Medium	Large	Small	Medium	Large
LRGV area										
71– 80	1	28	0	< 1	–	–	–	100.0	0	0
81– 90	2	133	2.8	< 1	73.0	27.0	0	40.0	44.0	17.0
111–120	6	100	4.0	4.7	4.3	51.1	44.6	18.9	41.5	39.6
121–130	2	5	1.0	4.0	0	0	100.0	75.0	25.0	0
Abasolo area										
81– 90	2	71	0.22	4.1	100.0	0	0	60.6	14.1	25.3
121–130	2	11	0	13.3	–	–	–	18.2	36.4	45.5
131–140	1	0	–	–	–	–	–	–	–	–

Table 8. Corn earworm (CEW) and fall armyworm (FAW) infestation of fruiting corn in northeastern Mexico and Lower Rio Grande Valley (LRGV) of Texas (spring-early summer observations)

Julian date	No. fields	Total larvae	Infestation level (%)		Frequency of larval stages (%)					
			CEW	FAW	CEW			FAW		
					Small	Medium	Large	Small	Medium	Large
LRGV area										
21– 30	1	16	13.0	3.0	– – – Not sized – – –			– – – Not sized – – –		
111–120	2	3	4.0	0	33.3	66.7	0	–	–	–
121–130	2	9	8.0	1.0	100.0	0	0	100.0	0	0
131–140	11	156	6.7	18.6	36.4	38.6	25.0	67.2	24.6	8.2
141–150	2	20	6.3	11.7	100.0	0	0	92.3	7.7	0
151–160	4	313	60.6	24.4	72.6	16.1	11.2	70.0	21.1	8.9
161–170	3	180	66.0	20.7	44.5	29.2	26.3	67.4	23.3	9.3
Abasolo area										
21– 30	1	34	29.0	5.0	– – – Not sized – – –			– – – Not sized – – –		
111–120	5	63	29.9	2.0	71.2	22.0	6.8	100.0	0	0
121–130	10	179	32.3	2.7	69.7	21.8	8.5	42.9	42.9	14.3
131–140	5	131	20.7	7.3	49.4	20.8	29.9	66.7	14.8	18.5
141–150	3	79	36.1	13.9	50.9	29.8	19.3	31.9	22.7	45.5

Table 9. Population dynamics of *Heliothis* spp. on cotton in Lower Rio Grande Valley, 1983

Julian date	No. fields	Infestation rate (%)		Frequency of corn earworm (%)
		Eggs	Larvae	
106–110	2	3.0	1.0	0
111–115	3	2.0	0	0
116–120	2	0	0	–
121–125	1	4.0	2.0	0
126–130	0	–	–	–
131–135	4	6.0	0.5	0
136–140	4	3.0	3.0	0
141–145	2	2.0	0	0
146–150	2	2.0	0	0
151–155	3	0.7	0	0
156–160	4	8.0	0.5	56.3
161–165	4	10.0	2.0	70.0
166–170	5	2.0	10.0	96.0
171–175	7	2.5	2.5	97.9
176–180	4	24.0	50.0	83.6
181–185	4	0	0	89.1[a]
186–190	5	6.5	1.5	52.2
191–195	3	1.3	0	0
196–200	1	42.0	4	17.6
201–205	3	0	0	–
206–210	0	–	–	–
211–215	1	–	–	7.2

[a] Corn earworm frequency calculated from insects collected outside of fields observed for infestation rate.

Heliothis spp. larval infestations were also observed on cotton at LRGV, and these data are presented in Table 9. Total *Heliothis* infestation in cotton ranged from 0–6.5% from days 106–155, and during this period, only TBW were observed. Between days 156–160, the infestation level increased to 8.5% of which 56% were CEW. The frequency of CEW larvae continued to increase through day 175, at which time it comprised 97.9% of the *Heliothis* population on cotton. This influx of CEW on cotton was associated with the adult emergence from the mature corn crop in the area. Thus, the deterioration of Habitat I (the mature-corn crop) resulted in a facultative dispersal into Habitat II (cotton) as described by Hughes (1979).

It is interesting to note that at the time when attractive corn was available, no CEW population development was occurring in cotton (Tables 7 and 8).

Following maturation of the corn crop in LRGV and Abasolo areas, observations were made on the frequency of larval exit holes from corn ears. In the LRGV area, 52% of the corn ears contained exit holes, while in the Abasolo area, 49% contained exit holes (Table 10). Based on CEW and FAW medium- and large-larvae frequencies presented in Table 8, 39.9% of the corn in LRGV produced mature-CEW larvae, and 12.1% produced mature-FAW larvae. Further, 39.5% of the corn at Abasolo produced mature CEW, and 9.5% produced mature FAW. An estimate of the magnitude of the CEW and FAW moth populations emerging from mature corn is also presented in Table 10. These estimates were based on larval frequencies on fruiting corn, moth emergence from mature corn, and subsequent soil-sample observations for qualitative and quantitative determination of pupal exuviae made in 1984 (J.R. Raulston, unpublished data). Based on these parameters, 16.7% of mature-CEW larvae and 84.2% of mature-FAW larvae which developed on fruiting corn reached adulthood. We estimate from these data that the 200000 ha corn crop in LRGV produced 1.04×10^9 and 1.58×10^9 CEW and FAW adults, respectively, and the Abasolo area produced 7.71×10^7 CEW adults and 9.33×10^7 FAW adults. The largest CEW trap-capture peaks of the year occurred between days 145–165 at LRGV and Abasolo which was associated with emergence from the mature-corn crops. A similar FAW peak occurred at LRGV during this period; however, no peak was observed at Abasolo related to the adult emergence from corn.

Radar observations on airborne insects made during the days 171–174 in 1984 (W.W. Wolf et al., unpublished data) showed that large numbers of airborne insects were being transported northward from the region. CEW moths were also captured in airplane-drawn aerial nets from insect layers at altitudes as high as 1767 m above ground level during the course of the above investigation.

During the peak-trap capture interval, we observed synoptic weather maps to determine the general direction airborne insects might be transported during this period. Between days 151 and 172, weather systems developed on three occasions capable of providing long-distance transport from our study region, primarily toward the north and northwest.

The first transport system (June 2–4) was associated with a cool front moving toward the southeast (Fig. 3A). This system would have resulted in an initial northwesterly displacement toward the Big Bend area of Texas, and a subsequent northerly displacement toward the Panhandle region of north Texas.

The second transport system (June 10–12) was initiated with a trough formation extending from North Dakota to west Texas, and the subsequent southeasterly move-

Table 10. Incidence of larval exit holes and/or damage to ears by full-grown larvae from mature corn fields in Lower Rio Grande Valley (LRGV) and Abasolo, 1983

Period of observation (Julian date)	No. fields observed	Total % damaged ears	% Damage attributed to [a]		Estimated adult emergence/ha [b]		Estimated No. adults produced/area [c]	
			Corn earworm	Fall armyworm	Corn earworm	Fall armyworm	Corn earworm	Fall armyworm
LRGV								
154–166	13	52.0	39.9	12.1	5.19×10^3	7.92×10^3	1.04×10^9	1.58×10^9
Abasolo								
145–152	9	49.0	39.5	9.5	5.14×10^3	6.22×10^3	7.71×10^7	9.33×10^7

[a] Based on frequency of medium and large larvae from ear-stage corn.

[b] Based on 77 854 ears ha^{-1}. Estimate assumes that 16.7% of mature corn-earworm larvae and 84.2% of mature fall-armyworm larvae reach adulthood. This assumption is based on observed larval frequencies vs pupae collected from soil samples in mature corn in 1984.

[c] Based on 200 000 ha of corn at LRGV and 15 000 ha at Abasolo.

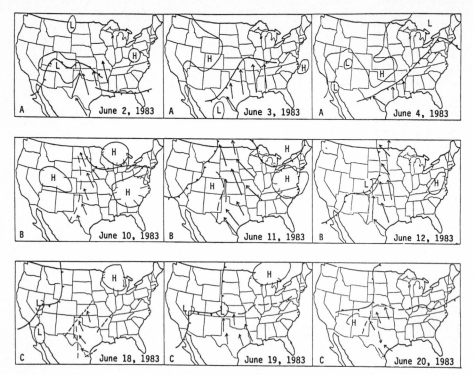

Fig. 3A–C. Synoptic-surface weather patterns illustrating favorable transport systems from north-eastern Mexico and Lower Rio Grande Valley of Texas during the emergence of *Heliothis zea* and *Spodoptera frugiperda* from fields of mature corn

ment of a cool front which developed over southwestern Canada and northwestern U.S. (Fig. 3B). Again the initial movement from our region was toward the Big Bend of Texas with a subsequent northerly direction occurring further to the north.

The third system (June 18–20) was also associated with frontal and trough systems moving across the U.S. toward the southeast with wind movement being funneled from our region toward the north (Fig. 3C). Interestingly, cotton producers in west Texas normally expect their initial *Heliothis* egg deposition to occur between days 167 and 177 (C. Allen, personal communication, 1984) which is well synchronized with the CEW emergence from our study region. Considering the magnitude of this emerging CEW population and the frequent development of aerial-transport systems similar to that shown in Fig. 3, we hypothesize that the *Heliothis* egg deposition occurring on cotton in west Texas is a likely result of moths migrating from southern Texas and northern Mexico as supported by radar observations of W.W. Wolf in 1984 (unpublished data).

5 Discussion

Although we have not attempted to physically track the movement of adult populations of the three subject moth species, our study illustrates the necessity and role of ecological studies in insect migration. Such studies define temporal and spatial trends in populations as well as areas where movements may originate and culminate. We have shown both a space and time relationship in development of populations within the study region with a progression occurring from south to north in the spring. We further implicate the corn-growing area in the northern section of the region as the source of populations subsequently migrating to the north and west in early summer. This implication is supported by the temporal synchrony of emerging populations from the corn in the region and the occurrence of initial infestations further north, as well as the occurrence of weather systems that develop in conjunction with this emergence that are conducive to aerial transport into those areas.

Acknowledgment. The authors gratefully acknowledge the efforts of Mr. John Norman, who provided data on *Heliothis* spp. populations occurring on cotton in the Lower Rio Grande Valley of Texas, and Dr. Alex Ortega (CYMMIT), who provided trap data from Poza Rica, Vera Cruz, Mexico. The authors further acknowledge Mr. James Houghtaling (USDA ARS, Brownsville, Texas), whose untiring efforts have made this study possible.

References

Barber GW (1941) Hibernation of the corn earworm in southeastern Georgia. US Dep Agric Tech Bull 791, 17pp

Blanchard RA (1942) Hibernation of the corn earworm in the central and northeastern parts of the United States. US Dep Agric Tech Bull 838, 14 pp

Eger JE Jr, Sterling WL, Hartstack AW (1983) Winter survival of *Heliothis virescens* and *Heliothis zea* (Lepidoptera: Noctuidae) in College Station, Texas. Environ Entomol 12:970–975

Graham HM, Hernandez NS, Llanes JR (1972) The role of host plants in the dynamics of populations of *Heliothis* spp. Environ Entomol 1:424–430

Graham HM, Robertson OT (1970) Host plants of *Heliothis virescens* and *H. zea* (Lepidoptera: Noctuidae) in the Lower Rio Grande Valley. Ann Entomol Soc Am 63:1261–1265

Hardwick DF (1965) The corn earworm complex. Mem Entomol Soc Can 40, Ottawa, Canada, 248 pp

Hartstack AW, Lopez JD, Muller RA, Sterling WL, King EG, Witz JA, Eversull AC (1982) Evidence of long range migration of *Heliothis zea* (Boddie) into Texas and Arkansas. Southwest Entomol 7:188–201

Hartstack AW Jr, Witz JA, Buck DR (1979) Moth traps for the tobacco budworm. J Econ Entomol 97:1077–1089

Hughes RD (1979) Movement in population dynamics. In: Rabb RL, Kennedy GG (eds) Movement of highly mobile insects: concepts on methodology in research. State University Press Raleigh, NC, pp 14–32

Joyce RJV (1982) A critical review of the role of chemical pesticides on *Heliothis* management. In: Proc Int Workshop Heliothis Management. International Crops Research Institute for the Semi-Arid Tropics (ICRISAT), 15–20 November 1981, ICRISAT Center, Patancheru, A.P., India, pp 173–188

Knipling EF (1982) The rational for areawide management of *Heliothis* populations. Bull Entomol Soc Am

Luginbill P (1928) The fall armyworm. US Dep Agric Tech Bull 34, 89 pp

Lukefahr MJ (1970) The tobacco budworm situation in the Lower Rio Grande Valley and Northern Mexico. In: Proc 2nd Am Tex Conf Insect Pl Dis, Weed and Brush Control. Texas A and M University, pp 140–145

Phillips WJ, Barber GW (1929) A study of hibernation of the corn earworm in Virginia. Va Agric Exp Sta Tech Bull 40, 24 pp

Phillips JR, Newson LD (1966) Diapause in *Heliothis zea* and *Heliothis virescens* (Lepidoptera: Noctuidae). Ann Entomol Soc Am 59:154–159

Raulston JR, Wolf WW, Lingren PD, Sparks AN (1982) Migration as a factor in *Heliothis* manage-
    ment. In: Proc Int Workshop *Heliothis* Management. International Crops Research Institute for
    the Semi-Arid Tropics (ICRISAT), 15–20 November 1981, ICRISAT Center, Patancheru, A.P.,
    India, pp 61–73
Raulston JR, Houghtaling JE (1984) Circumstantial ecological evidence for *Heliothis virescens* mi-
    gration into the Lower Rio Grande Valley of Texas from northeastern Mexico. In: Sparks AN
    (ed) Long range migration of moths of agronomic importance to the US and Canada. US Dep
    Agric Tech Bull (in press)
Shaver TN, Raulston JR (1971) A soybean-wheatgerm diet for rearing the tobacco budworm. Ann
    Entomol Soc Am 64:1077–1079
Slosser JE, Phillips JR, Herzog GA, Reynolds CR (1975) Overwinter survival and spring emergence
    of the bollworm in Arkansas. Environ Entomol 4:1015–1024
Snow JW, Copeland WW (1969) Fall armyworm: use of virgin female traps to detect males and to
    determine seasonal distribution. US Dep Agric Prod Res Rep no 110, 9 pp
Snow JW, Copeland WW (1971) Distribution and abundance of the corn earworm in the United
    States. US Dep Agric Plant Pest Control Div, Coop Econ Insect Rep 21:71–76
Stadelbacher EA, Martin DF (1980) Fall diapause, winter mortality, and spring emergence of the
    tobacco budworm in the delta of Mississippi. Environ Entomol 9:553–556
Stadelbacher EA, Pfrimmer TR (1972) Winter survival of the bollworm at Stoneville, Mississippi.
    J Econ Entomol 65:1030–1034

16 Radar Observations and Collections of Insects in the Gulf of Mexico

W. W. WOLF[1], A. N. SPARKS, S. D. PAIR, J. K. WESTBROOK, and F. M. TRUESDALE

1 Introduction

Infestations of insects such as corn earworm, *Heliothis zea* (Boddie), tobacco bud-worm, *H. virescens* (Fabricius), and fall armyworm, *Spodoptera frugiperda* (J.E. Smith), usually spread from southern to northern parts of the United States and Canada each year where they cause millions of dollars in economic losses. Since they do not survive the winter in these northern locations, they must disperse from their winter habitats each year. These winter habitats are assumed to be in the southern United States or Mexico. If the insects can cross the Gulf of Mexico, then Cuba and the Yucatan Penin-sula of Mexico may be additional sources of early-season infestations in the United States.

Many insects move long distances over land. Callahan et al. (1972) reported catching large numbers of corn-earworm moths and other Lepidoptera in traps on a television tower 317 m above the ground. The number of insects caught, the flight altitudes, and winds at those altitudes suggested mass movements for significant distances during a single night.

Radar observations near Brownsville, Texas (Wolf et al., in press) indicated that *Heliothis*-sized insects flew at altitudes where the southerly winds were maximum, and the flights lasted for several hours during the night. This type of insect behavior can result in long-distance transport of insects by synoptic winds. Climatic studies of syn-optic winds indicate that southerly winds occur frequently during the spring in the southwestern United States (Muller and Tucker, in press). These winds could transport insects from Mexico into the U.S. via Texas or across the western portions of the Gulf of Mexico. A high-pressure cell (often called the Bermuda high) develops each year off the southeastern coast of the U.S. The clockwise circulation of this synoptic system is favorable for assisting insect movement across even the widest portions of the Gulf of Mexico.

Insect traps operated aboard ships at sea (Gressitt and Nakata 1958; Yoshimoto and Gressitt 1959, 1960, 1961) caught insects at distances hundreds of kilometers from shore. Drake et al. (1981) reported that radar operated on a peninsula in northwestern Tasmania detected insects that apparently crossed the Bass Straight from Australia. The width of the Bass Straight is at least 350 km. The insect source areas in Australia and wind trajectories indicated that the flights covered much greater distances. Our

[1] Insect Biology and Population Management Research Laboratory, USDA, ARS, Tifton, GA 31793, USA.

Insect Flight: Dispersal and Migration
Edited by W. Danthanarayana
© Springer-Verlag Berlin Heidelberg 1986

radar observations on land, as well as those in Africa by Riley et al. (1983) and in Canada by Schaefer (1976), indicate that nocturnal insects stop flying before the sun rises. If these insects behave the same way over the ocean, then they must complete the crossing within one night or land in the ocean and die. Successful flight over the ocean would thus be limited by the wind speed during a single night. However, if nocturnal insects also flew during the day while over the ocean, the probability of their crossing the widest portions of the Gulf of Mexico would be dramatically increased.

This paper presents results of light-trap catches, insect collections on a ship, and radar observations made to assess the potential for insect movement across the Gulf of Mexico.

2 Methods

Initial observations were made during 1973 using blacklight traps mounted on unmanned oil platforms off the Texas-Louisiana shore (Sparks et al. 1975). During 1982 and 1983, radar observations and insect collections were made aboard a National Oceanic and Atmospheric Administration (NOAA) fisheries research vessel.

Blacklight traps were especially constructed for the marine environment of offshore oil platforms. With the cooperation of the U.S. Geological Survey, traps were mounted on unmanned platforms. A single, blacklight 15-W fluorescent lamp was mounted vertically in the center of four intersecting baffles and located above a funnel and collection container. One trap per platform was located ca. 32, 74, 106, and 160 km from shore. These traps were operated from September 11 to October 21, 1973.

Transportation was provided via the U.S. Geological Survey-leased helicopter service. Oil-field personnel on adjacent, manned platforms serviced the traps and changed collection containers.

Through cooperative efforts of the NOAA Southeast Fisheries Center at Pascagoula, Mississippi, a special radar for detecting insects was installed aboard the *Oregon II*. This vessel is used to sample marine organisms in the Gulf of Mexico. During each cruise the ship stops at predetermined sampling stations dictated by NOAA objectives. Radar data and insects were collected during cruise number 130, October 12–22, 1982, and during cruise number 134 from April 4–June 24, 1983.

The radar consisted of a 25-W peak power, X-band (9.45 GHz) marine transceiver connected to a 1.22-m diameter parabolic antenna. This antenna could be aimed vertically or at various angles above the horizon. The output from the radar receiver was displayed on an oscilloscope. This scope provided the altitude and signal strength of individual insects as they passed through the radar beam (A-scope).

Individual insects were counted as they passed through the vertically pointing radar beam. A vertical profile of relative insect density vs altitude was obtained by counting the number of insectlike signals observed on the A-scope during 2-min timed intervals. The counts were repeated for each of four altitude intervals: 300–450, 600–750, 900–1050, and 1200–1350 m, respectively, to complete a profile. Profile counts were repeated hourly during the first cruise (except when fatigue intervened). When insect activity was detected, the counts were repeated at ca. 30-min intervals. During the second cruise, profiles were measured every 2 h with increased frequency while insects were being detected.

The maximum-detection range of the insects was determined by pointing the antenna at various angles above the horizon and examining the A-scope for the maximum-target range. The maximum-detection range is related to the maximum size of a target passing through the center of the beam when the target is broadside to the beam.

The relative flight speeds of the targets were assessed by pointing the beam upwind or downwind (depending on the ship's heading). The displacement of individual targets was observed on the A-scope. The faster-flying targets approached or receded faster than the slower targets. Absolute measurements of air speed were not possible because wind vectors at target altitudes were not known.

Assessment of the proportion of the targets which were birds was made from the maximum-detection range, relative velocity of targets, and variation in amplitude of radar signals from individual targets. Intermittent variation of signal amplitude from individual targets was interpreted as intermittent wing beating and gliding of small birds. Steady fluctuations of signals at rates less than 10 to 20 Hz were also interpreted as birds as suggested by Schaefer (1976).

The number of insects per unit volume of air could not be determined due to the rolling of the ship and the unknown wind velocity above the ship. The motion of the ship caused the beam to move in a searchlight fashion so the sampling volume was unknown.

Insects were collected from the ship deck by the ship crew during their watches and by the radar operator after each set of radar observations. Collections were marked with time, date, location, and were frozen for later identification. During the second cruise, a 15-W blacklight trap and two pheromone traps were installed on the ship. A trap baited with fall-armyworm pheromone was located near the front of the ship and a corn-earworm trap near the center. These traps were checked at least four times per day.

One of the NOAA experiments included sampling the ocean surface for ichthyo-plankton with a neuston net. This net was 1 m high by 2 m wide and was dragged with the lower lip of the net 0.5 m below the ocean surface for ca. 600 m. Collections from this net were preserved for later identification. The actual sea-surface area sampled was calculated from the ship's speed, time in the water, and net width. This area was used to calculate insects per hectare of ocean.

The ship's log provided air temperature, wind speed, and direction. Additional meteorological data were obtained from the National Weather Service buoys stationed in the center of the Gulf. Upper-level wind data were obtained from stations along the Gulf Coast of the U.S. and were used to construct wind trajectories (Reap 1972) depicting wind displacement prior to arrival at the ship's location. These trajectories were calculated for a 100-mb pressure altitude (ca. 150 m above sea level).

3 Results

In 1973, Hurricane Delia passed through the Gulf before traps were placed on the offshore platforms. Subsequently, at least two cool fronts, accompanied by northerly winds, passed the platforms while the traps were operating. Trap collections contained 9 orders, 66 families, and 169 species of insects. The family Noctuidae and the order Diptera represented 50 and 27% of the total collected, respectively (Table 1). Some of

Table 1. Orders and families of insects collected in light traps on unmanned oil platforms located in the Gulf of Mexico (9 orders, 66 families, 169 species), September 11–October 7, 1973

Order	Family	No. of species	Total catch	Distance from shore in km[a]			
				32	74	106	160
Coleoptera:	Anthicidae	1	1	x			
	Anthribidae	1	3		x	x	
	Bostrichidae	1	1	x			
	Carabidae	17	71+	x	x	x	x
	Coccinellidae	2	77	x		x	
	Curculionidae	2	5	x			
	Dytiscidae	1	1				x
	Elateridae	2	2	x	x		
	Heteroceridae	1	6	x	x	x	
	Limnichidae	1	15	x			
	Platypodidae	1	1	x			
	Scarabaeidae	4	30	x	x	x	x
	Scolytidae	1	1			x	
	Staphylinidae	?	58	x	x		x
	Tenebrionidae	1	1	x			
Diptera:	Anthomyiidae	1	1			x	
	Calliphoridae	2	120	x	x		x
	Ceratopogonidae	6	38	x		x	
	Chironomidae	3	567	x	x	x	
	Chloropidae	3	11	x	x	x	
	Culicidae	6	9	x	x		
	Dolichopodidae	3	4	x	x		
	Ephydridae	3	8	x	x		
	Micropezidae	1	1	x			
	Milichiidae	1	2		x		
	Muscidae	2	117	x	x	x	x
	Mycetophilidae	1	1		x		
	Otitidae	2	5	x	x		
	Phoridae	1	1		x		
	Sarcophagidae	1	1	x	x	x	x
	Sciomyzidae	1	1		x		
	Sepsidae	1	1		x		
	Simuliidae	1	1		x	x	
	Sphaeroceridae	1	4	x	x		
	Stratiomyidae	1	61	x	x	x	x
	Syrphidae	3	53	x	x	x	x
	Tabanidae	3	4	x			
	Tachinidae	2	12	x		x	
	Tephritidae	2	3	x		x	x
Ephemeroptera:	Ephemeridae	2	16	x	x		
Hemiptera-Homoptera:	Aphididae	4	6	x	x	x	
	Cicadellidae	7	32	x	x	x	x
	Delphacidae	9	306	x	x	x	x
	Pentatomidae	3	25	x	x	x	x
	Psyllidae	1	1			x	

Table 1 (continued)

Order	Family	No. of species	Total catch	Distance from shore in km[a]			
				32	74	106	160
Hymenoptera:	Braconidae	1	1	x			
	Ichneumonidae	2	6		x	x	
	Scelionidae	1	1	x			
Lepidoptera:	Arctiidae	3	12	x	x	x	
	Ctenuchidae	1	1	x			
	Gelechiidae	1	1			x	
	Hyblaeidae	1	76	x	x		
	Noctuidae	23	1917	x	x	x	x
	Olethreutidae	2	6		x	x	
	Pyralidae	6+	58	x	x	x	x
	Sphingidae	4	11	x	x		
	Tortricidae	2	6	x	x	x	x
	Yponomeutidae	1	23	x	x	x	x
Odonata:	Aeschnidae	1	8		x	x	x
	Coenagrionidae	1	9	x	x	x	
	Libellulidae	2	2	x			x
Orthoptera:	Gryllidae	1	18	x			x
	Tettigoniidae	1	2			x	x
Trichoptera:	Hydropsychidae	1	1		x		
	Leptoceridae	1	1	x			
	Psychomyiidae	1	1	x			

[a] An x indicates location of insect catches; actual numbers were not available.

Table 2. Number of economic species of Lepidoptera caught in light traps on oil platforms located in the Gulf of Mexico, September 11–October 7, 1973 (previously reported by Sparks et al. 1975)

Scientific name	Common name	Distance from shore (km)			
		32	74	106	160
Agrotis ipsilon (Hufnagel)	Black cutworm	2	1	3	2
Anticarsia gemmatalis (Hübner)	Velvetbean caterpillar	16	3	1	4
Spilosoma virginica (Fabricius)	Yellow woollybear	0	1	0	0
Agrotis subterranea (Fabricius)	Granulate cutworm	4	28	12	9
Heliothis virescens (Fabricius)	Tobacco budworm	0	1	0	1
Heliothis zea (Boddie)	Corn earworm	27	16	8	3
Hyphantria cunea (Drury)	Fall webworm	5	0	0	0
Peridroma saucia (Hübner)	Variegated cutworm	0	1	1	0
Pseudaletia unipuncta (Haworth)	Armyworm	22	2	2	0
Pseudoplusia includens (Walker)	Soybean looper	3	0	6	0
Spodoptera dolichos (Fabricius)	Large cotton cutworm	0	0	0	2
Spodoptera exigua (Hübner)	Beet armyworm	16	0	30	31
Spodoptera frugiperda (J.E. Smith)	Fall armyworm	120	24	31	41
Spodoptera ornithogalli (Guenée)	Yellow-striped armyworm	1	0	0	0
Trichoplusia ni (Hübner)	Cabbage looper	163	89	526	631

the most economically important lepidopteran species in the U.S. were collected as far as 160 km from shore, and the number of most species decreased with increased distance from the shore (Table 2). Two exceptions were *Spodoptera exigua* (Hübner) and *Trichoplusia ni* (Hübner).

The cruise track of the *Oregon II* during October 1982 stayed within 25 km of the Alabama-Florida coast for the first 3 days. Starting October 15, the ship traveled as far as 320 km from the shore and by October 19 had returned to within 25 km of the mouth of the Mississippi River (Fig. 1A). The winds at the ship were initially from the north, shifted to the northwest, back to the north, east, south, and back to the northeast. The curvature and direction of wind trajectories indicated that on some days insects traveled significantly further than the shortest distance from ship to shore (Fig. 1B). From October 12–16, the wind direction was from the Mississippi-Alabama

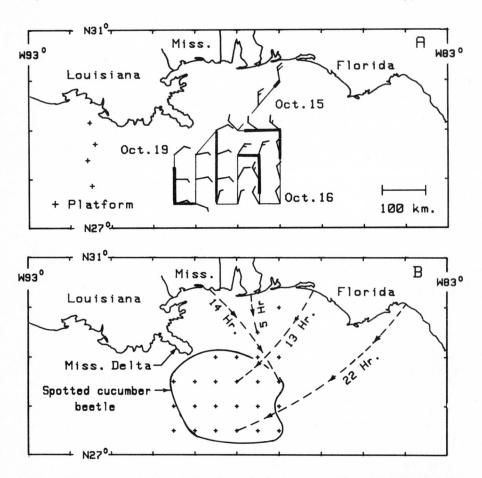

Fig. 1. A Offshore oil-platform locations in 1973 and the cruise track of *Oregon II* during Oct. 1982. The *shafts* of the wind flags denote direction, and *large* and *small barbs* denote 5- and 10-knot wind velocity, respectively. *Narrow* and *wide cruise tracks* denote day and night. **B** *Dashed lines* denote variation in wind trajectories from Oct. 15–17. Neuston nets collected spotted cucumber beetles within the area shown by the *solid line*

Table 3. Insect collections from neuston nets in the Gulf of Mexico, October 1982

Sample identification (order/family/species)	No. samples	Nearest mainland (km)	No. insects
Coleoptera:			
Anthicidae:			
Notoxus	1	250	1
Carabidae:			
Undetermined spp.	7	100–200	8
Chrysomelidae:			
Diabrotica u. howardi Barber	16	100–320	43[a]
Systena blanda Melsheimer	1	140	1
Coccinellidae:			
Cycloneda sanguinea (L.)	2	40–190	2
Curculionidae:			
Undetermined species	1	130	1
Dermestidae:			
Undetermined species	1	200	1
Platypodidae:			
Platypus	1	200	1
Diptera:			
Agromyzidae:			
Liriomyza	1	200	2
Chironomidae:			
Undetermined spp.	5	150–320	8
Culicidae:			
Undetermined spp.	2	40–200	2
Drosophilidae:			
Drosophila	12	130–240	165
Undetermined spp.	1	250	3
Phoridae:			
Undetermined spp.	1	200	3
Simuliidae:			
Undetermined spp.	1	200	1
Syrphidae:			
Eristalis	1	150	1
Undetermined spp.	1	200	1
Tipulidae:			
Undetermined spp.	9	100–300	20
Undetermined family:	8	110–300	27
Hemiptera:			
Coreidae:			
Undetermined spp.	1	40	1[a]
Gerridae:			
Undetermined spp.	2	240–280	2
Lygaeidae:			
Undetermined spp.	1	210	2
Paromius longulus (Dallas)	4	140–300	5
Undetermined spp.	2	140–320	2
Miridae:			
Undetermined spp.	5	200–320	11
Nabidae:			
Nabis	1	320	2

Table 3 (continued)

Sample identification (order/family/species)	No. samples	Nearest mainland (km)	No. insects
Homoptera:			
Aphididae:			
Undetermined spp.	5	100–280	147
Cicadellidae:			
Craminella	1	280	15
Undetermined spp.	1	210	3
Undetermined spp.	7	40–250	24
Delphacidae:			
Delphacodes	5	100–300	7
Liburniella ornata (Stal)	2	100–320	2
Sogatella kolophon (Kirkaldy)	12	40–320	45
Hymenoptera:			
Formicidae:			
Solenopsis	1	40	1
Ichneumonidae:			
Enicospilus	3	40–300	3
Undetermined family:	1	210	1
Lepidoptera:			
Noctuidae:			
Anticarsia gemmatalis (Hübner)	1	240	1[a]
Mocis	1	280	1[a]
Plathypena scabra (Fabricius)	2	200–210	2[a]
Undetermined spp.	2	130–200	2[a]
Undetermined spp.	4	210–320	6
Tineidae:			
Undetermined spp.	1	250	1
Neuroptera:			
Chrysopidae:			
Chrysopa	4	100–200	5
Undetermined spp.	1	300	4
Hemerobiidae:			
Undetermined spp.	16	40–320	61
Undetermined family:	1	170	1

[a] Dissected, no eggs in ovaries of female; gonads and fat bodies characteristic of migrating or diapausing populations.

coast. By October 17, the wind had shifted so that the probable insect sources were from Florida. From early October 18 to the middle of October 19, the wind trajectories were easterly to southerly and the nearest upwind shore was beyond the limits of our meteorological data set (600 km). No trajectories were computed after October 19, because the proximity and shape of the Mississippi River delta made interpretation of source areas ambiguous.

The neuston nets collected 5 orders and 13 families of insects from the ocean surface (Table 3). Presence of live insects could not be determined due to neuston-net, standard-operating procedures. The largest number were Diptera and other small insects typical of insects collected from the sea surface (Zaitsev 1970). Twenty-three of 24 neuston net samples contained insects. The number of insects per hectare of ocean reached a peak of 290 insects per ha on October 16 near noon and again on October 17 before midnight (Table 4). The greatest surface density, 1350 insects per ha, occurred during the night of October 18 while the ship was in a zone of convergent ocean currents. Although the collection time of the neuston nets was known, the amount of time the insects may have floated on the water was unknown. Zaitsev (1970) reported that the Colorado potato beetle, *Leptinotarsa decemlineata* (Say), could survive for several days in the Black Sea. If the insects floated for 1 day, they could have drifted more than 30 km with the ocean currents before being caught in our nets. Because of the unknown arrival times and unknown drift with ocean currents, wind trajectories corresponding to neuston-net collections were not calculated. The area from which spotted cucumber beetles, *Diabrotica undecimpunctata howardi* Barber, were collected is circled in Fig. 1B. These beetles had no eggs in the ovaries of females, and gonads and fat bodies were characteristic of migrating or diapausing populations (Table 3).

Insects collected from the deck of the *Oregon II* during the first cruise included 6 orders and 12 families (Table 5). Locating these insects against the white paint of the

Table 4. Number of insects per hectare sampled with neuston nets in the Gulf of Mexico during October 1982

Date	Time (h)	Position		Total insects collected	Insects per hectare
		N. Lat.	W. Long.		
October 15	0220	30.0	87.0	4	35
	0648	29.5	87.0	4	32
	1116	29.5	87.5	0	0
	1723	29.0	88.0	10	81
	2112	29.0	87.5	3	24
October 16	0039	29.0	87.0	3	26
	0453	28.5	87.0	13	110
	1233	27.5	87.0	33	290
	1551	27.5	87.5	23	190
	2018	28.0	87.5	11	89
October 17	0041	28.5	87.5	21	200
	0504	28.5	88.0	9	73
	0932	28.0	88.0	21	190
	1419	27.5	88.0	21	190
	1736	27.5	88.5	14	110
	2120	28.0	88.5	36	290
October 18	0110	28.5	88.5	20	180
	0514	29.0	88.5	18	150
	1014	28.5	89.0	22	180
	1413	28.0	89.0	6	49
	1806	27.5	89.0	66	530
	2151	27.5	89.5	25	200
October 19	0155	28.0	89.5	101	890
	0613	28.5	89.5	167	1350

Table 5. Insects collected from the decks of the *Oregon II* in the Gulf of Mexico during October 1982

Insect	No. samples	Upwind[a] mainland (km)	No. insects
Coleoptera:			
Chrysomelidae:			
Diabrotica u. howardi Barber	5	500	74
Hymenoptera:			
Ichneumonidae: undetermined spp.	1	340	1
Neuroptera:			
Hemerobiidae: undetermined spp.	1	320	1
Odonata:			
Libellulidae: undetermined spp.	1	< 40	1
Orthoptera:			
Tettigoniidae: undetermined spp.	1	< 40	1
Lepidoptera:			
Danaidae:			
Danaus plexippus (Linnaeus)	2	>500	2
Hesperiidae:			
Urbanus proteus (Linnaeus)	1	< 40	1
Noctuidae:			
Anticarsia gemmatalis (Hübner)	4	300	6
Heliothis zea (Boddie)	1	280	1
Pseudoplusia includens (Walker)	3	450	3
Undetermined spp.	2	450	2
Nymphalidae: undetermined spp.	3	< 40	5
Pyralidae: undetermined spp.	2	40	2
Sphingidae:			
Agrius cingulata (Fabricius)	2	< 40	2
Undetermined spp.	2	< 40	2
Microlepidoptera	4	40	5

[a] Maximum distance upwind along wind trajectory.

ship was relatively easy. The most numerous single species collected was spotted cucumber beetles (74 specimens). Unlike the neuston-net collections, these were collected only during the day of October 17 and ca. one-third were collected from the screen covering the ship's air ventilation inlet duct. The wind trajectories for October 17 indicated that the upwind distance to the Florida mainland was 340 km and 500 km with an over-water travel time of 16 h and 25 h for beetles collected in the morning and evening, respectively.

Most of the other insects collected aboard ship were found resting on various surfaces, usually protected from bright light and wind. The upwind distances to the mainland for these other insects are shown in Table 5. Usually, when insects were found on the ship's deck, the radar was also detecting insects. One exception was a monarch butterfly, *Danaus plexippus* (Linnaeus), collected at 1014 h on October 18 with an easterly wind. Another monarch butterfly was collected near shore when the radar was not operating. If the monarchs or other species were flying below 300 m, they would not have been detected by radar.

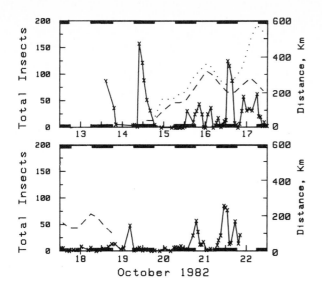

Fig. 2. Total insects detected by radar for each profile (*solid line*), distance to nearest land (*dashed line*), and distance upwind along wind trajectory to shore (*dotted line*) during cruise of *Oregon II* Oct. 1982. *Narrow* and *wide* portions of the *horizontal axis* indicate day and night

Most of the insects detected by the radar were flying less than 750 m above the water; therefore, the number observed at each of the four altitudes was summed for each profile (Fig. 2). The insects passed the ship in discrete flights lasting from a few hours to more than 10 h. While the ship was near shore, the insect activity started shortly after sundown and then decreased during the next few hours. We had previously observed similar patterns with land-based radar. As the ship moved further from shore, the insects arrived at the ship later during the night and flights even occurred during the day.

The radar detected insects during two sunsets and four sunrises. The distance from shore indicated that these insects initiated their flight earlier and continued flying during sunset and sunrise. The authors have never observed similar insect activity at similar altitudes during sunset and sunrise while operating radar on land. Schaefer (1976) and Riley et al. (1983) reported that their land-based radars showed no significant insect activity during sunset and sunrise as compared to other times of the day or night. Drake et al. (1981) reported that insects coming across the Bass Strait from Australia continued to arrive during morning hours, but did not specifically mention activity at sunrise. Thus, the flight activity we detected over the ocean during sunset and sunrise, as well as Drake's observations, may indicate a different behavior over water as compared to over land.

With only one radar, the total duration or displacement of an insect flight could not be determined. Our radar provided only the altitude, relative density, and time of day of insects passing the ship. We suggest that the arrival and duration of the individual flights of insects (Fig. 2) resulted from the timing of emigration at the source, flight behavior enroute, and the wind velocity and direction. For example, the number of insects passing the ship-based radar tended to stop more abruptly than noted in land-based radar observations. This sharp cutoff at sea may have resulted when a portion of the insects of a particular flight terminated their flight on land at the normal time. Those insects inadvertently over water (at the normal, flight-termination time) continued flying and eventually passed our ship.

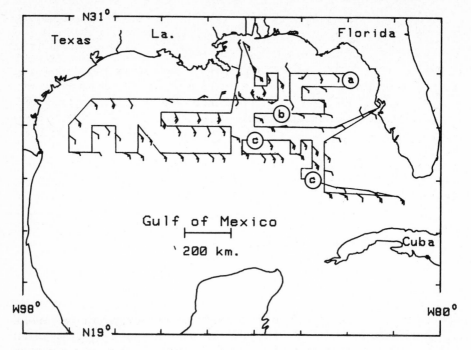

Fig. 3. Cruise track of *Oregon II* during April and May 1983. *Wind flags* indicate wind direction at the ship. Insects collected more than 1 day from port were: *a* Nessus sphinx; *b* houseflies; and *c* fall armyworms

During the second cruise, the *Oregon II* visited stations in the northern half of the Gulf of Mexico between April 21 and May 24, 1983 (Fig. 3). Brief port calls were made on May 5, 13, and 17 for repairs and supplies. The wind during this cruise was almost always from the southeastern quadrant. Wind trajectories were not computed for this trip due to lack of appropriate upper-air reporting stations. Neuston-net samples from this second trip have not been identified.

No insects were detected by the radar or collected from the ship's deck in the western Gulf of Mexico though occasionally birdlike signals were detected on radar.

During the eastern portion of the cruise, the number of insects detected by the radar was usually less than five per profile. When the ship was within 80 km of St. Petersburg and 60 km of Key West, Florida, the radar detected 18 and 29 small insects per profile, respectively. These insects were mostly below 450 m altitude.

On May 19 at 0336 h, the radar indicated 18 insects per 2-min counting period at an altitude of 675 m. At 0503 h, only four insects were counted at this altitude. There were 14 insects per 2-min period at 975 m altitude at 0633 h, and by 0818 h the insect counts had decreased to three at the 675 m altitude. During this period, the ship was at least 390 km west of the Florida coast and 440 km northwest of Cuba (between 25.5 N, 85.5 W and 26.0 N, 85.5 W, latitude and longitude). The surface wind was about 28 knots and no insects were collected from the traps or decks. This was the only significant flight of insects detected while the ship was more than 100 km from shore.

The blacklight trap on the ship caught four houseflies on May 11 between 0001 and 0600 h ca. 410 km east of Florida (27.5 N, 85.8 W). One sphingid, *Amphion floridensis* B.P. Clark, was caught between 1800 and 2400 h on May 18, ca. 110 km east of the Florida coast (28.3 N, 84.0 W). Other miscellaneous, small, noneconomic insects were caught the first night after our departure from the ports of St. Petersburg and Key West, Florida. While the ship was docked overnight at Key West, the light trap caught 224 miscellaneous insects. In addition, this collection included one velvetbean caterpillar, *Anticarsia gemmatalis* Hübner, and two male and two mated female fall armyworms.

One male fall armyworm was resting on the fall armyworm pheromone trap on May 18 at 1215 h when the ship was 390 km west of Florida and 440 km northwest of Cuba (25.0 N, 83.7 W). A second male fall armyworm was collected inside the pilot house (doors and windows open and within 10 m of pheromone trap) on May 20 at 2015 h 610 km west of Florida and 610 km northwest of Cuba (26.5 N, 88.0 W). No insects were collected from the corn earworm pheromone trap.

4 Discussion

The first trip aboard the *Oregon II* was scheduled for the fall because we expected that the large source areas on the U.S. mainland, relatively high insect densities in the fall, and frequent north winds would supply enough insects over the Gulf of Mexico to obtain useful radar data. The data collected on this trip confirmed earlier observations of Sparks (1979) and Baust et al. (1981) that insects could move significant distances over the ocean. The radar observations showed that most of the insects were flying below 450 m altitude during the fall, probably due to the cold air associated with the northerly winds. The radar and insect collections also showed that insects could travel at least 500 km over water and suggest that they have the ability to fly for longer periods of time over water. This longer flight time would enable them to cross wide bodies of water, especially when the winds are strong.

A second trip aboard the *Oregon II* was scheduled during the spring because early season insect movement across the Gulf of Mexico is economically important to the U.S. Not many insects were caught or detected on this occasion, presumably because of the relatively small source areas and lower populations in the source areas.

The collection of fall armyworm moths near the center of the Gulf of Mexico and the detection of a layer of insects on the morning of May 19 provide further evidence that economic insects can cross the widest parts of the Gulf of Mexico with favorable winds. The persistent southeast wind was favorable for transporting insects from Florida and Cuba to the southern U.S. during April and May.

Insects collected on the ship deck or in light traps on offshore platforms may represent those individuals unlikely to cross the ocean. The ship or oil platforms may have provided a refuge for those insects seeking to land. Interpretations of the temporal and spatial patterns obtained from neuston-net collections are difficult because the amount of time the specimens floated on the water is not known.

Acknowledgments. The authors are grateful to the NOAA Southeast Fisheries Center at Pascagoula, Mississippi, for allowing our radar to be installed on the *Oregon II*, to the ship's crew for their assistance, and to Dr. Dale Newsom of Louisiana State University for identifying the insects collected in the neuston nets.

References

Baust JG, Benton AH, Aumann GD (1981) The influence of off-shore platforms on insect dispersal and migration. J Econ Entomol 27:23–25

Callahan PS, Sparks AN, Snow JW, Copeland WW (1972) Corn earworm moth: vertical distribution in nocturnal flight. Environ Entomol 1:497–503

Drake VM, Helm KF, Readshaw JL, Reid DG (1981) Insect migration across Bass Strait during spring: a radar study. Bull Entomol Res 71:449–466

Gressitt LJ, Nakata S (1958) Trapping of air-borne insects on ships in the Pacific. Proc Hawaii Entomol Soc 16(3):363–365

Muller RA, Tucker NL (1985) Climatic opportunities for the long-range migration of moths. In: Sparks AN (ed) Long-range migration of moths of agronomic importance to the United States and Canada: specific examples of occurrence and synoptic weather patterns conducive to migration. USDA Misc Pub (in press)

Reap RM (1972) An operational three-dimensional trajectory model. J Appl Meteorol 11:1193–1202

Riley JR, Reynolds DR, Farmery MJ (1983) Observations of the flight behavior of the armyworm moth, *Spodoptera exempta*, at an emergence site using radar and infra-red optical techniques. Ecol Entomol 8:395–418

Schaefer GW (1976) Radar observations of insect flight. In: Rainey RC (ed) Insect flight. Oxford, Blackwell, p 287

Sparks AN (1979) An introduction to the status, current knowledge, and research on movement of selected Lepidoptera in southeastern United States. In: Rabb RL, Kennedy GG (ed) Movement of highly mobile insects: concepts and methodology in research. Conference Proc. North Carolina State University, Raleigh, NC

Sparks AN, Jackson RD, Allen CL (1975) Corn earworms: capture of adults in light traps on unmanned oil platforms in the Gulf of Mexico. J Econ Entomol 68:431–432

Wolf WW, Westbrook JK, Sparks AN (1985) Relationship between radar entomological measurements and atmospheric structure in south Texas during March and April, 1982. In: Sparks AN (ed) Long-range migration of moths of agronomic importance to the United States and Canada: specific examples of occurrence and synoptic weather patterns conductive to migration. USDA Misc Pub (in press)

Yoshimoto CM, Gressitt JL (1959) Trapping of air-borne insects on ships on the Pacific II. Proc Hawaii Entomol Soc 17:150–155

Yoshimoto CM, Gressitt JL (1960) Trapping of air-borne insects on ships on the Pacific III. Pac Insects 2:239–243

Yoshimoto CM, Gressitt JL (1961) Trapping of air-borne insects on ships on the Pacific IV. Pac Insects 3:556–558

Zaitsev YP (1970) Marine neustonology. Academy of Sciences of the Ukranian SSR. Translated from Russian by Israel Program for Scientific Translations, Israel, pp 9–11

17 Modeling of Agricultural Pest Displacement

R. E. STINNER [1], M. SAKS, and L. DOHSE

1 Introduction

One can view movement at many different levels and from many perspectives (see Stinner et al. 1983). Much work in foraging theory, for example, relies on data dealing with plant-to-plant or small patch movement over minutes or hours. At the other extreme, one can study range expansion over years or milennia, encompassing both movement and evolutionary change. Between these two extremes lie studies of local resource-to-resource movement occurring over a few hours or days, and studies of interregional migration. Many of the talks you will hear today deal with this latter aspect of movement. From an agricultural perspective, it is imperative that we be able to predict when migrations into a region will occur. However, it is equally imperative that we also know to what resources (e.g., crop species) migrant (or nonmigrant) insects will go when they are in a given region. Thus, our work has focused on the relatively local movement of several species of pest insects among fields and overwintering habitats within a region.

Since we work with insects of agricultural importance, our efforts may be biased toward species that tend to be relatively mobile. Our focus is further constrained to studies from a population-dynamics perspective. We recognize that population dynamics are ultimately driven by the evolutionary processes that lead to the observed behaviors, but one must also be cognizant that the evolutionary pathways taken are in large part determined by the population dynamics.

Since our interests are in understanding and predicting pest dynamics, our research has concentrated on the proximal factors involved in movement and the interaction of movement processes with other population processes, using computer simulations.

In order to be more complete, we should like to briefly discuss a number of different approaches to quantifying local movement, several of which are presently being pursued by a number of outstanding young scientists.

It is obvious that insects for the most part are not distributed at random. This, however, does not preclude random movement with varying "rules" that serve to collect individuals in patches. Rules can take the form of insect-density effects, impacts of host distributions, or simply varying the distances that individuals can move.

One approach to examining local movement is to determine how the distribution of hosts within a region affect the distribution of pests whose movement pathways

[1] Department of Entomology, North Carolina State University, Raleigh, N.C. 27695, USA.

Insect Flight: Dispersal and Migration
Edited by W. Danthanarayana
© Springer-Verlag Berlin Heidelberg 1986

(trajectories) are themselves random. Thus, the first type of displacement model that we would like to briefly address is the diffusion model, which in its basic form assumes that insects behave like gas particles. The density (N), at a given point (x, y) from the initial starting point at time (t) is given as:

$$\frac{\partial N(x,y,t)}{\partial t} = D \frac{\partial^2 N}{\partial x^2} \frac{\partial^2 N}{\partial y^2} \, ,$$
(1)

where D is the diffusion coefficient. Kareiva (1983) provides an excellent discussion of this model. He points out that if D is a constant, then the solution to this equation at time t is a bivariate normal distribution, with a variance in any dimension equal to 2 Dt.

He also points out, however, with many examples, that D is often not constant, but rather can vary with density, or, in a heterogenous environment, may vary with position. He further discusses many cases where leptokurtosis in the population distribution from a release point has been observed and reiterates Okubo's (1980) explanations. In spite of the mathematical difficulties in working with diffusion equations with non-constant coefficients, Kareiva (1982) has made considerable progress, both experimentally and theoretically, by examining a discrete-closed Markov model of interpatch movement of *Phyllotreta cruciferae* (Goeze) and *P. striolata* (F.) in collards, *Brassica oleracea*.

A related approach has been to use this discrete "form" of these continuous diffusion equations – random-walk theory. In this approach, as has been used by Wellington et al. (1975), Sawyer (1978), and Kareiva and Shigesada (1983), individuals are allowed to move with specific step lengths and turning angles (e.g., making turning angles correlated with each other, or changing angles when a border is reached), one is able to examine models ranging from pure-random walk (i.e., a first-order Markov chain) to more complicated, correlated-random walk. Again, however, problems of density- or patch quality-related movements are poorly described and Kareiva and Shigesada (1983) suggest investigation of higher order Markov chains to describe such effects.

2 Heliothis zea

Our own work has involved a somewhat different approach to quantifying movement. Rather than setting up a series of possible rules for movement and then selecting a species to test, we started with a particular species and developed the model (rules) from observations of actual behavior. Our first efforts dealt with an extremely mobile species, *Heliothis zea* (Boddie), in a highly patchy environment. Both the mobility of this species and its use of numerous crop species are quite important in defining the species dynamics and our approach. *H. zea* is a polyphagous species that oviposits on maize, soybean, tobacco, cotton, peanuts, tomato, and sorghum, as well as many other minor crops and numerous wild hosts. It exhibits distinct preferences in its oviposition (and thus implicitly in its movements), not only among these crops, but also among phenological stages within a given host crop. Since *H. zea* has a full complement of eggs upon emergence, total fecundity does not appear related to any individual crop species upon which adults oviposit.

Agriculture in North Carolina is comprised primarily of small fields (2–3 ha average) of randomly dispersed crops. In addition, plantings for an individual crop can often occur with up to a 2-month difference in first and last planting. Thus, the adult *Heliothis* population is presented with a virtual smorgasbord of fields with varying degrees of "attractiveness."

Our model of oviposition (movement implicit) thus is based on this system. We (Johnson et al. 1975) assumed that each crop in each phenological stage has a basic attraction, ϕ, but that the stimulus to a given moth, ϕ', is affected by the distance, d, of the moth to that field:

$$\phi' = \phi/d^2 \ . \tag{2}$$

Thus, the total stimulus, Φ, received by a moth is:

$$\Phi = \sum_{\text{crop}} \ \sum_{\text{stage}} \ (\phi/d^2) \ . \tag{3}$$

Assuming a mobile, randomly distributed adult population and proportional response, the proportion of adults, P_{ij}, in a given field i of stage j is:

$$P_{ij} = (\phi_{ij}/d_{ij}^2)/\Phi \ . \tag{4}$$

Fields in much of North Carolina are randomly distributed with respect to each other. Thus, through a modification of the nearest-neighbor index, the average distance to the nearest field of a given crop and stage can be calculated and is proportional to the density of fields (this information is readily available from our crop surveys).

So far we have discussed movement as an independent process. In order to fully utilize this information, it is necessary to place movement in the context of the total-population dynamics (e.g., including fecundity and mortality processes, given colonization). We (Stinner et al. 1974, 1977) have used this *Heliothis*-movement model as part of a larger regional simulator. Recently, in work with G. Kennedy (unpubl. data), we examined the impact of different types of host-plant resistance in maize on the population). We (Stinner et al. 1974, 1977) have used this *Heliothis*-movement model as part certain types of resistance (e.g., antibiosis), other types of resistance (e.g., nonpreference) can greatly increase the *Heliothis* populations, not only in other crops, but also in maize. Studies of movement or of mortality taken separately would not have allowed us to quantitatively develop this scenario.

3 *Epilachna varivestis*

At this point we would like to shift our discussion from the highly mobile *H. zea* to more recent work that we and several colleagues have been conducting on a species with much more limited mobility: the Mexican bean beetle, *Epilachna varivestis* Mulsant, a pest of *Phaseolus* spp. and soybean.

Our first studies of this species involved examining host-plant/physical-environment effects on this beetle's development, reproduction, and survival. However, it quickly became apparent that we would not be able to understand the dynamics of this species without considering movement in connection with the other population processes. Our most recent work has involved laboratory, greenhouse, and field studies of *Epilachna* movement.

Unlike *Heliothis*, *E. varivestis* does not move freely and often from field to field. In addition, its oviposition is closely tied to its adult host. Thus, the results governing *Epilachna*-population dynamics differ markedly from those of *H. zea*. The following discussion is drawn from Dohse (1982).

In estimating movements of these beetles among fields, the first step is to calculate the conditional probability of reaching field i from another field j (PR), given that a beetle will migrate from field j. (Fields i and j are any two fields in the system). This probability will be the product of two probability functions of two (assumed) independent-random variables. The first of these (Pr) governs the beetle's flight direction and orientation, the second (Pd) governs the maximum distance a beetle can fly.

$$PR \text{ (from j to i)} = Pr \text{ (orientation)} \times Pd \text{ (distance)} . \tag{5}$$

The problem of calculating the above-mentioned conditional probability has been broken into two parts. The first of these probability functions is assumed to be uniform over the interval 0 to 2π. The second is assumed to be a shifted-negative exponential [see Eq. (6)].

The uniform distribution was chosen because no evidence for directionality was found. Siniff and Jesson (1969) suggested a polar-coordinate, normal distribution defined over the unit circle, but this would be applicable only if the beetles seemed to have a preferred direction. Field observations, however, are presently too few and have too large a variance to assume directionality (Blau, personal communication). If future observations should prove that there is a preferred direction, the model is constructed in such a way that only one equation needs to be replaced and one parameter added.

The probability-density function governing the maximum-time duration a beetle can fly was taken from the literature. Freeman (1977) and Jones (1977) suggested a negative-exponential function for the distance of one flight. Tethered flight experiments, though inconclusive, also suggest this type of curve for the duration of flight (Blau, personal communication).

The function that is used in the model to calculate the chance of a beetle reaching a specified distance x is:

$$Pd(x) = \begin{cases} 1 & \text{for } x \leq d_m \\ \exp[-u(x-d_m)] & \text{for } x > d_m \end{cases}, \tag{6}$$

where:

Pd(x) = the probability that a beetle can fly at least x kilometers,
u = relative death rate per kilometer,
d_m = minimum distance before death rate is considered.

To simplify the calculations and reduce the number of parameters, the fields are viewed as points in a plane. The entire beetle population is viewed as being in the center of the field, i.e., on that point. Each field is surrounded by a circular "aura" of a specified radius. A beetle is assumed to immigrate into a field if it encounters this aura, i.e., if its flight path intersects this circular region. The equation for the probability of going in the direction of field j is given by [Eq. (7)].

$$Pr(x,r_j) - [\arcsin(r_j/x)]/\pi , \tag{7}$$

where:

r_j = the radius of field j (includes the "aura").

x = distance between point of origin and center of field j.

The above functions are strictly dependent on the geographic parameters, field size, and location. The probabilities of migrations, on the other hand, depend on the organisms (beetles and crops) of the model. With respect to the probability of migration initiation, there are some data available. Blau and Stinner (1983) observed the earliest age of migration for the Mexican bean beetle. Their results showed the same characteristic skewed curve seen in Dingle's (1980) data. Most of the newly emerged beetles were observed to leave the field within 4 to 9 days after eclosion. After comparing the results of two different hosts (soy and lima), the indication was that beetles will stay longer in a lima bean field than in a soybean field, i.e., the expected time until departure is longer in limas. This difference can be simulated with appropriate values for the probabilities of migration from a given field type.

In this model it is assumed that an adult beetle can be in one of four migratory states. The first of these is a preadult (premigratory) state, s_0, where no flight behavior is possible. This is followed by a "ready to migrate state," s_1.

The beetle leaves state s_1 only if it has migrated. Since a beetle could be in a different physiological state during migration, the probability of migration could depend on whether a beetle has or has not migrated on a given previous day. The model is flexible enough to incorporate this dependency (memory). However, in order to keep the Markovian assumption, two other migration states are needed. A beetle is in the migrating state, s_2, if it migrated on the previous day. (Probability for flight in this state is high.) A beetle is in the "colonizing" state, s_3, if it has migrated, but did not migrate on the previous day. Probability of migration in this state is assumed low.

Since the insect is initially in a flightless state (s_0) followed by a state in which flight is likely (s_1), there should be an underlying probability-density function that governs the chance of maturing (going from s_0 to s_1) at a given age. This random variable, age of beetle at first flight, is assumed to be distributed by a shifted-binomial distribution:

$$f(age) = \begin{cases} 0 & \text{if } age < m \\ \binom{M-m}{age-m} \times p^{(age-m)} \times (1-p)^{(M-age)} & \text{if } m \leq age \leq M , \\ 0 & \text{if } M < age \end{cases} \tag{8}$$

where:

f(age) = probability of maturing (dependent on the age of the beetle),

m = minimum age for maturation,

M = maximum age for maturation,

p = estimated mean age for maturation/(M−m) (age is measured in number of days).

This distribution was chosen for its simplicity in calculation and its goodness of fit (Dohse 1982). The only parameters needed are the earliest, latest, and expected age of maturation. These parameters depend on the type and condition of the field. Together with the probability of leaving, the above distribution, f(age), is flexible enough to fit most unimodal distributions that could be expected for this process.

Once a beetle has matured into a migratory state, the probability of migration on any given day will depend on the state of the beetle, the field's condition, and crop type. The more "preferred" the field, the less likely a beetle is to emigrate from that field. Note that it is assumed, for each migration state, the probabilities of migrating are constant.

Thus, the probability of leaving a field depends on the type of field and the state of the beetle. There is no density-dependent behavior assumed, with one exception: if the population reaches a specified maximum, the field will be destroyed and all the beetles will leave. On the other hand, the probability of reaching any given field is dependent on the location (distribution) of emigrating beetles and the size of the given field. Thus, the chance of leaving a given habitat is dependent on biological parameters and the chance of reaching a given habitat is dependent on the geometry (spatial structure) of the system. Although the actual probability of leaving on a given day is affected by physical factors (Blau and Stinner 1983), weather records indicate that there are relatively few days where physical factors are such that movement is prohibited.

4 Discussion

It is of interest to note that sensitivity analyses suggest that the *Epilachna* model is quite insensitive to the minimum and maximum distances that beetles are capable of flying, at least under our agricultural system. In addition, when allowed to reach equilibrium, relative beetle densities in fields of the same host are proportional to the field *perimeters*. In contrast, with *Heliothis* such densities are proportional to field *areas*. Thus, if these models are correct, they demonstrate a basic difference in potential agroecosystem-structure effect on highly mobile versus less mobile species' dynamics. [For a complete discussion of model sensitivities, see Dohse (1982).]

The present model for *Epilachna* is, however, far from complete. From a field experiment just completed (Saks and Stinner, unpubl. data), we believe there are a number of misconceptions in the present movement model.

1. Although adults do move at random, their residency times appear much shorter than presently estimated in the model, at least for males.
2. Males appear to move more readily, and are at least locally attracted to feeding damage, thus requiring a separate model for male movement with not only different parameters, but also including a density effect.
3. Beetles appear more hesitant to cross a border than to move from one part of the field to another.

Given a set of movement behaviors (rules) and a description of those environmental attributes that affect these behaviors, to which environmental and behavioral parameters is the particular system most sensitive? Answering this question should be one of the major objectives in modelling movement. All too often we spend a great deal of time and funds in an attempt to measure parameters to which the system is relatively insensitive. For example, as mentioned earlier, the *Epilachna* system appears relatively insensitive to the maximum and minimum distances a beetle can fly *within the North Carolina* agroecosystem where small fields are randomly distributed and relatively close to each other. Under this set of environmental conditions, changes in the probabilities of leaving a given field yield far greater differences in the dynamics that change in these potential flight distances. Under another set of conditions (e.g., large monocultures, widely separated as in the midwestern United States), however, the sensitivities could easily be reversed.

It, thus, behooves us to critically examine our system (both behavior and environment) with initial models before large-scale experiments are conducted. This is particularly true in studies of movement, where many of the necessary parameters are difficult and expensive to estimate.

References

Blau WS, Stinner RE (1983) Temporal flight patterns in the Mexican bean beetle (Coleoptera: Coccinellidae) and the relation to weather. Environ Entomol 12:1047–1054

Dingle H (1980) Ecology and evolution of migration. In: Gauthreaux SA Jr (ed) Animal migration, orientation and navigation. Academic Press, New York

Dohse L (1982) A discrete model simulating the interfield movement of a multihost phytophagous beetle. PhD Thesis, Biomath Program, North Carolina State University, Raleigh, 107 pp

Freeman GH (1977) A model relating numbers of dispersing insects to distance and time. J Appl Ecol 14:477–487

Johnson MW, Stinner RE, Rabb RL (1975) Ovipositional response of *Heliothis zea* (Boddie) to its major hosts in North Carolina. Environ Entomol 4:291–297

Jones R (1977) Movement patterns and egg distribution in cabbage butterflies. J Anim Ecol 46: 195–212

Kareiva P (1982) Experimental and mathematical analysis of herbivore movement: quantifying the influence of plant spacing and quality of foraging discrimination. Ecol Monogr 52:261–282

Kareiva P (1983) Local movement in herbivorous insects: applying a passive diffusion model to mark-recapture field experiments. Oecologia (Berl) 57:322–327

Kareiva P, Shigesada N (1983) Analyzing insect movement as a correlated random walk. Oecologia (Berl) 56:234–238

Okubo A (1980) Diffusion and ecological problems: mathematical models. Springer, Berlin Heidelberg New York

Sawyer AJ (1978) A model for the distribution and abundance of the cereal leaf beetle in a regional crop system. Michigan State University, Pest Mngt Tech Rep 22, 213 pp

Siniff DB, Jesson CR (1969) A simulation model of animal movement patterns. Adv Ecol Res 6: 185–217

Stinner RE, Rabb RL, Bradley JR Jr (1974) Population dynamics of *Heliothis zea* (Boddie) and *H. virescens* (F.) in North Carolina: a simulation model. Environ Entomol 3:163–168

Stinner RE, Rabb RL, Bradley JR Jr (1977) Natural factors operating in the population dynamics of *Heliothis zea* in North Carolina. Proc 15th Int Congr Entomol Washington DC, 622–642

Stinner RE, Regniere J, Wilson KG (1982) The differential effects of agroecosystem structure on the dynamics of three soybean herbivores. Environ Entomol 11:538–543

Stinner RE, Barfield CS, Stimac JL, Dohse L (1983) Dispersal and movement of insect pests. Annu Rev Entomol 28:319–335

Wellington WG, Cameron PG, Thompson WA, Vertinsky IB, Landsburg AS (1975) A stochastic model for assessing the effects of external and internal heterogeneity on an insect population. Res Popul Ecol (Kyoto) 17:1–28

18 Dispersal of Insects of Public Health Importance[1]

W. STEIN[2]

1 Introduction

Dispersal, irrespective of its method of occurrence, is an integral part of the life-history patterns of all living organisms. It is generally accepted that dispersal has three main consequences that are significant in the survival of species: avoidance of crowding, expansion of the range of distribution and recolonization of appropriate biotopes, and the flow of genetic material within and between the colonized biotopes. In adaptive dispersal, also referred to as migration (Johnson 1969), a species changes its biotope or moves into another geographical area. Migration also enables a species to escape unfavourable conditions (Kennedy 1961), and some insect species migrate in search of hibernation sites (Schneider 1962, Stein 1971).

It is not too difficult to envisage that dispersal (including migration) of insects of public health importance may result in enormous health problems, for over 50% of all infectious diseases, among which are some of the most common diseases in the world, are insectborne (Davidson 1976). A knowledge of the dispersal behaviour of insects will, therefore, be of immense value in any vector control or disease-prevention programme. There are some insect species which are primarily harmful to animals, but also affect humans. These, as well as the acarine pests, are considered in this review.

2 Mode of Dispersal

Fundamentally, two types of dispersal may be distinguished: active and passive dispersal. The latter mode is particularly useful to small insects. Active dispersal is represented by flight, crawling or jumping. Dispersal by flight is by far the most important since both short- and long-distance displacement can be achieved by this type of movement. In terms of public health pest problems, short-range dispersal is more relevant.

Insect movement is often classified into two categories, appetitive and migratory, which have distinct behavioural and physiological attributes. During appetitive movements insects are attracted to the hosts or to certain habitat components for the purposes of feeding, mating, oviposition, and shelter. An insect in a migratory phase is usually not distracted by the above vegetative activities (Kennedy 1961). Migration

---

[1] Dedicated to Professor Dr. B. Rensch on his 85th birthday.
[2] Institut für Phytopathologie und Angewandte Zoologie, Justus-Liebig-Universität Gießen, 6300 Gießen, FRG.

Insect Flight: Dispersal and Migration
Edited by W. Danthanarayana
© Springer-Verlag Berlin Heidelberg 1986

movements, on the other hand, are a temporary syndrome of persistent locomotion that may lead, for example, also to the colonization of new habitats (Dingle 1972; Johnson 1963; Kennedy 1961). As a rule, the migratory phase occurs only once in the lifetime of an insect (Johnson 1969). Subsequent flights are usually of the appetitive type. Also insect flight and other movements are negatively or positively influenced by factors outside the animal such as temperature, speed and duration of wind and precipitation.

Passive dispersal in insects treated here occurs when an insect already adapted to the human environment makes use of the host itself or items associated with it (e.g. clothes, foodstuffs, waste, household rubbish, transport vehicles) for displacement. If the pest has alternative animal hosts, then the latter may also act as carriers of the pest enabling passive dispersal. Other modes of passive dispersal include abiotic means such as wind transportation, floating, and rafting on water, carriage in water currents – particularly those species with aquatic life stages. It is quite conceivable that many insects dangerous to public health attained worldwide distribution by passive means. In this context the ability to survive and adapt in a new environment is crucial for a species, and it is logical to suppose that survival and adaptation may be enhanced by mutation of the genetic material, leading ultimately to the buildup of high-density populations in the new environments.

3 Influence of Individual Environmental Factors

Many of these aspects have been previously reviewed in detail by Johnson (1969), and for this reason only brief accounts, relevant to this discussion, are presented here.

3.1 Temperature

It is well-established that activities of poikilothermic animals are largely governed by the prevailing temperature. In general, the lower and upper temperature thresholds for locomotory activities are $10°$ and $35°C$, respectively, though there are exceptions to this in some species. The modifications to normal room temperatures induced by central heating and air conditioning would certainly improve the dispersal and settlement efficiency of many household pests such as cockroaches, ants, fleas, and flies.

3.2 Humidity and Rainfall

The influence of humidity on dispersal is less pronounced than that of temperature. According to Rowley and Graham (1968), relative-humidity values ranging from 30% to 90% have no influence on the flight performance of *Aedes aegypti* (L.). Rainfall normally has an inhibitory effect upon insect dispersal, but in brachyceran flies there is often an increase in flight activity immediately after rainfall (Chamberlain 1984; Haschemi 1981).

3.3 Light

Distinct values of light intensity can affect insect dispersal in a pronounced manner. Brachyceran Diptera, under normal circumstances, do not fly during darkness. A light intensity of 10 lx induces the flight activity of *Calliphora erythrocephala* (Meig.) (Digby 1958a), and also of mosquitoes (Nielson and Greve 1950). In some species which are active at dusk and/or dawn, low light values are essential for the initiation of flight activities. An inhibition of flight activity may occur under very strong solar radiation, as with *Lucilia* spp. (Haschemi 1981), but in this case the associated increase in temperature may also be an influencing factor. Also the normal influence of olfactory stimuli on orientation movements can be affected by optic components as in mosquitoes (Bidlingmayer and Hem 1979) and in the flies *Musca domestica* L. and *Hydrotaea irritans* (Fall.) (Berlyn 1978).

3.4 Time of Day

Circadian rhythms of insect activity have been intensely studied and well-documented (e.g. Lewis and Taylor 1965; Saunders 1982). Thus, only several examples are discussed here. Dispersal activities of insects of health importance may occur at various times during the day or night, depending on the species, but such activities are often subject to changes induced by other physical-environmental factors. Some species such as *Haematobia irritans* De Meijere and *Anopheles stephensi mysorensis* Sweet and Rao are known to disperse at night (Hoelscher et al. 1968; Quraishi et al. 1966), whereas that of others such as *Hydrotaea irritans* occurs from 1000-1800 h, peaking at 1300 h (Berlyn 1978).

3.5 Position of the Sun

The position of the sun in the sky, independent of the light intensity, is known to influence the dispersal of insects, particularly those species which use skylight for navigation (see Wehner 1984) or initiate flight and dispersal activities using sunlight as a cue (Johnson 1969). Among public health pests it has been shown that *Haematobia irritans* flies up-sun, at least during the morning hours (Chamberlain 1984). Also the direction during the dispersal phase of *Blattella germanica* in the field is strongly dependent on the position of the sun (Haschemi and Stein, in prep.).

3.6 Air Pressure

Very little is known of any significance about the effect of air pressure on insect dispersal. According to Burnett and Hays (1974) air pressure may influence the flight activity of *Tabanus* spp.

3.7 Chemical Factors

It is well-known that the direction of movement of insects can be governed by chemical components. Carbon dioxide, fatty acids, and other animal products play an important role in dispersal and orientation of blood-sucking parasites. The odours of food and breeding materials may enable insects to locate their hosts and this occurs in association with single hosts and also on a very large scale as demonstrated by the attraction of insects to urban refuse tips for breeding purposes (Stein and Danthanarayana 1974) and attraction to small towns (Shura-Bura 1955; Shura-Bura et al. 1956). On the other hand, the odour of an appropriate breeding substrate may inhibit emigration as has been shown to happen with certain flies such as *Lucilia sericata* (Mg.) and *Ophyra aenescens* (Wied.) (El-Dessouki and Stein 1978).

3.8 Wind

Of all the physical environmental factors, wind undoubtedly has the greatest effect on insect dispersal, provided other factors, particularly the temperature, allow flight activity. Passive transport as well as migration, of small insects in particular, could be dependent on wind velocity (see Taylor 1974). Lower wind speeds enhance carriage of odour, and thereby determine the direction of dispersal. Wind velocity itself may initiate and terminate flight activity. For example, wind speeds of up to $0.7 \, \mathrm{m s^{-1}}$ enhance the flight activity of *Calliphora erythrocephala*, and speeds in excess of this value have an inhibitory effect (Digby 1958b). *Musca domestica* is known to fly at velocities between $0.89–3.1 \, \mathrm{m s^{-1}}$ against the wind (Pickens et al. 1967) and across the wind (Hindle and Merriman 1914; Parker 1916). Monsoon winds are responsible for the long-range dispersal of *Simulium damnosum* Theo., an important element in the extension of onchocerciasis in the Volta-River area in West Africa (Garms et al. 1979). The mosquito *Aedes taeniorhynchus* (Wied.) flies strictly downwind (Provost 1957), whereas the fly *Hippelates collusor* (Tns.) flies upwind at speeds of up to $0.89 \, \mathrm{m s^{-1}}$ and downwind at higher wind speeds (Dorner and Mulla 1962). Winds with velocities less than $2 \, \mathrm{m s^{-1}}$ have no important effect upon the flight direction of *Haematobia irritans* (Chamberlain 1981).

3.9 Topography

Landscape formations may influence the short- and long-range dispersal of insects. For instance, high vegetation surrounding a marshland may, in certain cases, reduce or even inhibit the dispersal of *Tabanus nigrovittatus* Macq. (Rockel and Hansens 1970). Certain types of distinct demarcations, such as shores and rivers, have been shown to direct the flights of *Aedes taeniorhynchus* and other insects (Provost 1952). Also the velocity of dispersal may be influenced by the topography. Uncultivated areas are crossed more quickly than cultivated ones by calliphorid and muscid flies; but flight distances can be decreased over cultivated land and small townships, because of the availability of a multitude of appetitive stimuli (Parker 1916; Shura-Bura et al. 1958). Nevertheless, the dispersal of these and other flies is not affected by even larger urbanized areas

(Dow 1959; Greenberg and Bornstein 1964; Schoof and Mail 1953; Schoof et al. 1952). Pronounced topographical features such as steep slopes with an ascent of 45° and larger water surfaces are no barriers for the dispersal of calliphorids and muscids (MacLeod and Donnelley 1958).

4 Dispersal Within Buildings

Pest insects can considerably reduce the living qualities in dwelling houses, in business buildings, in factories and processing plants, in hotels, eating houses, and other similar situations where people live, work, and eat. In addition to the deterioration of these premises, damage to equipment and/or contamination of products are known to occur. This obviously is of enormous economic and health importance especially in the food industry. Also insects and mites can cause problems in hospitals by causing or accentuating allergies, besides their potential of being disease vectors. Unfortunately, not very much is known about many of the aspects mentioned above, and this has led to the difficulty in proposing adequate prophylactic measures to prevent the spread of insects and other arthropods within buildings. Modern buildings provide close to the optimum temperature for insect development and movement all year-round. Piping, ducting passages associated with water, gas, electricity, heating, sewerage, and other facilities in buildings provide excellent routes for horizontal and vertical dispersal of insects. Other features in buildings such as sandwich-walls, ceilings and floor coverings provide nesting, hiding, and dispersal facilities to insects and arachnids despite their primary functions as amenity and insulation materials. In air-conditioned buildings small insects and mites are known to be carried and spread by the streams of circulating air.

5 Sex and Age of Insects

The dispersal abilities of the two sexes may be different. For example, in the long-range dispersal of *Simulium damnosum* mainly the older females take part (Garms et al. 1979), whereas in most other species the newly emerged females are predominantly dispersive (Johnson 1969). The distance travelled by *Aedes taeniorhynchus* females is greater than that of the males (Provost 1952). Wellington (1944) established that the females of *Culex* sp. have a lower temperature threshold for flight activity than the males. This would mean that in these mosquitoes, females have a longer dispersive phase, and may also result in some separation of the sexes during the swarming period. In many species age, too, may influence the dispersal activity. So, the beginning of the development of the gonads may bring the activity to an end or, as this is known in brachyceran flies, the increasing damage of the wings of older insects may reduce or even inhibit the dispersal ability by flight.

6 Food

Freshly obtained carbohydrates and the fat body provide the main source of energy necessary for dispersal. For long-range dispersal fat-body development is a prerequisite, for usually no food is consumed during this phase. If the food source becomes rare, as can happen with blood-sucking insects, then the availability of food determines the dispersal capabilities. In flight-mill studies, it has been shown that *Aedes aegypti* flies until the glycogen reserve is fully exhausted (Rowley and Graham 1968a,b). In principle, unavailability of food, particularly of the right quality, may lead to a reduction of dispersal activity. Nevertheless, this could also lead to increased activity and consequently to dispersal when the insect searches for suitable food, as it is common with blood-sucking parasites, and this is followed by a reduction in dispersal activity as the insect (e.g. female mosquitoes), after a full-blood meal, is too heavy for immediate long-range displacement.

7 Population Density

It is a common phenomenon in many insect species that an increase in population density enhances displacement activities. This is often associated with or caused by changes in the physiology and behaviour or production of morphs adapted for dispersal. With respect to public health pests, increased flight activity correlated with increased population density has been demonstrated for *Aedes taeniorhynchus* and this is associated with phase polymorphism (Nayar and Sauerman 1969). Under field conditions *Blattella germanica* L. exhibits higher dispersive abilities at very high densities (Haschemi and Stein, in prep.).

8 Dispersal Distances and Flight Velocities

The determinations of distances travelled and flight velocities are difficult and the results obtained are usually approximate values. Nevertheless, such information is essential from a practical point of view for the elucidation of the biology and ecology as well as the control of a pest species. The most common methods used in these studies are mark-release-recapture techniques in the field and wind tunnel (flight chamber) and flight-mill studies in the laboratory. More recently, optical and remote-sensing methods have been advanced for agricultural pests (see Farrow, Riley, this volume; Schaefer and Bent 1984). The main difficulty encountered under field conditions is the thinning out (diffusion) of the mark-released insects with increasing distance. Also the direction of movement whether by flight or by walking may be undirectional, with no guarantee of results for the research effort. With remote-sensing methods it is impossible, at this present time, to determine anything else than information on takeoff, flight speeds and the flying altitude and orientation. With these constraints taken into account, examples of maximum distances travelled and the flight speeds of the various species are presented in Tables 1 and 2. As long distances can be covered by many species in a relatively short time period (Table 2), dispersal of vectors could lead to a corresponding dispersal of the diseases they transmit. However, it is known, too, that pathogens adapted to higher temperatures of their warmblooded hosts may be inactivated during the transport by insect vectors (Radvan 1960).

Table 1. Some maximum distances of active and/or passive dispersal of arthropods of public health importance (F = female, M = male)

Species		Distance	References
Ixodidae			
Amblyomma americanum (L.)		29 m	Smittle et al. (1967)
Simuliidae			
Simulium damnosum Theo.		300 km	Garms et al. (1979)
S. metallicum Bellardi		11.9 km	Dalmat (1950)
Culicidae			
Aedes sticticus (Mg.)		11 km	Brust (1980)
A. flavescens (Mull.)	F	10.6 km	Shemanchuk et al. (1955)
	M	1.3 km	Shemanchuk et al. (1955)
A. taeniorhynchus (Wied.)	F	40.2 km	Provost (1957)
	M	4.8 km	Provost (1957)
A. vexans Meigen		48.3 km	Horsfall (1954)
Anopheles stephensi List.		4.5 km	Quraishi et al. (1966)
Muscidae			
Musca domestica L.		33.2 km	Schoof (1959)
M. autumnalis Deg.		6.4 km	Killough et al. (1965)
Muscina stabulans (Fall.)		15 km	Shura-Bura et al. (1958)
M. assimilis (Fall.)		4.5 km	Shura-Bura et al. (1958)
Fannia canicularis (L.)		4.5 km	Shura-Bura et al. (1958)
F. pusio (Wied.)		5.6 km	Quarterman et al. (1954)
Ophyra leucostoma (Wied.)		11.3 km	Bishopp and Laake (1921)
O. aenescens (Wied.)		11.3 km	Schoof (1959)
Haematobia irritans L.		11.8 km	Kinzer and Reeves (1974)
Hydrotaea dentipes (F.)		15 km	Shura-Bura et al. (1958)
Stomoxys calcitrans L.		> 3.2 km	Bailey et al. (1973)
Calliphoridae			
Sarcophaga sueta (Wulp.)		3.4 km	Quarterman et al. (1954)
Lucilia caesar (L.)		6.2 km	Shura-Bura et al. (1958)
L. sericata (Mg.)		6.4 km	Schoof (1959)
L. cuprina (Wied.)		8.1 km	Schoof (1959)
Phormia terrae-novae R.-D.		10.7 km	Shura-Bura et al. (1958)
P. regina (Mg.)		45.1 km	Schoof (1959)
Calliphora erythrocephala (Meig.)		5 km	Shura-Bura et al. (1958)
C. uralensis Villeneuve		8.2 km	Shura-Bura et al. (1958)
Callitroga macellaria (F.)		24.1 km	Schoof (1959)
Cochliomyia hominivorax (Coquerel)		289.6 km	Hightower et al. (1965)

Table 2. Some data on dispersal velocity of arthropods of public health importance (F = female, M = male)

Species	Dispersal distance	Needed time	References
Ixodidae			
Amblyomma americanum (L.)	23 m	3 days	Smittle et al. (1967)
	29 m	11–12 weeks	
Simuliidae			
Simulium metallicum Bellardi	6.1 km	1 day	Dalmat (1950)

Table 2 (continued)

Species		Dispersal distance	Needed time	References
Culicidae				
Aedes flavescens (Mull.)		10.6 km	11−27 days	Shemanchuk et al. (1955)
Chloropidae				
Hippelates pusio Lw.		1.6 km	3.5 h	Dow (1959)
Oestridae				
Hypoderma sp.	F	8−9 m	1 s	Nogge and Staack (1969)
Muscidae				
Musca domestica L.		10.1 km	1 day	Schoof (1959)
		732 m	35−45 min	Parker (1916)
Musca autumnalis Deg.		3.2 km	1 day	Killough et al. (1965)
		6.4 km	5 days	
Haematobia irritans L.		1.4 km	4 h	Tugwell et al. (1966)
		11.7 km	10 h	Kinzer and Reeves (1974)
Stomoxys calcitrans L.	F	29.1 km	1 day	Bailey et al. (1973)
	M	28.9 km	1 day	
Calliphoridae				
Sarcophaga bullata Parker		2.9 km	1 day	Schoof (1959)
Lucilia sericata (Mg.)		6.4 km	1 day	Lindquist et al. (1951)
L. cuprina (Wied.)		8.0 km	2 days	Schoof (1959)
Phormia terrae-novae R.-D.		1 km	1 day	Shura-Bura et al. (1956)
		3 km	3 days	
		5 km	10 days	
P. regina (Meig.)		6.4 km	1 day	Lindquist et al. (1951)
		16.5 km	2 days	Schoof (1959)
Callitroga macellaria (Fab.)		12.9 km	1 day	Schoof (1959)
		16.1 km	2 days	

9 Dispersal of Some Pest Species

There is very little published work on the dispersal of health pests, particularly of
Acarina, Siphonaptera, and Hemiptera. The most widely studied group is Diptera, and
is very well reviewed by Johnson (1969). Diptera, with mosquitoes, tsetse flies, calliphorids, muscids, and simuliids, well-known vectors of the majority of communicable
diseases, can also be of mere nuisance value or cause myiasis. Presumably for these
reasons more is known on the dispersal of Nematocera and Brachycera than any other
group. Many species of mosquitoes and flies are easily bred, large in size and are recover-
able in the field and it is not surprising that many more investigations have been carried
out on these than on any other group. In comparison, ticks and mites, which are obligate
parasites on humans and animals, are transported passively from host to host, and are
therefore less 'suitable' for experimentation (on dispersal) than those species which
have obvious migratory phases.

Among other groups of some importance are the cockroaches which have increased
in numbers during the last decades as a result of the changes in human life-styles and
the parallel development of buildings. The German cockroach *Blattella germanica* is,
perhaps, the best example in this respect. In geographical areas with moderate climatic

conditions, passive transport is the mode of dispersal in this species. But recent investigations suggest that active dispersal occurs in the field when, for example, extensive movement out of a refuse tip, followed by hibernation in the open field, ultimately leads to an extension of its geographical range in a new habitat (Haschemi and Stein, in prep.). Spread within buildings has been already discussed in Sect. 4. Certain ant species also disperse in a manner similar to that of cockroaches. The pharaoh ant, *Monomorium pharaonis* (L.) is an important pest, especially in hospitals, as it transmits pathogens, contaminates medical equipment and disturbs patients. Transportation is usually by passive means, but establishment of new colonies occurs only under certain conditions (Petersen and Buschinger 1971).

It seems clear from these few examples that the dispersal of health pests is complex, with a variety of mechanisms involved, but is often aided by human intervention. As passive transport is a significant form of dispersal, measures to reduce this type of spread should contribute to the reduction of the pest status of some of these insects. It needs to be reiterated that more research is required on all aspects of dispersal and migration of health pests.

References

Bailey DL, Whitfield TL, Smittle BJ (1973) Flight and dispersal of the stable fly. J Econ Entomol 66:410–411

Berlyn AD (1978) The flight activity of the sheep headfly, *Hydrotaea irritans* (Fallen) (Diptera: Muscidae). Bull Entomol Res 68:219–228

Bidlingmayer WL, Hem DG (1979) Mosquito (Diptera:Culicidae) flight behaviour near conspicuous objects. Bull Entomol Res 69:691–700

Bishopp FC, Laake EW (1921) Dispersion of flies by flight. J Agric Res 21:729–766

Brust RA (1980) Dispersal behavior of adult *Aedes sticticus* and *Aedes vexans* (Diptera:Culicidae) in Manitoba. Can Entomol 112:31–42

Burnett AM, Hays KL (1974) Some influences of meteorological factors on flight activity of female horse flies (Diptera:Tabanidae). Environ Entomol 3:515–521

Chamberlain WF (1981) Dispersal of horn flies: Effect of host proximity. Southwest Entomol 6: 316–325

Chamberlain WF (1984) Dispersal of horn flies III. Effect of environmental factors. Southwest Entomol 9:73–78

Dalmat HT (1950) Studies on the flight range of certain Simuliidae, with the use of aniline dye marker. Ann Entomol Soc Am 43:537–545

Davidson G (1976) Vector-borne diseases and need to control them. In: Gunn DL, Stevens JGR (eds) Pesticides and human welfare. Oxford University Press, pp 29–41

Digby PSB (1958a) Flight activity in the blowfly *Calliphora erythrocephala*, in relation to light and radiant heat, with special reference to adaptation. J Exp Biol 35:1–19

Digby PSB (1958b) Flight activity in the blowfly, *Calliphora erythrocephala*, in relation to wind speed, with special reference to adaptation. J Exp Biol 35:776–795

Dingle H (1972) Migration strategies of insects. Science 175:1327–1335

Dorner RW, Mulla MS (1962) Laboratory study of wind velocity and temperature preference of *Hippelates* eye gnats. Ann Entomol Soc Am 55:36–39

Dow RP (1959) Dispersal of adult *Hippelates pusio*, the eye gnat. Ann Entomol Soc Am 52:372–383

El-Dessouki S, Stein W (1978) Untersuchungen über die Insektenfauna von Mülldeponien III. Die Ausbreitung von Fliegen einer Rottedeponie (Dipt., Muscidae und Calliphoridae). Z Angew Zool 65:367–375

Garms R, Walsh JF, Davies JB (1979) Studies on the reinvasion of the onchocerciasis control programme in the Volta River Basin by *Simulium damnosum* s.l. with emphasis on the South-Western areas. Tropenmed Parasitol 30:345–362

Greenberg B, Bornstein AA (1964) Fly dispersion from a rural Mexican slaughterhouse. Am J Trop Med Hyg 13:881–886

Haschemi H (1981) Untersuchungen zur Biotopbindung von *Lucilia*-Arten (Dipt., Calliphoridae). Thesis, J-Liebig-University, Gießen

Hightower BG, Adams AL, Alley DA (1965) Dispersal of released irradiated laboratory-reared screw-worm flies. J Econ Entomol 58:373–374

Hindle E, Merriman G (1914) The range of flight of *Musca domestica*. J Hyg 14:23–45

Hoelscher CE, Combs RL Jr, Brazzel JR (1968) Horn fly dispersal. J Econ Entomol 61:370–373

Horsfall WR (1954) A migration of *Aedes vexans* Meigen. J Econ Entomol 47:544

Johnson CG (1963) Physiological factors in insect migration by flight. Nature 28:26–27

Johnson CG (1969) Migration and dispersal of insects by flight. Methuen, London

Kennedy JS (1961) A turning point in the study of insect migration. Nature 189:785–791

Killough RA, Hartsock JG, Wolf WW, Smith JW (1965) Face fly dispersal, nocturnal resting places, and activity during sunset as observed in 1963. J Econ Entomol 58:711–715

Kinzer HG, Reeves JM (1974) Dispersal and host location of the horn fly. Environ Entomol 3:107–111

Lewis T, Taylor LR (1965) Diurnal periodicity of flight of insects. Trans R Entomol Soc Lond 116:393–479

Lindquist AE, Yates WW, Hoffman RA (1951) Studies on the flight habits of three species of flies tagged with radioactive phosphorus. J Econ Entomol 44:397–400

MacLeod J, Donnelly J (1958) Local distribution and dispersal paths of blowflies in hill country. J Anim Ecol 27:349–374

Nayar JK, Sauerman DM Jr (1969) Flight behaviour and phase polymorphism in the mosquito *Aedes taeniorhynchus*. Entomol Exp Appl 12:365–375

Nielsen ET, Greve H (1950) Studies on the swarming habits of mosquitoes and other Nematocera. Bull Entomol Res 41:227–258

Nogge G, Staack W (1969) Das Flugverhalten der Dasselfliege (*Hypoderma* Latreille) (Diptera, Hypodermatidae) und das Biesen der Rinder. Behaviour 35:200–211

Parker RR (1916) Dispersion of *Musca domestica* Linnaeus under city conditions in Montana. J Econ Entomol 9:325–362

Petersen M, Buschinger A (1971) Untersuchungen zur Koloniegründung der Pharaoameise *Monomorium pharaonis* (L.). Anz Schädlingskd Pflanzenschutz 44:121–127

Pickens LG, Morgan NO, Hartsock JG, Smith JW (1967) Dispersal patterns and populations of the housefly affected by sanitation and weather in rural Maryland. J Econ Entomol 60:1250–1255

Provost MW (1952) The dispersal of *Aedes taeniorhynchus* I. Preliminary studies. Mosq News 12:174–190

Provost MW (1957) The dispersal of *Aedes taeniorhynchus* II. The second experiment. Mosq News 17:233–247

Quarterman KD, Mathis W, Kilpatrick JW (1954) Urban fly dispersal in the area of Savannah, Georgia. J Econ Entomol 47:405–412

Quraishi MS, Faghih MA, Esghi N (1966) Flight range, length of gonotrophic cycles, and longevity of P[32] – labelled *Anopheles stephensi mysorensis*. J Econ Entomol 59:50–55

Radvan R (1960) Epidemiological importance of spread and transmission of bacteria surviving during the development of flies. Cesk Epidemiol Mikrobiol Immunol 9:497–500

Rockel EG, Hansens EJ (1970) Emergence and flight activity of salt-marsh horse flies and deer flies. Ann Entomol Soc Am 63:27–31

Rowley WA, Graham CL (1968a) The effect of age on the flight performance of females *Aedes aegypti* mosquitoes. J Insect Physiol 14:719–728

Rowley WA, Graham CL (1968b) The effect of temperature and relative humidity on the flight performance of female *Aedes aegypti*. J Insect Physiol 14:1251–1257

Saunders DS (1982) Insect clocks, 2nd edn. Pergamon, Oxford

Schaefer GW, Bent GA (1984) An infra-red remote sensing system for the active detection and automatic determination of insect flight trajectories (IRADIT). Bull Entomol Res 74:261–278

Schneider F (1962) Dispersal and migration. Annu Rev Entomol 7:223–242

Schoof HF (1959) How far do flies fly? Pest Control 27:16–22

Schoof HF, Mail GA (1953) Dispersal habits of *Phormia regina* in Charleston, West Virginia. J Econ Entomol 46:258–262

Schoof HF, Siverly RE, Jensen JA (1952) House fly dispersion studies in metropolitan areas. J Econ Entomol 45:675–683

Shemanchuk JA, Fredeen FJH, Kristjanson AM (1955) Studies on flight range and dispersal habits of *Aedes flavescens* (Muller) (Diptera: Culicidae) tagged with radio-phosphorus. Can Entomol 87:376–379

Shura-Bura BL (1955) Untersuchungen über die Migration von Fliegen von einem Müllplatz aus mittels Markierung (russ) Gig Sanit 9:12–15

Shura-Bura BL, Ivanova EV, Onutshin AN, Glazunova AJ, Shaikov AD (1956) Migration of flies of medical importance (Diptera, Muscidae, Calliphoridae, Sarcophagidae) in Leningrad district. Entomol Obozr 35:334–346

Shura-Bura BL, Shaikov AD, Ivanova EV, Glazunova AI, Mitriukova MS, Fedorova KG (1958) The character of dispersion from the point of release in certain species of flies of medical importance. Entomol Obozr 37:282–290

Simmonds M, Stein W (1981) Ein Beitrag zur Hygienesituation auf Autobahnparkplätzen I. Das Vorkommen hygienisch bedenklicher Fliegen. Forum Städte-Hyg 32:174–180

Smittle BJ, Hill SO, Philips FM (1967) Migration and dispersal patterns of the Fe^{59}-labeled lone star ticks. J Econ Entomol 60:1029–1031

Stein W (1971) Das Ausbreitungs- und Wanderverhalten von Curculioniden und seine Bedeutung für die Besiedlung neuer Lebensräume. In: den Boer PJ (ed) Dispersal and dispersal power of carabid beetles. Misc Pap 8, Landbouwhogeschool Wageningen, pp 111–118

Stein W, Danthanarayana W (1974) Untersuchungen über die Insekten-Fauna von Mülldeponien I. Freilassung von markierten *Lucilia sericata* (Meig.) (Dipt., Calliphoridae) außerhalb einer Deponie. Z Angew Zool 61:407–417

Taylor LR (1974) Insect migration, flight periodicity and the boundary layer. J Anim Ecol 43:225–238

Tugwell P, Burns EC, Witherspoon B (1966) Notes on the flight behavior of the horn fly, *Haematobia irritans* (L.) (Diptera: Muscidae). J Kans Entomol Soc 39:561–565

Wellington WG (1944) The effect of ground temperature inversions upon the flight activity of *Culex* sp. (Diptera, Culicidae). Can Entomol 76:223

Wehner R (1984) Astronavigation in insects. Annu Rev Entomol 29:277–298

19 Night-Vision Equipment, Reproductive Biology, and Nocturnal Behavior: Importance to Studies of Insect Flight, Dispersal, and Migration[1]

P. D. LINGREN[2], J. R. RAULSTON[3], T. J. HENNEBERRY[2], and A. N. SPARKS[4]

1 Introduction

The adults of well over one-half of our insect pests are active during dark or crepuscular periods and most species are highly mobile. Their mobility, along with our limited night vision, has resulted in a large void in our knowledge of their nocturnal behavior. Nevertheless, in recent years, a number of technologies and techniques have emerged that are adaptable for studying nocturnal behavior of insects in their natural environment. The technologies include the use of various forms of visible light such as flashlights, lanterns, Aldis lamps, and 6-V headlamps (Lingren et al. 1977; Rose and Dewhurst 1979); night-vision devices using natural-light intensification and amplification (Lingren et al. 1978; Greenbank et al. 1980); infrared-optical devices including video (Conner and Master 1978; Lingren et al. 1982; Riley et al. 1983; Schaefer and Bent 1984); and radar (Schaefer 1976; Wolf 1978; Riley 1980).

A primary purpose of this paper is to acquaint the reader with some of the equipment currently used to visually locate and observe activities of nocturnal insects. Additionally, we will attempt to point out reproductive biology and nocturnal activities considered important to dispersal of nocturnal adult forms. In general, we will concentrate primarily on our experience with *Heliothis* spp. [corn earworm, *H. zea* (Boddie); and tobacco budworm, *H. virescens* (F.)] located in cotton and corn fields.

2 Equipment

2.1 Headlamps

For initial field observations, we normally equip ourselves with a 6-V battery-powered headlamp fitted with a 6.35-cm polished reflector. A carpenter's apron is used to carry the battery and various types of catch and collection containers. The lamps are very useful for locating and observing certain types of activity in the field such as mating pairs, feeding, oviposition, sex pheromone-secreting females, short-range flight, emerging adults, nocturnal predation, etc. We know that the insects will sometimes react to the

---

[1] Mention of a proprietary product does not constitute endorsement by the U.S. Department of Agriculture.
[2] USDA, ARS, 4135 East Broadway, Phoenix, AZ 85040, USA.
[3] USDA, ARS, P.O. Box 1033, Brownsville, TX 78520, USA; Current address of senior author: USDA, ARS, P.O. Box 157, Gone OK 74555, USA.
[4] USDA, ARS, Southern Grain Research Laboratory, Tifton, GA 31793, USA.

Insect Flight: Dispersal and Migration
Edited by W. Danthanarayana
© Springer-Verlag Berlin Heidelberg 1986

light emitted by the headlamp. In most situations, however, adult insects that are programmed for and are well into a specific type of activity pattern will momentarily react to the light, but then continue their original activity as the light intensity is decreased. Therefore, when an activity is located, the headlamp is positioned so that the target is at the outer fringes of the beam, or various types of filters are placed over the lamp to decrease the intensity and/or quality of light being emitted.

Headlamps and batteries should be available in most countries. In the United States, they are found in most hardware stores and numerous other types of general merchandise retail outlets. Lamps and batteries may also be purchased from primary manufacturers: (1) Ray-o-vac Division, E.S.B., Inc., Madison, Wisconsin 53703; or (2) Eveready Battery Division, Union Carbide Corp., Danbury, Connecticut 06817. The headlamp with a heavy-duty battery and carrying case can be purchased for about 11 U.S. dollars; batteries range from 2 to 3.5 U.S. dollars. The functional life of a heavy-duty battery for field use is about 6 h, while that of a regular battery is only about 3 h.

2.2 Night-Vision Goggles

Night vision goggles are very useful for observing insect activity under situations of low-light levels (10^{-6} to 1 ft. candle) ranging from an overcast night to twilight. The goggles are equipped with an electro-optical system that has good sensitivity to light radiation in the visible range as well as the NIR (near-infrared) range of the electro-magnetic spectrum up to about 860 nm. Consequently, on dark nights or in other very dark situations, auxillary lighting in the form of headlamps or other light sources provided with NIR filters can be used to provide greater scene illumination to which the insects are not sensitive.

Currently, there are two types of night-vision goggles (first and second generation devices) available to the commercial market (Slusher 1978). Both types are passive image intensification devices that convert photons to electrons which are amplified and focused on a phosphor screen that absorbs the energy and radiates it as a visible image. First generation devices utilize a three-stage fiber-optic, electron-acceleration system, and second-generation devices utilize a microchannel-plate, image-intensification system. The first-generation device provides a light gain of about 50000 times, whereas the second-generation goggles provide a light gain of about 40000 times.

Generally speaking, the first-generation goggles are less expensive, larger, more subject to flare and screen burn from bright-light sources, and create more image distortion than second-generation devices which usually weigh about 30 oz. and provide a 40° field of view at unity magnification with less than 2% distortion. The image-intensifier tubes have an operating life of about 2000 h and they are available in 18- and 25-mm microchannel-plate wafers. The goggles are fitted with a headstrap assembly, as shown in Fig. 1, that allows for viewing while freeing the hands for other uses. On a typical starlight night, one can see a moving insect with a wing span of 3–4 mm up to about 20 m without using auxillary lighting. Second generation goggles can be purchased for 4000 to 6000 U.S. dollars.

Goggles are typically powered by a 2.7 V mercury battery made by Mallory Battery Co., South Broadway, Tarrytown, N.Y., USA 10591. Their batteries provide about 20 h

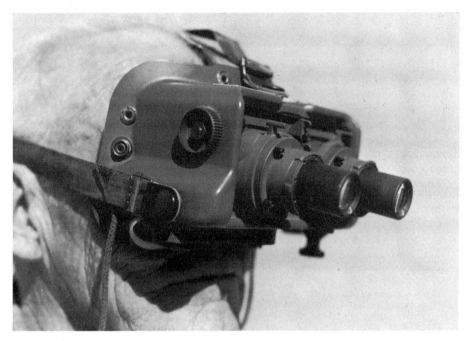

Fig. 1. Typical night-vision goggles showing headstrap

of operating life, cost about 5-6 U.S. dollars, and may be purchased only in lots of 50. Lots of this size are expensive and not needed for most commercial (as compared to military use) uses since only three to five batteries are normally used during a year. A solution to this problem now exists in the form of a face-mask assembly recently released by ITT Electro-Optical Products Div., Roanoke, Virginia, USA 24019, that utilizes universally available double-A batteries. The assembly is adaptable to most goggles and can be purchased for about 180 U.S. dollars.

A third-generation goggle has now been developed by ITT that includes a photocathode which provides much greater sensitivity in the 580-920 nm range of the electromagnetic spectrum than existing devices. These goggles provide much greater scene illumination than first- and second-generation models under very dark conditions. Unfortunately, the third-generation goggles are not available for commercial use at this time, but they are expected to be available in the near future.

2.3 Monocular Night-Viewing Devices

Monocular devices are available with 18- and 25-mm image-intensification tubes (first- and second-generation tubes). There are numerous systems adaptable to various closed-circuit television video (CCTV) and 35-mm single-lens reflex cameras. Table 1 provides a partial list of companies involved in the sale of monocular and/or binocular night-viewing devices. Because of the variety of accessories and options with monocular systems, we suggest that the reader contact several companies before purchasing such a system.

Table 1. A partial list of companies involved in the manufacturing and supply of night-vision devices[a,b]

1. ITT Electro-Optical Products Division 7635 Plantation Road Roanoke, Virginia 24019 USA Tel. (703) 563-0371 Jeff Slusher, Manager, Night-Vision Marketing	7. AEG-Telefunken Industriestraße 29 2000 Wedel, West Germany
2. Javelin Electronics, Inc. P.O. Box 2033 Torrance, California 90510 USA Tel. (213) 327-7440 Chris Cladis, Sales Manager	8. M.E.S. 00156 Roma Vannina 78 Rome, Italy
	9. Odelft (N V Optische Industrie De Ovde Delft) P.O. Box 72 2600 MD Delft, Netherlands
3. Litton Industries Electron Tube Division 1215 South 52nd Street Tempe, Arizona 85281 USA Tel. (602) 968-4471 John Wagner, Sales Manager	10. Bofors Aerotronics AB 18181 Lidingo Sweden
	11. Pilkington P.E. Ltd. Optical Division Glascoed Rd., St. Asaph Clwdy LL 17 OLL, England
4. Ni-Tec Inc. 5600 West Jarvis Niles, Illinois 60648 USA Tel. (312) 647-7702 Mike Nolan, International Sales Manager	12. Rank Pullin Controls Langston Road Debden, Loughton Essex, England
5. Varo, Inc. Integrated Systems Division 1919 South Shiloh Garland, Texas 75042 USA Tel. (214) 487-1436 Mervyn Simpson	13. United Scientific Instruments 10 Fitzroy Square London WIP 6AB, England
	14. Avimo Ltd. 140 Tottenham Court Rd. London WIP OJD, England
6. Sopelem 10 rue Chaptal 92306 Levallois-Perret France	15. Genzor Optics Corp. 1006 W 15th St. Riviera Beach, Florida 33404 USA

[a] For information only; not meant to be totally inclusive.
[b] ITT, Litton, and Varo are primary manufacturers of night-vision goggles in the USA. Ni-Tech and Javelin concentrate primarily on monocular devices. Varo and Litton also produce monocular devices.

Our use of the monocular system has been mostly associated with delayed-video viewing of partially stationary targets such as the entrance orifices of sex-pheromone traps, oviposition sites, point sources of attractants, adult emergence from pupal cells, etc. The system we have used consists of a low-light level TV camera equipped with a 20-mm Newvicon® tube with extended sensitivity down to 0.05 ft. candles and an electronic viewfinder. The camera is attached to a second-generation 18-mm night-viewing tube via an adjustable 25-mm fl. 4 C-mount relay lens. A 135-mm fl. 8 objective telephoto lens with mechanical-stop adjustment is then attached to the tube via a C-mount adapter. The camera is connected to a 10-mm portable video tape recorder adapted for

AC or DC usage. A homemade, lightweight frame houses the camera system and attaches to an adjustable tripod. The frame provides adequate stability to the camera system and allows for good maneuverability. Minimum resolution of the system is 36 line pairs per millimeter and the overall system weighs about 3 kg. Various photographic techniques are adaptable to the night-viewing devices (Boogher and Slusher 1978).

We have experienced several problems with this system because of excess electronic noise and poor editing capabilities of the recording unit. The system has recently been updated by the companies involved and there are numerous companies that make low-light level CCTV systems with C-mounts adaptable to first- and second-generation image-intensifying devices of various weights and configurations with numerous accessories. However, we feel it is essential to obtain a good professional-type portable recorder in order to obtain quality video from such systems under field conditions. One has many choices for procuring an adequate CCTV night-viewing system with the costs ranging from 4000–6000 U.S. dollars. The costs of such systems may be high by most standards, but the returns should be very great. Therefore, we strongly urge those researchers interested in nocturnal-insect activities to procure a good CCTV night-viewing system.

3 Uses of Equipment

A list of some major ways we have applied the information gathered while utilizing night-vision aids is provided in Table 2. We recognize that the process of gathering such information is highly labor-intensive, the observations are of a subjective nature, and in many cases the data are density-dependent. Nevertheless, in our opinion, a good understanding of the activity of nocturnal species is an essential aid for unraveling the key parameters involved in their migratory activity.

3.1 Characterization of Nocturnal-Insect Activity

One of the primary applications we have made of night-vision aids is to characterize nocturnal-activity patterns for several species of adults (Lingren et al. 1979; Lingren et al. 1980; Lingren and Wolf 1982; Lingren et al. 1982; Lingren 1983) in the context

Table 2. A list of some uses made from information gathered by using night-vision equipment

1. Establish what's, where's, and when's of nocturnal activity
2. Support radar observations of insect movement
3. Observe responses of insects to stimulants under natural conditions
4. Improve sex-pheromone trap design and efficiency
5. Relationship of trap capture to in-field mating or other activities
6. Evaluate insect-control programs
7. Establish population emergence, density, aging, and dynamics
8. Nocturnal predation
9. Study low-level insect movement
10. Design of new and/or more effective control technology

of defining the temporal and spatial relationships of such activity. Such information may be important for understanding and reporting those activities that could be used to identify migratory and nonmigratory individuals or populations. In general, activity patterns are sequential and of a circadian nature that can be modulated by both physical and biological parameters. Temperature extremes tend to shorten the duration of nocturnal activity, but the sequence of activity remains the same. Cool temperatures generally create earlier activity, while warm temperatures delay activity. Biological parameters such as the age structure of the insect population and the phenology of available host plants also result in temporal and spatial shifts in activity patterns.

A good understanding of nocturnal-activity patterns is considered a useful adjunct in studies on migration of nocturnal species. An example we wish to present here is that of the tobacco budworm, representing a lepidopteran pest species. The nocturnal activity in tobacco budworm populations is initiated with a flurry of flight activity that appears to be oriented primarily downwind. Similar activity has been reported on the fall armyworm, *Spodoptera frugiperda*, (Smith) (Sparks 1979). This activity occurs about 30 min after sundown and it takes place over a short duration of 5–15 min. Heavy feeding interspersed with oviposition then begins and continues for about 2 h. Both feeding and oviposition continue sporadically and at low levels until about 30 min before daylight, but the peak time of these activities occurs during early evening. A "fast-flight activity" of males that we have associated with mate-searching behavior begins near the end of the primary-feeding and oviposition period. The flight activity during this period appears to be oriented diagonally to the wind, whereas the feeding and ovipositing adults generally orient upwind when approaching feeding and oviposition sites.

Mate-searching activity of males is closely synchronized with female calling (sex-pheromone secretion) activity, but this behavior is not triggered by the presence of the female's sex pheromone(s) (Lingren et al. 1979). When laboratory-reared adults of both sexes are released into the field, we begin to see calling females at about the same time that fast-flying males are observed. However, calling females are seldom seen in natural populations unless their habitat has been saturated with their synthetic sex pheromone and closely-related mimics, or when temperatures have reached the lower threshold of activity. Calling females in natural populations are observed occasionally during the late afternoon and early evening when temperatures have decreased from an optimum range of $21°$–$28 °C$ to about $13 °C$. At this temperature the males become inactive, but females may continue their nocturnal activities to about $10 °C$.

Mating is first observed shortly after what appears to be a short lull in activity following the intense feeding and oviposition activities. Virgin females mate earlier than previously mated females (Raulston et al. 1975). Observations of the first occurrences of so-called fast-flight activity and calling of females is closely correlated with the initiation of mating. The mating activity generally begins about 3–5 h after sundown and continues for 2–4 h into the early morning hours. After the cessation of mating activity, the population appears to become inactive for a few minutes and then a short flurry of activity takes place about 30 min before daylight. This activity lasts for about 30 min. Light feeding, oviposition, and even mating activities sometimes occur during this period.

3.2 Establishing Population Emergence, Aging, and Generation Dynamics

Presumedly, because many of the nocturnal species that we are concerned with are short-lived, their migratory flights occur early in adult life. Consequently, a knowledge of the age structure of a population should be of paramount importance in determining its potential for migratory activity. A sound understanding of the reproductive biology of a species and its relationship to nocturnal-activity patterns can provide clues to the population age structure. For instance, the tobacco budworm emerges primarily at night (Lingren et al. 1979). Males are sexually mature about 6 h after emergence and sperm is available in the duplex at about 8 h after emergence (Henneberry and Clayton 1984). Although males are capable of mating each night after the night of emergence, the incidence of mating decreases as the males age (Henneberry and Clayton 1985). Males are not sexually mature on the night of emergence until after the nocturnal mating cycle. Collections and dissections of newly emerged males and mating pairs from both laboratory and field populations show that the males normally do not mate on the night of emergence.

A dark-red, gellike substance forms in the simplex of virgin males at 10-24 h after emergence and remains identifiable throughout the life of unmated males (Henneberry and Clayton 1984). This material is passed to females during the first mating and is not regenerated in the male. Therefore, any male possessing the dyelike substance is a virgin, that is, at least 10 h of age. A similar dyelike substance has been reported from the simplex of virgin-male fall armyworms (Snow and Carlyle 1967). The simplex of numerous other species contain a granular material that exists in various stages of pigmentation. Possibly these materials can be differentiated in virgin and mated males or the materials may change structurally with age, sexual activity, or both. If so, and if these changes can be measured, then one could utilize the procedure to aid in determining the age of males in other species. Such a capability would greatly enhance our ability to determine a more exact male-age structure in populations.

Female tobacco budworms emerge at night, but do not mate on the night of emergence. Eggs are immature upon emergence and become mature 40 h later. Virgins can mate on the first night after emergence, but mating occurs after the primary oviposition activity in mated individuals. Oviposition begins on the second night after emergence. A delay in mating of just 5 days from emergence (4 days from normal mating date) can result in a 44% reduction in the number of eggs deposited (Proshold et al. 1982).

Raulston et al. (1975) showed a direct relationship between the age of female tobacco budworm and the number of times they have mated. Further, Raulston et al. (1979) indirectly described the dynamics of an emerging generation of tobacco budworm through measurements of mating activity, male response to pheromone traps, and age structure of females over time (Fig. 2). A majority of the population emerged (established by the numbers of virgin females within the population) over a relatively short period of time (5-7 days). A majority of the mating occurred sharply in conjunction with the emergence. During this period, the virgin females were competing with the pheromone traps for males. As a result, trap capture was low during the major emergence and mating of the population and it did not increase until a preponderance of the newly emerged females had mated.

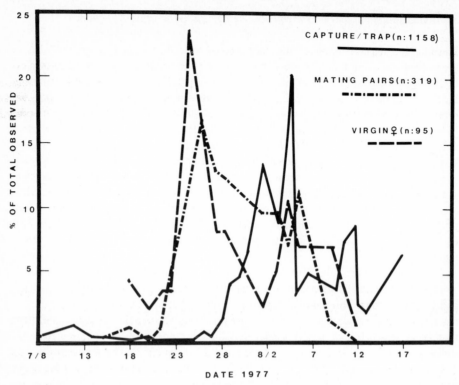

Fig. 2. Comparison of emergence of native tobacco budworm adults in cotton field (No. of virgin females) and mating to capture of males in direction-live pheromone traps (from Raulston et al. 1979)

The study of the emergence, generation dynamics, and flight behavior of source populations certainly provides another approach for obtaining a better understanding of the migratory activity in a population. An excellent example of such studies is given for the African armyworm, *Spodoptera exempta* (Walker), by Rose and Dewhurst (1979) and Riley et al. (1981, 1983). Here we provide a short summary of the emergence activity of a population of corn earworms from corn fields (Lingren et al., in press). In this study, corn plants were cut at about 38 cm above the ground just after 91% of the corn earworm larvae had vacated the ears. Plants were cut to provide a clear area for observations of emergence. Adults began to emerge from pupal cells in the soil with unexpanded wing pads. They rapidly moved to the corn stubble to assume a vertical position on the stalks about 15 cm above the soil with their heads oriented upward (Table 3). There they began inflating their wings. The wing-inflation process took about 18 min. After remaining motionless for about 2 min, they extended the wings directly overhead and held them in that position for about 25 min, after which the wings were again lowered to the normal resting position. The adults remained on the stalks in a very docile condition for another 101 min before taking their first flight. The first flight was usually a short-range flight of less than 20 m and occurred about 2.5 h after emergence.

Table 3. Sequence and duration of activities of corn earworms from adult eclosion to first flight in the corn field[a,b,c] (From Lingren et al., in press)

Sequence of activities	Cm above soil			Time (min)		
	Mean	(SE)	Range	Mean	(SE)	Range
Location of adults on plants	15.4	(2.2)	6–30	–	–	–
Emergence to fully-inflated wings	–	–	–	18.3	(2.2)	10– 38
Inflated to overhead extension of wings	–	–	–	2.3	(0.6)	1– 8
Wing extension (overhead) to normal resting position	–	–	–	24.9	(0.9)	19– 31
Normal resting position of wings to first flight	–	–	–	101.3	(5.2)	77–129
Total time (eclosion to first flight)	–	–	–	146.8	–	–

[a] Eleven to 15 observations of each type on August 9 and 10, 1983.
[b] Observation area = 242 m² per night.
[c] Intermittent rain throughout observation period on night of August 9.

Moth emergence began just prior to 2100 h and peaked about 2300 h with very little emergence occurring after 0100 h (Fig. 3). There were two peaks of flight activity of the newly emerged moths. The first peak took place between midnight and 0100 h about 2 h after peak emergence and the second peak occurred between 0400–0500 h.

There were also two peaks in collections of newly emerged moths with fully-formed wings. The first peak occurred at 2400 h just after peak emergence. The numbers collected then declined until a second peak of about equal size occurred at 0400 h. Recovery in total collections of newly emerged moths with fully-formed wings in large numbers between 0300–0400 h in comparison to that at 2300–2400 h following the cessation of emergence and first-flight activity strongly suggests that the moths did not leave the field during their first peak in flight activity. Likewise, the early morning (0400–0600 h) flight activity leads to the emigration of the moths from the field because only one nonflying moth was collected after 0500 h and its wings were deformed.

The nightly emergence and flight activity observed in the studies of Lingren et al. (in press) are quite similar to observations made on the African armyworm by Rose and Dewhurst (1979) with two exceptions. The former investigators found no dusk flights or concentration of moths in trees. The lack of a dusk flight was likely associated with the fact that the crop had been removed from the field and that the newly emerged moths left the field during the predawn flight. However, dusk flights of *Heliothis* spp. have been observed on numerous other occasions by visual observations (Lingren et al. 1979) and by radar (Schaefer 1976; Lingren and Wolf 1982). In radar studies the latter two authors have also observed plumes of insects (probably *Heliothis* spp.) leaving a cotton field in the early evening during the peak feeding period. The dusk, flight-activity period has also been reported for the African armyworm and appears to be the beginning of takeoff for migratory flight (Riley et al. 1983).

Observations of a few nearby trees by Lingren et al. (in press) revealed no bollworm moths following each period of flight activity. The predawn flight appeared to be oriented downwind, but information was not collected pertaining to moth destination. The predawn flight of bollworm moths must be a nonmigratory flight since visual observations

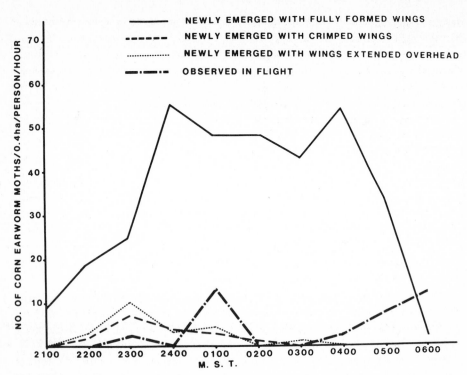

Fig. 3. Chronological sequence of wing condition and flight activity of corn earworm moths emerging from 10.2 ha corn field; sample size equals average numbers of moths observed per person per hour of search from 0.4 ha; one to five observers recorded wing condition and activity from 2 to 14 rows per hour of search on August 8, 9, 10, 14, and 17, 1983; different portion of field observed on each sample-time interval each night. (From Lingren et al., in press)

(Lingren et al. 1977) and radar observations (Schaefer 1976; Lingren and Wolf 1982; Riley et al. 1983) have shown a cessation in flight activity of several lepidopteran species shortly after daylight. If the population was in a migratory mode, then that migratory activity would most likely have taken place the following night during the early evening takeoff. In bollworm populations, the early evening takeoff occurs at about the same time as intense feeding and oviposition. It is not known whether the insects feed before migratory takeoff. However, feeding on the night of emergence could have some influence on the time of migratory flight because Rose and Dewhurst (1979) observed newly emerged moths feeding in trees and Lingren et al. (in press) observed numerous newly emerged bollworm moths feeding on exudates from corn stalks between 18 and 24 min after emergence while their wings were still extended overhead. Certainly, it would seem that feeding prior to a major migratory flight would enhance such flight activity.

The generation-emergence dynamics of *Heliothis* spp. populations reported by Lingren et al. (in press) and Raulston et al. (1979) show that most of the emergence occurs over a relatively short period (5–7 days) in nondiapausing generations. Likewise, somatic-cell development takes place at a faster rate in females than males (Fye and

McAdo 1972; Roome 1979) and germ-cell development occurs at a faster rate in males than females (Callahan 1961; Henneberry and Clayton 1984). Therefore, there appears to be a preponderance of females early in the emergence cycle of a given generation which changes to a preponderance of males as the population continues its emergence cycle. Generally, the sex ratio of the population begins to equalize near the end of a generation emergence cycle, but overall sex ratios can vary from field to field.

4 Summary

The important points presented here are that populations emerge over a relatively short period of time, they can be observed, and rough age determinations can be made over time. Obviously, there must be some as yet unknown physiological difference(s) between migratory and nonmigratory individuals. Knowledge of the nocturnal activities of the species under study provides some basis for studying the migratory activity of source populations. Nocturnal observations and collections from emerging source populations, along with radar determinations of flight activity and associated aerial and migratory-fallout collections (see Greenbank et al. 1980), offer a means of providing source material for determinations of physiological and behavioral differences in migratory and nonmigratory individuals or populations. If a valid and usable physiological or behavioral difference can be established between such populations, then adult control of source populations as suggested by Joyce (1982) could become a reality.

References

Boogher B, Slusher JA (1978) Successful photographic techniques through night vision devices. In: Lingren PD (conv) ESA Symp: night vision equipment for studying nocturnal behavior of insects. Bull Entomol Soc Am 24:203–206

Callahan PS (1961) Relationship of the crop capacity to the fat body and egg development in the corn earworm, *Heliothis zea* and the fall armyworm, *Laphygma frugiperda* (Lepidoptera: Noctuidae). Ann Entomol Soc Am 54:819–827

Conner WE, Master WM (1978) Infrared video viewing. Science 199:1004–1005

Fye RE, McAdo WC (1972) Laboratory studies on the development, longevity, and fecundity of six lepidopterous pests of cotton in Arizona. USDA Tech Bull 1554, p 73

Greenbank DO, Schaefer GW, Rainey RC (1980) Spruce budworm (Lepidoptera:Tortricidae), moth flight and dispersal: new understanding from canopy observations, radar, and aircraft. Mem Entomol Soc Can no 110, p 49

Henneberry TJ, Clayton TE (1984) Time of emergence, mating, sperm movement, and transfer of ejaculatory duct secretory fluid by *Heliothis virescens* (F.) (Lepidoptera:Noctuidae) under reversed light-dark cycle laboratory conditions. Ann Entomol Soc Am 77:301–305

Henneberry TJ, Clayton TE (1985) Tobacco Budworm Moths (Lepidoptera:Noctuidae): effect of time and emergence, male age, and frequency of mating on sperm transfer and egg viability. J Econ Entomol 78:379–382

Joyce RJV (1982) A critical review of the role of chemical pesticides in *Heliothis* management, pp 173–188. In: Reed W (ed) Proc Int Workshop on *Heliothis* Management, 15–20 November 1981. (ICRISAT) International Crops Research Institute of the Semi-Arid Tropics, Patancheru, A.P., India, p 418

Lingren PD (1983) Behavior of Pink bollworm (Lepidoptera:Gelechiidae) adults during eclosion to departure from site of emergence. Ann Entomol Soc Am 76:657–660

Lingren PD, Wolf WW (1982) Nocturnal activity of the tobacco budworm and other insects, pp 211–228. In: Hatfield JL, Thomason IJ (eds) Biometerology in integrated pest management. Academic Press, New York, p 491

Lingren PD, Greene GL, Davis DR, Baumhover AH, Henneberry TJ (1977) Nocturnal behavior of four lepidopteran pests that attack tobacco and other crops. Ann Entomol Soc Am 70:161–167

Lingren PD, Sparks AN, Raulston JR, Wolf WW (1978) Applications for nocturnal studies of insects. In: Lingren PD (conv) ESA Symp: night vision equipment for studying nocturnal behavior of insects. Bull Entomol Soc Am 24:206–212

Lingren PD, Raulston JR, Sparks AN, Proshold FI (1979) Tobacco Budworm: nocturnal behavior of laboratory-reared irradiated and native adults in the field. USDA/SEA, ARR-W-5, p 17

Lingren PD, Henneberry TJ, Bariola LA (1980) Nocturnal behavior of adult cotton leafperforators in cotton. Ann Entomol Soc Am 73:44–48

Lingren PD, Raulston JR, Sparks AN, Wolf WW (1982) Insect monitoring technology for evaluation of suppression via pheromone systems, pp 171–193. In: Kydonieus AF, Beroza M (eds) Insect suppression using controlled release pheromone systems, vol 1. CRC, Boca Raton, Florida, p 274

Lingren PD, Warner B, Kehat M, Henneberry TJ, Zvirgzdens A, Gillespie JM (1986) In field emergence of feral populations of corn earworm, *Heliothis zea* (Boddie). Environ Entomol (in press)

Proshold FI, Karpenko CP, Graham CK (1982) Egg production and oviposition in the Tobacco Budworm: effect of age at mating. Ann Entomol Soc Am 75:51–55

Raulston JR, Snow JW, Graham HM, Lingren PD (1975) Tobacco Budworm: effect of prior mating and sperm content on the mating behavior of females. Ann Entomol Soc Am 68:701–704

Raulston JR, Lingren PD, Sparks AN, Martin DF (1979) Mating interaction between native tobacco budworms and released backcross adults. Environ Entomol 8:349–353

Riley JR (1980) Radar as an aid to the study of insect flight, pp 134–140. In: Amlaner CJ Jr, MacDonald DW (eds) A handbook on biotelemetry and radio tracking. Permagon, Oxford, p 804

Riley JR, Reynolds DR, Farmery MJ (1981) Radar observations of *Spodoptera exempta* Kenya, March-April 1979. Centre for Overseas Pest Research Misc Rep, no 54, p 43

Riley JR, Reynolds DR, Farmery MJ (1983) Observations of the flight behavior of the armyworm moth, *Spodoptera exempta*, at an emergence site using radar and infra-red optical techniques. Ecol Entomol 8:395–418

Roome RE (1979) Pupal diapause in *Heliothis armigera* (Hübner) (Lepidoptera:Noctuidae) in Botswana: its regulation by environmental factors. Bull Entomol Res 69:149–160

Rose DJW, Dewhurst CF (1979) The African armyworm, *Spodoptera exempta* Congregation of moths in trees before flight. Entomol Exp Appl 26:346–348

Schaefer GW (1976) Radar observations of insect flight, pp 157–197. In: Rainey RC (ed) Insect Flight Symp R Entomol Soc no 7, p 287

Schaefer GW, Bent GA (1984) An infra-red remote sensing system for the active detection and automatic determination of insect flight trajectories (IRADIT). Bull Entomol Res 74:261–278

Slusher JA (1978) Night vision equipment developments: where we've been, where we are today, and where we're going. In: Lingren PD (conv) ESA Symp: night vision equipment for studying nocturnal behavior of insects. Bull Entomol Soc Am 24:197–200

Snow JW, Carlysle TC (1967) A characteristic indicating the mating status of male fall armyworm moths. Ann Entomol Soc Am 60:1071–1074

Sparks AN (1979) A review of the biology of the fall armyworm. Fl Entomol 62:82–87

Wolf WW (1978) Entomological radar studies in the United States, pp 263–266. In: Rabb RL, Kennedy GG (eds) Movement of highly mobile insects: concepts and methodology in research. University Graphics, North Carolina State University, Raleigh, North Carolina, pp 456

20 The Four Kinds of Migration

L. R. TAYLOR [1, 2]

1 Introduction

The papers presented here cover a wide range of aspects of insect migration, from genetical and biochemical through physiological and behavioural to ecological, aerodynamical, and agricultural. This range makes fascinating general reading. It is a topical survey and statement of how the subject of migration has developed in entomology, especially over the last 30 years and how it now appears to entomologists. It brings Johnson's (1969) great compendium up to date in many fields. Much of it is recognizably descended from that impressive work, even when the authors may not be aware of it.

This polymath approach yields a round view which is appropriately initiated in the Introductory Chapter by Danthanarayana and continued in an evolutionary and genetic dissertation by Dingle (Chap. 1, this Vol.) that sets the mark of a mature scientific discipline on what was, only a few decades ago, the barely noticed specialist province of a very few authors.

Dingle reminds us that the dominant force in insect migration was the spectacular; locust swarms were a historical plague; mass butterfly migration was brought to ecological attention, largely by C.B. Williams (1930). Outside entomology, concepts of migration projected birds, lemmings, salmon, and eels. Elton's (1927) rather quiet humdrum process was not widely associated with migration. The arrival of colonists was regarded as an unusual event and, perhaps because of this, the known artificial introductions of insects that produced spectacular damage, especially in agriculture, tended to be regarded as evidence that humdrum movements were of slight concern except when Man interfered. In population dynamics the idea of emigration as a major factor was dismissed without due consideration, except as a component of mortality; whilst immigration comprised so small a component of natality, with which it was equated except by geneticists, as to be neglected altogether.

A part of this neglect was due to the laboratory experimental approach to biology. Migration is not easy to create indoors. When so created, the result is often a travesty of reality. This distortion is now manipulated to advantage, but initially it often gave a misleading picture of migration. It needed field techniques and a patient, flexible but quantitative, observational approach to see what actually happened before the experiments had meaning; and especially it needed the development of quantitative techniques to measure the scale of it. The latest radar developments are now confirming the

[1] 12 Carisbrooke Road, Harpenden, Hertfordshire, AL5 5QT, United Kingdom.
[2] L.R. Taylor holds an Emeritus Research Fellowship from the Liverhulme Trust and wishes to acknowledge their support.

Insect Flight: Dispersal and Migration
Edited by W. Danthanarayana
© Springer-Verlag Berlin Heidelberg 1986

remarkable mobility of some insect populations and, with field work, confirms the surprising navigational achievements of some species. Ecology is again turning from the
quantitative observational long-term study to the illustrative experiment. The value of
this is undoubted – when results can be adequately measured and replicated in space
and time without too much unwitting interference in the process. Ecology is concerned
with the unfettered response of organisms to environment; the subtle and crucial behavioural link with an environment is often indefinable and never perfectly replicated.
Migration is an exceptionally elusive property in this intricate association. The danger
of experimental tunnel vision remains, because it is too easy to misinterpret the real life
function of an ecological response when the emphasis is on the mechanics of it.

Fortunately that does not appear to have seriously affected the papers here.

What holds back the study of migration now is that because of this functional un-
certainty, migration means different things to different people, especially across major
taxonomic divides. Much of the confusion is caused by people discussing different
"migratory" functions without recognizing those differences because the language used
is not precisely defined; they are talking at cross purposes.

Perhaps that has become clear as we perused the contents of this book. To be sure
of this, it may still be worthwhile to clarify the several meanings attached to migration
by different groups of people in this final chapter.

2 The Meaning of Migration

In entomology an extremely wide range of life-styles and life histories, combined with
the quantitative facility offered by large samples and good experimental material, has
helped to develop an awareness of the problems of the interpretation and classification
of migration and dispersal. The classic *Migration and Dispersal of Insects by Flight* by
C.G. Johnson in 1969 brought together a vast amount of information and interpretation that began with J.W. Tutt's *The Migration and Dispersal of Insects* in 1902. These
titles show that the functional distinction between migration and dispersal had long
been recognized in entomology, although this is not so for all taxa, and also the need
to integrate individual migratory behaviour and population dispersal, even before this
could be accomplished in practice. Especially in the migratory locusts and the Aphididae initially, and later in other insect groups, there have been extensive studies of migratory genetics, morphology, physiology, and behaviour and quantitative comparative
analyses of movement, population structure and redistribution. In the Lepidoptera
there is a vast literature on field observation and experiment.

This experience is more difficult to obtain in the vertebrates where samples are
relatively small and where quantitative, replicated experiment in laboratory and field,
especially on the creation of, as well as the behaviour and function of, migratory morphs,
is limited. Long-term changes in migratory distribution, often over hundreds of generations of field populations over Great Britain, have now been monitored in hundreds of
species of aphid and, in some of these, migrant morphs can be produced, or inhibited,
indefinitely in laboratory and field experiments (Taylor 1986a,b).

In a discussion of migratory behaviour Rankin and Singer (1984) show how the
oogenesis-flight syndrome of C.G. Johnson, based in the "alternate flight and reproduc-

tion" concepts of J.S. Kennedy, is a good description of some, but not all, migratory conditions in insects; this physiological-behaviour experimentation, combined with endless hours of field-work, generated a population-dynamics approach to the subject that was for long lacking in vertebrate studies. In other words, migration is not merely an amazing seasonal transportation of millions of birds from one part of the globe to another, with another return migration for which the navigational abilities are even more remarkable. Migration in insects is seen as vital for colonization in a constantly moving environment; the maximization of reproductive output, given the spatial dynamics of real populations, is secondary to the ability to place that potential in a viable situation; migratory viability comes first, chronologically and behaviourally (Rankin et al., Chap. 2 this Vol.).

Many insects are highly mobile and much of the mobility is associated with reproduction, so that insect populations are also often mobile; for this reason the study of migration has run a parallel course with population dynamics. In contrast, bird migration, which is highly visible, and fish migration, which is less visible but economically more important, provide a second kind of migratory movement that accentuates the concept of static populations. In these taxonomic groups the major migratory event does not interfere with the reproductive sequence which takes place in the same place year after year. Migration appears as a refuelling spatial side-cycle to the main population temporal cycle and can be ignored in the dynamic progression. Population dynamics and migration are separated.

For many years the continuing discussion about the meaning of migration, with which the names of C.B. Williams, J.S. Kennedy, and C.G. Johnson were closely linked, was thought by many entomologists to be merely semantic; not worthy of serious scientific consideration. Southwood (1962) was thence regarded as having resolved the matter by introducing the evolutionary component, and interest died as the three main protagonists turned to other issues, leaving an impasse more evident from outside entomology than from inside. Here we are concerned with the wider implications of migration, defined precisely, so that it can be used with equal understanding inside and outside entomology. Again the issue is not *merely* semantic; it has a crucial purpose in interpretation of function.

When Tutt (1902) wrote extensively about insect migration, he touched every aspect and remained aloof from critical comment so that no ground rules were laid. Williams (1930, 1958) wrote about the visual observation of butterfly migration, in essence, with some additional insect families as was opportune. Kennedy (1956, 1985) wrote about a behavioural concept of migration based on an interpretation of physiological mechanism and was concerned very largely with this behavioural mechanism or orientation in migratory flight. Johnson (1954, 1969) added a physiological, ontogenetic component that concerned insect life histories. Southwood (1962) encased it all in an evolutionary explanation.

All these authors had a different interpretation of migration. My personal interpretation of their conceptual image at that time, was of four sporting competitors, all highly successful, shooting at different targets and being occasionally frustrated by not being able to get all four targets into one perspective.

This perspective is examined historically elsewhere (Taylor and Taylor 1983). The issue here is that these entomologists confined their attention to insects. At a later date

Baker (1978) attempted to expand these concepts and to combine all animal migration under one umbrella. In this he failed because he tried to incorporate in a single all-embracing explanation, too many different kinds of migration. Taylor and Taylor (1977) tried to account for the dynamics of migratory motion, using the word in the Southwood sense, with a simple conceptual model but also ignored the completely different concepts invoked by the word migration in different zoologists – quite apart from botanists. It is now essential to clarify this terminology and define what we mean by migration.

The essays presented here consider some of the behaviour associated with migration of a particular kind in insects, as pointed out by Danthanarayana in the Introduction. Because other zoologists already have pre-empted the word migration for clearly defined processes, these different usages must be accounted for. The most familiar and well-established usage is in bird migration and I take it first because it facilitates the introduction of others.

2.1 Bird Migration

This cyclic phenomenon, a pair of out-and-home migrations, repeated annually during the lifetime of a single individual, does not exist in the complete form in insects if only because no insect lives as an adult for more than 1 year. It may occur in some insects in an abbreviated form, e.g. the monarch butterfly (Gibo, Chap. 12, this Vol.) and the armyworm (Gatehouse, Chap. 9, this Vol.) but there is no evidence that the same insect ever returns after the second, return, migration, to precisely the same site from which it emigrated the first time. If it does not, then two kinds of migration have been elicited in the single life process. The out-and-home migrations of the kind familiar in birds, have an additional geometrical component. The whole set of three migrations is represented, not by a single line between two sets of spatial coordinates, but by a triangle between three sets (Fig. 1). In this simple geometrical construction the real definitions of migrations are made explicit.

The differences between the many definitions of migration in the literature, and the confusion resulting from them, often result from failure to define the organizational level at which the definition is made; in many cases definitions were sought that would cover several levels, notably physiological, behavioural and ecological simultaneously. This is not only overly optimistic; it is impossible. Migration is a life function of whole individuals, not of their parts or elements, and can only be defined ecologically.

The ecological definition of migration must cover a point of emigration, a process of migration and a point of immigration. These points may be conceived as being in two different populations, although the migrating individual remains the same. Since we have, as yet, no adequate definition for "population" this is not an acceptable ground for a definition and it is part of the cause of the past confusion. The only possible alternative points of reference are coordinate points in space, and since a large part of the insect literature on migration is concerned essentially with spatial-point observations or real-space vector observations, this must be the definitive frame.

Emigration then becomes the movement from $C(x_1, y_1)$ to $D(x_2, y_2)$ as seen from C; immigration is the same movement seen from D; and (passively) migration is the whole

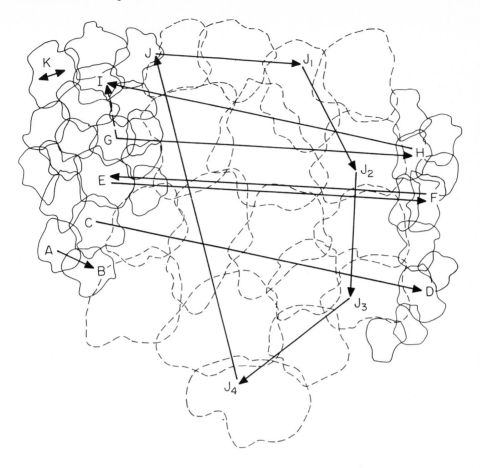

Fig. 1. Two regions of permanent territories (*solid outlines*) are separated by an empty or ephemeral region (*dashed outlines*). $E-F-E$ is characteristic two-way migration of the third kind ($E \rightarrow F \rightarrow E$). $C-D$ is one-way long-distance migration of the second kind ($C \rightarrow D$). $A-B$ is short-distance migration of the second kind ($A \rightarrow B$) between occupied territories. $G-H-I$ is combined third ($G \rightarrow H \rightarrow G$) and second ($G \rightarrow I$) kinds of migration. K is internal, fourth kind, of migration. $J-J_1-J_2...J$ is long-distance, circular development of migration of the third kind ($E \rightarrow F \rightarrow E$) (not usually so recognized) in, say, albatross, that always breed at J, but at other times travel continuously ($J \rightarrow J_1 \rightarrow J_2 \rightarrow ...J$) (after Taylor 1986b)

transition from C to D including take-off and landing, seen from outside, the actual travelling sometimes being called transmigration; whilst migration is also (actively) the motion that produces the transition, e.g. migratory flight = migrating.

The ornithologists have some difficulty with this concept. Bird migration is considered essentially as a seasonal movement by which individuals benefit from the different spatial distributions of summer and winter on a spinning, circumambulating globe with a corrugated surface. Migration may be, roughly, north and south due to the axis; or up- and down-hill; surface irregularities may also introduce an east-west component through climatic distortions in relation to continental masses, and the self-centred rotary motion of the globe. In all cases the migration demands a return to the origin. In Fig. 1 it requires motion from $E \to F \to E$, i.e. two migrations. At the second site, the arrival of birds is an immigration and their leaving an emigration, just as at the first site. Otherwise confusion reigns. All this is independent of population, either spatially or on the biological scale of levels of organization. Such migration is, and essentially must be capable of individual definition because it is an active verb – to migrate is what an *individual* does – as well as a passive observation – 'immigrants arrived'. In ornithology, the individual behaviour and the dynamic population function are not in conflict so long as the individual returns to precisely the original site after the second 'return migration'; $(E_1 \to F \to E_2)$ where E_2 is coincident with E_1.

When E_2 is not coincident with E_1, as in $G \to H \to I$ (Fig. 1), a secondary problem arises that has usually been ignored both in entomology and ornithology. Birds are commonly treated, for convenience, as returning to the 'same' habitat or locality. More recently within-locality movement has been investigated in greater depth, but mainly for 'non-migratory' birds, and for the purpose of common migratory processes such small deviation have been overlooked. They are now classified as natal and reproductive dispersal (Greenwood and Harvey 1982).

The same is true for many ocean-going freshwater fish; return migration is regarded as a return to the point of departure.

2.2 Insect Migration

In insects, no such confidence can be maintained. There are no reliable records of individuals returning to the identical place after migration. The Monarch butterfly, which is and is likely to remain for some time, the only species with sufficient visual presence to make the tracking of marked individuals feasible, has never been known to return precisely to its birth place. Nor is this likely. The ability to home in on the 'butterfly trees' that form its over-wintering roost in California or Mexico is remarkable, but there it has a large fixed site for an 'inherited memory', of whatever form, to orientate to. The final stages of migration undoubtedly owe much to the same aggregatory roosting behaviour that is evident in this species at lower densities, at the evening roosts during the transmigration. For each individual to return to its source and find its own *unique* birthplace, demands a precision of navigation of a different order of magnitude. It also assumes that the host plant is perennial and had survived, that being the only relevant identifiable source, and this is uncertain.

This being so, for the best authenticated migrant insect species the seasonal ground track is not linear, $E \to F \to E$, but triangular $G \to H \to I$ (Fig. 1). In most insects the third side is relatively longer, as compared with birds, than is acceptable in terms of a permanently static 'population' from which individuals emigrate and to which they return. The 'returned emigrants' from a single site in an insect population are immigrants to a whole series of different sites, when these are defined geographically, i.e. by spatial coordinates. The spatial continuity of insect ground tracks is broken in each generation, and this introduces the third side of the triangle $(G \to I)$ as a migratory component of great significance in population dynamics. 'The population' is no longer static;

it has an amoebic motion of its own that is the sum of the *migrations of the second kind* of the individuals of the population.

The third side of the triangle also represents "dispersal" when the product of all individual movements changes, not only the spatial coordinates of the population centre of gravity, but also changes density. The relevance of this component in many insects' lives is overwhelming. Only this third side of the triangle of migration exists for many species, aphids for example; however, it is still called migration in contrast to the usage in the small mammal literature, for example. In that literature migration is rarely mentioned and 'dispersal' is considered a major dynamic factor. In birds it is 'natal dispersal' on the first occasion and 'reproductive dispersal' on subsequent occasions that fulfils this function.

In the small mammal literature local movement is referred to almost exclusively as dispersal; the word is used synonymously apparently with 'exodus' or 'emigration' because it seems to be concerned mainly with factors affecting the activation of emigration. Factors affecting termination of migration and immigration to the new site, are treated less thoroughly. This may be because of the preoccupation with population density cycling in *time*, which has a large literature of its own, rather than in *space*.

The continuous flights of seabirds such as the albatross, between breeding periods, are an extension of the two-way migration of the third kind ($E \rightarrow F \rightarrow E$) in which F is a continuously resumed journey until the spatial cycle is completed by a return to the original nesting site ($J \rightarrow J_1 \rightarrow J_2 \rightarrow J_3 \rightarrow ... J$). This is the evolutionary end-point, achieved with incredible navigational skill, of the near-random wanderings of the desert locust swarm that eventually returns to its earlier source many generations later, wind-borne by cyclic weather patterns (Rainey 1963) over which its only "control" is massive reproductive rates with the resultant ability to wander for generations.

Migration of the second kind is epitomized by the Insecta. It is the means for dynamic change in spatial co-ordinates, usually between generations. It is hardly ever a random process but behaviourally controlled. Nor is its outcome geometrically randomizing in space; even in the apparently most uniform environments, species distribution patterns differ as a consequence of migration of the second kind. This specific behavioural component is the product of evolutionary selection operating on migratory movement to maximize survival of the present generation during transit and, after arrival, to protect the next generation by the meticulous selection of a place for it to live; paying special regard to where other individuals of the same species are located.

Despite the convincing evidence in linear, oriented flight for some species, control of track is not possible for many species for much of the time; the wind is too powerful. Nevertheless orientation mechanisms are abundant and sophisticated and evolutionary precepts demand that they have a function. I am firmly convinced that the behavioural component of migration is pre-eminent and vital. Its function is dynamic; to maximize environmental search efficiency for *new* optimal sites for colonization; not to return to *old* ones for breeding. The spatial distributions resulting from this species-specific spatially-dynamic search behaviour are equally species-specific and are the determining factor in subsequent temporal population growth. Random approaches to the analysis of spatial distribution are contrary to evolutionary principles.

3 Introduction to Function: Definitions and Classification

There is no immediately obvious single definition of migration that can include, say, typical aphid or locust migration and the familiar bird migration – depending as they do on apparently different principles – and also apply to, say, voles or whales. This problem can perhaps be overcome by accepting the dictionary definitions for migration which acknowledge four different kinds (Oxford English Dictionary, 1971; Websters Universal, 1968) and relaxing, where necessary, the constraints on the classes to which they are applied in common usage.

1. The action of moving from one place to another ... chiefly of *things*.
2. Removal from one place of residence to another, especially of *persons or tribes*.
3. The periodical departure from and return to a region at a particular time of year, especially of some *birds and fishes*.
4. Alteration of position within the *body*; of an *organ*.

The first kind, being largely inanimate, is the only one that leads to random distributions. The second and the third kind correspond to one-way migration as in aphids and locusts as well as to 'persons'; and two-way migration occurs in gnu and caribou as well as in birds and fish. The fourth kind does not, as yet, present a clear-cut concept, although for completeness I include it; with the organ represented by an individual in the (body) population. It opens a new door to epideictic behaviour and social migration.

It is not surprising that these ecological distinctions are found in dictionaries, since they are readily recognizable by common observation, some of them in human populations. For ecological purposes they need slight modification and rather more precision. In particular, the definitions need criteria independent of the organisms they are applied to. For ecological purposes the degree of control of movement required for a given kind of migration provides such an independent criterion (Table 1).

Table 1 is basically a linear series in time for the individual, or its descendants, to recover its original spatial coordinates with a segregating mechanism. As Table 1 progresses, the return time to the original coordinates becomes shorter. The segregation is simple, based on evolutionary sequence, not visualimpact, i.e. on the movement control required.

Reciprocating, two-way migrations require navigation to recover the original coordinates (even approximately). It may take more than one lifetime, but the mechanism functions on coordinates in space (sensory navigation or memory).

Redistributing, one-way migration requires no navigation: it may have control of emigration to maximize prospects for travelling conditions; it may also have immigration control to maximize host-finding, but it does not have true navigation and there is a larger element of risk, hence a longer time if ever, for the organism to return to its origin. The function, however, is much more primitive and hence fundamental; it is not concerned (as 3) with maximizing energy by increasing feeding time during the year. It is concerned with tracking the environment; optimizing the survival of the next generation by correctly placing it in space. Coordinate navigation is not necessary to find, by simple search, an equivalent environment; it is required to find an exact site.

Table 1. The degree of control over movement exerted by organisms during different kinds of migration

1. No migration control
   Random distribution
   Chance return to origin
 	
   First kind
   [Dust: some spores etc.]
	4. Direction control weak
	   Non-random distribution
	   At least two generations to return;
	   partially defined
	
	   Second and third kinds
	   [Monarch: armyworm etc.]

2. Emigration control only
   Non-random distribution
   Chance return to origin
 	
   Second kind
   [Basidiospores etc.]
	5. Direction control
	   Non-random distribution
	   One generation to return; precisely
	   defined in space
	
	   Third kind
	   [Salmon: eels]

3. Em- and immigration control
   Non-random distribution
   Prospects of return increased, but not defined
 	
   Second kind
   [Aphids: locusts]
	6. Direction control
	   Non-random distribution
	   Annual return; precisely defined in space
	   and time
	
	   Third kind
	   [Amphibians: reptiles: birds: large mammals]

Migration of the fourth kind, internal to the population remains, at present, something of a mystery: see Table 2.

Type 3 migration ranges from the mass flights between two predetermined sets of coordinates of some song birds to the continuous circuit migrations of some human nomad bands and the highly individual continuous migrations of some colonially-breeding seabirds, although these are not commonly classed as migrants in ornithology.

It should be noted that both functions, to return to original spatial coordinates and to explore new ones, are vital to survival in a changing environment. They occur on different scales of space and time for, say, a swallow and a woodlouse. When these migratory functions fail to operate effectively, or the environment changes faster than 'expected', fragments of populations become detached and go extinct. Many butterfly species of Great Britain, for example, are now in that situation (Thomas 1984).

Migration is an ecological phenomenon that expands life histories and adapts them to the environment. Navigation is a physiological phenomenon concerned with the location of individuals in space. It is made use of by the animal's behaviour and that, in turn, services the animal's ecology. Navigation is two levels of organization removed from migration, so that it cannot be used to define migration. It does, however, help to classify it. The main determining characters for migration are size of the organism and time; navigation is a secondary segregating character.

Table 1 shows two main categories of migration, based on the presence or absence of a return to an abode at fixed latitude x longitude coordinates in space; that depends upon an ability to navigate, by whatever means, which in turn depends on size. Size also governs the sequence for progressive increase in environmental control. First and

Table 2. The four kinds of migration

First kind: Passive Migration: One way; projected emigration; transported migration without volition or control over the immigration end point. Usually inanimate, but some animate migration of spores, minute seeds, etc. in this class.
Function: random distribution (not necessarily dispersal); *no navigation; no return.*

Second kind: Dynamic Migration: One-way; single migration; actively initiated emigration; volition continued through transmigration to immigration; may use extrinsic power and direction (wind, tides, etc.). *Navigation not essential; no return.*
Function: dynamic, non-random re-distribution between generations.

Third kind: Homeostatic Migration: Two-way, double migrations; actively controlled at all stages — emigration, transmigration, and immigration; extrinsic power made use of but with no loss of control of navigation, except by default, (e.g. wind-assisted). *Second, return migration to breeding site. Navigation vital.*
Function: maintenance of energy input over seasonal cycle.

Fourth kind: Social Migration: Movement of individuals within population structure as members of a body.
Function: sublimated migration within the pack, herd, etc. in relation to each other, apparently not to coordinate space; behaviour not yet defined in detail, but includes change of dominance, rank, etc. with age, size, learning, etc. Leads to epideictic behaviour, and true "social migration", and possibly to dispersal or aggregation.

---

smallest are inert particles; then timing of take-off is controlled (emigration) as in spore discharge; then alighting is controlled (immigration), as in most insects; this allows some increase in the prospect of a return to source, but no direct navigational control over it.

Finally control of direction (navigation) appears in the larger animals. Within this category of navigational control, the time sequence is progressively shortened from at least two generations for the full return cycle of migration, through one generation, to annual cycles by the same individual. The classification stops short at diel cycles which are not usually treated as migratory, except in plankton; I am not sure how that can be justified.

The resultant classes are grouped together in the formal nomenclature of migration as defined by the O.E.D. Since migration is a word in common usage, although with several meanings in human life cycles that are slightly different from its usage for birds and fish, the familiar O.E.D. categorization has been retained in Table 2 by merely changing the body (physical) for the body (political); we can use the fourth kind for migration within the population.

The common component in migration of the first three kinds is that it is an individual activity changing spatial coordinates of latitude and longitude (and possibly altitude). Unlike migration of the fourth kind, most migration presumes nothing about the other members of the species or community, although it may incidentally be performed en masse or in isolation. Migration of the fourth kind is the activity that results in aggregation or dispersal, which in turn changes population density. Whilst other kinds of migration change the positions of individuals in space, and so may move the population centre if the migrations are not random, the fourth kind affects the relative spacing of individuals, i.e. a response to each other. Migration and dispersal may be concurrent, but they are not interdependent. Dispersal is the complement of aggregation; the first diminishes local population density, and the second increases it. Both conditions are caused by or result from migration which is the active processes by which individuals disperse, aggregate or move, collectively or alone, to change spatial coordinates or local density.

The different processes described as migration for birds, other animals and plants, require qualification as to the ecological function performed by the migration. This is clear from the dictionary definitions. These functions depend on the following conditions: movement, active or passive; direction and time, predetermined; migration, mono- or bi-directional (i.e. single or double migration); completed annually, in one lifetime, in two lifetimes, or more.

These components form a simple classification tree leading to the dictionary classes of animate and inanimate migration.

The primary division lies in the possession of directional control of movement and the second in navigation equipment. For a reciprocating migratory system of the third kind, moving to specified coordinates at a particular time of year whether annual, lifetime or possibly over two lifetimes, navigation is essential. This divides directional migration into two classes.

The reciprocating system of migration of the third kind includes:

1. Two-generation (or more) monarch butterflies (Gibo, Chap. 12) and armyworms (Gatehouse, Chap. 9), probably many of the butterfly migrants, and similar powerful insect fliers. Despite their size, locusts do not conform to this criterion. Their prospect of return to base are little or no better than for aphids, for the swarm structural mechanism prevents individual navigation by presenting the swarm to the maximum downwind displacement as an intact unit. Evolved adaptation to wind or water systems then provides the only likelihood of return to base.

2. One-generation, return-to-breed-and-die double migrations in fish such as eels and salmon, which have feeding and breeding grounds widely separated by a sea since they evolved their life cycle, require navigation but only once each way in a lifetime.

3. Annual return migrants, such as large mammals and birds, must have highly effective navigation, either by unknown extrinsic sensory signals, sky signals or imprinted memory of local ground systems. They must complete a round trip each seasonal cycle to benefit from a different season at each end of their journey, often thousands of kilometers. This may be two clear-cut migrations, or a continued series. These sophisticated organisms use collective, inherited or individual memory, as well as true navigation, to make the repeated search for new sites for reproduction unnecessary. They effectively avoid the limitations of a single site, with its seasonal cycle of temperature and food by their migrations between two or more sites, hence here referred to as homeostatic migrations, avoiding dynamic change.

The alternative migratory system of the second kind (one-way) is a move from 'one's country of habitation or residence' involving permanent change in the spatial coordinates of the abode in the constant search for a better place to reproduce, i.e. the function of this migration is dynamic, not homeostatic. It is a basic function. Lacking either the ability to navigate, or the physical power to steer a course against the elements, or both, the individual migrant cannot return to the spatial coordinates of its abode, its own birthplace. However, with or without that ability, the search for a new abode is positive and universal and is the ecological criterion. Only by the slow change of weather systems (in locusts) or by chance over many generations (aphids) may the offspring arrive back at the same point in space, so that any homeostatic component is indirect and largely by chance.

This classification is, of necessity, greatly simplified in its details; only a tiny fraction of the possible taxonomic types are included. It may not be adequate for all purposes and does not pretend to be complete; but, because the ground rules are simple, designed to cover several major taxa and not narrowly for one group only, it may be capable of expansion or adaptation. The definitions are therefore given in more detail in an Appendix, to be tested and developed and especially to be tried out on other species.

The definitions are part of an attempt to persuade ecologists who deal with widely different taxa to look at their own concepts in the light of the life-styles outside their immediate concern. The design is to separate, so far as is reasonable, the plant seed and spore, one-way migration from the equivalent animal, one-way migration, and both from the two-way seasonal, conventional bird and large mammal migrations, and to relate all these to elements of human migration that have so many not-quite-equivalent terms. The terminology is important here because most of the attention given to bird and insect migration (for example) has been directed to almost diametrically opposite functions. Hence the disagreement.

For the present purpose, successful migration of the *second kind* is defined as follows: *movement made by a living organism between its own birthplace* (set of spatial coordinates) *and those of its offspring*.

Most insect migration falls into this category of migration of the second kind. Category one includes movement without 'volition' and is exemplified by seeds, spores, and dust and may include some insect eggs and inactive individuals or stages. Category three is the return, feeding-and-breeding cycle of birds, large mammals, migratory river fish and may include some butterflies. Category four is novel in that it treats the individual as an element in the population and its function in insects is unexplored.

Human migration covers all categories. Ship-wrecked mariners are in category one; migration to the New World is category two; cyclic working migrations, regular nomadic seasonal migrations and the return of the prodigal son are category three; and social migration is category four.

The logical distinctions are quite specific, although one kind may change into another with age or environmental change. The only exclusions are daily rhythms and for these, whatever their function, it seems to me commuting is a more suitable term, with its connotation of regular, daily return journeys concerned primarily with food and reproduction, but achieving no progressive change of coordinates in space. This then includes the daily cycles of plankton and insect diel cycles of activity in the tree canopy, for example.

This listing is not comprehensive, but it effectively isolates most of those movements of short-lived, mobile (as adults) insects on the basis of the ecological function of movement, *not on the associated behaviour* which cannot *define* an ecological function, although it may be essential to perform it. Behaviour has been the root of most of the misunderstanding in the past.

The seasonal element and the concentration of some insect migrants into swarms or masses, in static swarms, in transit or at the end of a migration, is another distracting feature more likely concerned with feeding and reproduction than with redistribution.

It is perhaps significant that such aggregations are usually at an unfavourable season and that the migrants to the aggregation are often sexually immature, whilst those leaving are sexually mature. Could these aggregations serve a population function by resisting the constant risk of disintegration for populations of small flying organisms and, by this seasonal congregation, maintain coehsion? The ultimate effect of intraspecific competition is to generate a repulsion between individuals of the same species. Given free rein this could only lead to irrevocable decrease in population density – once animals had acquired sufficient mobility to indulge this repulsion. Only a common interest in food, and more strongly in sex overcomes this; the main ecological function of sex could be to retain population integrity. Hence, dispersal and aggregation are functions of population density. The individual behaviour involved is in response directly to population density, i.e. to other individuals or their artifacts. One individual cannot disperse. The behaviour involved in dispersal, as in aggregation, is migratory.

Appendix

Definitions for Four Kinds of Migration

The term migration has its origins in common usage and is correspondingly lacking in ecological precision. Even so, it is an indispensible tool in general ecological terminology; nobody has succeeded in avoiding its use despite it being pre-empted for activities peculiar to specific taxa. Therefore, it seems wise to state the dictionary definitions biologically, imprecise though these are; it is not helpful to have specialist use that directly contravenes common usage. The Oxford English Dictionary (1971) gives four definitions and Websters (1968) also; they differ only marginally.

I take them in the O.E.D. sequence. MIGRATION (1) *Chiefly of things*; e.g. centre of gravity, accidents, elements, souls – to pass from one place to another, to be transported – of some elements from primary growth which ... constitute the centres of secondary formation. (2) *Chiefly of persons*; e.g. moving from one country, locality, *place of residence*, to another. (3) *Chiefly of animals*; e.g. movement in flocks or shoals, esp. of birds and fishes, *periodical, seasonal departure and return* from one region to another. (4) *Chiefly of bodily organs*; alteration of position *within the body*.

1. *Migration of the first kind* refers initially to inactive particles, and applies only to biological entities in their totally inactive resistant spore or seed stage when time is stopped until the right environment comes along. However, the 'centres of secondary' growth category seems possibly suited to vegetative reproduction.

Emigration and immigration are not relevant to migration of the *first kind* since it is not intrinsically activated. However, its outcome is not, by dictionary definition, random; e.g. nuts 'migrated by' squirrels become aggregated, but they have *intrinsic* behavioural initiation of motion and control of termination, only in evolutionary terms. Similarly, shedding of spores and pollen may be timed to take advantage of wind transport in a particular direction or for a particular distance.

Migration of the first kind is a randomizing process which includes all inanimate objects, i.e. with no volition, and hence covers logs in a river, sand in the tide and dust in the air. Whether or not a projected spore is of the *first kind* depends upon whether the resultant distribution is random. This may require an evolutionary justification. If a plant projects its own seeds or spores, I would classify this as migration of the second kind as I would tumbleweed, where the plant becomes a machine for seed migration. If a nut uses a squirrel to acquire non-random distribution, that is also migration of the second kind.

2. *Migration of the second kind*, applied to humans, has direct equivalence with migration in other animals. Its essence is a one-way, intrinsically motivated and controlled movement to a *new place to live*; and, by implication, to stay and reproduce there. The description is immediately recognizable as the migrations of people from the Old World to the New, either collectively or individually.

Migration of the second kind is ecologically specific and *undertaken by almost all organisms*. It is the spatial journey, of no specified length, from the residence of one generation to that of the next; from birthplace of parent to that of offspring. Only rarely do these coincide. The true Autochthone is a mythical beast that never migrates; it is an interesting (and futile) exercise to work out its life cycle; it must replace its only parent, precisely, in every generation, otherwise the population can drift and is no longer autochthonous.

That drift is the population expression of the sum of all individual migrations of the second kind seen in classic form in many insects. In aggregations of the higher, and larger, animals it may be commuted into dominance hierarchies and territories and these collective residences (see the fourth kind) may be the nearest to authochthonous.

Migration of the second kind involves emigration behaviour from the locus of the parents birth, transmigration, and evolved immigration behaviour into the other locus where the offspring is to be produced. The loci and their real spatial coordinates are an essential feature. It is related to humans by the dictionary definition, but such migration is not peculiar to man. Clearly, specially evolved organs of dissemination in plants are intrinsically activated and belong in this class.

Migration of the second kind is the category most relevant to *spatial dynamics* and most often implicated in redistribution, as in natal and breeding "dispersal" (Greenwood and Harvey 1982). Unlike the third kind, it is not a repetitive activity. A second migration leads to a third set of coordinates. If no arbitrary limits are imposed with respect to mode of transport or distance travelled, all active, living organisms that reproduce elsewhere, other than where they were born, are migrants of the second kind by definition; e.g. tumbleweeds, fur-carried and bird-eaten seeds, since seeds carried, but not digested by animals, have an improved prospect of immigrating into another environment like that from which they emigrated. Hence, their ultimate distribution is not random and they are migrants of the second, not first, kind.

3. *Migration of the third kind* is referred initially to animals meaning vertebrates, specifically to birds and fishes, but it could include other vertebrates. It is again immediately identifiable from the description, but of limited taxonomic extent, being concerned initially with the transition between sites of *reproduction and of feeding*. It is a periodic, reciprocating activity usually annual, or once in each generation. It is the migration most concerned with *temporal dynamics*, being a source of energy without change of place of reproduction.

Migration of the third kind is again quite specific. It is concerned with the highly specialized and controlled movements of a taxonomically small elite of large species. Where it occurs, it so obscures the more primitive function of migration of the second kind that the two become obfuscated. By given example, it is applied to the many familiar migrant birds and to diadromous fishes, including specifically salmon. The common element is essentially the return migration to the source *within a generation,* to reproduce. This double migration is not primarily a change of residence, but a feeding cycle, that is usually seasonal, but also may be a life-span in species that usually breed only once. These cases are highly and specifically, intrinsically activated and require navigation to complete the return.

4. *Migration of the fourth kind* is apparently at the wrong level or organization for ecology, dealing as it does with parts of bodies; unless the bodies are equated with 'populations', and the organs with individuals.

Migration of the fourth kind then becomes internal migration, within the population or local community (single species). This may have an increasingly important role as migratory function becomes clearer. It can be considered as social migration resulting from epideictic behaviour, related to the positions of other individuals rather than spatial coordinates, and derived from the second kind, especially in complex societies like ants and primates.

All of these are categories of true migration, being definitive of *individual* behavioural motion and capable of being arranged in a sequence of active *individual* contribution.

These four definitions from the Oxford English Dictionary and Webster's are quite clear in intent, and each may be adapted without undue distortion into an ecological function.

References

Baker RR (1978) The evolutionary ecology of animal migration. Hodder & Stoughton, London, pp 1012

Elton C (1927) Animal ecology. Sigwick & Jackson, London, pp 207

Greenwood PJ, Harvey PH (1982) The natal and breeding dispersal of birds. Annu Rev Ecol System 13:1–21

Johnson CG (1954) Aphid migration in relation to weather. Biol Rev 29:87–118

Johnson CG (1969) Migration and dispersal of insects by flight. Methuen, London, pp 763

Kennedy JS (1956) Phase transformation in locust biology. Biol Rev 31:349–370

Kennedy JS (1985) Migration, behavioural and ecological. In: Rankin MA (ed) Migration: mechanisms and adaptive significance. Contributions in Marine Science, 27 (suppl), pp 1–20

Oxford English Dictionary (1971) Oxford University Press, Oxford

Rainey RC (1963) Meteorology and the migration of desert locusts. Tech Notes Wld Met Org (54), pp 115

Rankin MA, Singer M (1984) Insect movement: mechanisms and effects. In: Huffaker CB, Rabb RL (eds) Ecological entomology, pp 185–216. John Wiley and Sons, New York, pp 844

Southwood TRE (1962) Migration of terrestrial arthropods in relation to habitat. Biol Rev 37: 171–214

Taylor LR (1986a) An international standard for synoptic monitoring and dynamic mapping of migrant insect populations. In: MacKenzie DR, Barfield C, Kennedy GG, Berger RD (eds) Movement and dispersal of biotic agents. Claitor's Law Books and Publishing Division, Baton Rouge

Taylor LR (1986b) Synoptic ecology, migration of the second kind, and the Rothamsted Insect Survey. Presidential address. J Anim Ecol 55:1–38

Taylor LR, Taylor RAJ (1977) Aggregation, migration and population mechanics. Nature (Lond) 265:415–421

Taylor LR, Taylor RAJ (1983) Insect migration as a paradigm for survival by movement. In: Swingland IR, Greenwood PJ (eds) The ecology of animal movement pp 181–214. Clarendon Press, Oxford, pp 311

Tutt JW (1902) The migration and dispersal of insects. Elliott Stock, London

Thomas JA (1984) The conservation of butterflies in temperate countries: past efforts and lessons for the future. In: Vane-Wright RI, Ackery PR (eds) The biology of butterflies. Roy Entom Soc, London, pp 429

Websters Universal Dictionary (1968) Unabridged international edition. Wyld HC, Partridge EH (eds) Haver Publishing Inc, New York

Williams CB (1930) Migration of butterflies. Oliver and Boyd, Edinburgh, pp 473

Williams CB (1958) Insect migration. Collins, London, pp 235

Species Index

Subject Index